MANAGING HEALTH AND SAFETY IN THE UK

A course book for the NEBOSH National General Certificate in Occupational Health and Safety

Series Editor:
Paul Randall

Authors:
Dr Luise Vassie
Tony Morriss
Dr Mark Cooper
Caroline Copson
Dr David Towlson

Editorial and project management by Haremi Ltd. (www.haremi.co.uk)
Typesetting by York Publishing Solutions Pvt. Ltd., INDIA
Cover design by Kamae Design

Edition: 1
Version: 1

References to legislation in this course book are as at
15 August 2022. For the most up-to-date versions of legislation,
please see legislation.gov.uk.

Contains public sector information published by the Health and
Safety Executive and licensed under the Open Government Licence

Every effort has been made to trace copyright material and obtain
permission to reproduce it. If there are any errors or omissions,
NEBOSH would welcome notification so that corrections may be
incorporated in future reprints or editions of this course book.

© 2022 NEBOSH
All rights reserved.
No part of this publication may be reproduced, stored in a retrieval system, or
transmitted in any form, or by any means, electronic, electrostatic, mechanical
photocopied or otherwise, without the express permission in writing from NEBOSH.

CONTENTS

Introduction	**5**
Element 1: Why we should manage workplace health and safety	**7**
1.1: Morals and money	7
1.2: The force of law – punishment and compensation	12
1.3: The most important legal duties for employers and workers	34
1.4: Managing contractors effectively	42
Element 2: How health and safety management systems work and what they look like	**52**
2.1: What they are and the benefits they bring	52
2.2: What good health and safety management systems look like	66
Element 3: Managing risk – understanding people and processes	**78**
3.1: Health and safety culture	78
3.2: Improving health and safety culture	86
3.3: How human factors influence behaviour positively or negatively	102
3.4: Assessing risk	114
3.5: Management of change	135
3.6: Safe systems of work for general work activities	142
3.7: Permit-to-work systems	152
3.8: Emergency procedures	161
Element 4: Health and safety monitoring and measuring	**170**
4.1: Active and reactive monitoring	170
4.2: Investigating incidents	183
4.3: Health and safety auditing	193
4.4: Review of health and safety performance	203
Element 5: Physical and psychological health	**209**
5.1: Noise	209
5.2: Vibration	222
5.3: Radiation	229
5.4: Mental ill-health	243
5.5: Violence at work	254
5.6: Substance abuse at work	260

Element 6: Musculoskeletal health — 267

6.1: Work-related upper limb disorders — 267
6.2: Manual handling — 281
6.3: Load-handling equipment — 292

Element 7: Chemical and biological agents — 305

7.1: Hazardous substances — 305
7.2: Assessment of health risks — 316
7.3: Occupational exposure limits — 327
7.4: Control measures — 332
7.5: Specific agents — 348

Element 8: General workplace issues — 363

8.1: Health, welfare and work environment — 363
8.2: Working at height — 374
8.3: Safe working in confined spaces — 392
8.4: Lone working — 406
8.5: Slips and trips — 417
8.6: Safe movement of people and vehicles in the workplace — 429
8.7: Work-related driving — 445

Element 9: Work equipment — 462

9.1: General requirements — 462
9.2: Hand-held tools — 472
9.3: Machinery hazards — 479
9.4: Control measures for machinery — 493

Element 10: Fire — 502

10.1: Fire principles — 502
10.2: Preventing fire and fire spread — 511
10.3: Fire alarms and fire-fighting — 525
10.4: Fire evacuation — 535

Element 11: Electricity — 544

11.1: Hazards and risks — 544
11.2: Control measures — 552

INTRODUCTION

This National General Certificate in Occupational Health and Safety course book has been designed to support learner development. A variety of topics will be covered, of which risk assessment is central.

The work undertaken in recent decades by various UK, European and International professional, standards and trade bodies has provided a solid, logical foundation for the practice and teaching of occupational health and safety all around the world and the National General Certificate qualification runs in close parallel with its International equivalent.

As you will learn, the harmonisation of scientifically based exposure limits for noise and hazardous chemicals, and the worldwide adoption of the Plan, Do, Check, Act health and safety management system, have all contributed to our core body of workplace health and safety knowledge. Consistent use of the terminology 'Risk', 'Hazard', 'Danger', 'Practicable' and 'Reasonable' is now well established.

This introduction has been written by someone who has spent the last 40 years teaching and promoting workplace health and safety. I have never been on the staff of NEBOSH but have, over the years, done much work for them, and with them. My background is in materials research and, for me, workplace health and safety provides a rich combination of engineering and human behaviour. Over the years, the syllabus has seen the balance between these two evolve, partly to reflect the changing nature of workplaces. It is now widely recognised that issues such as bullying, exhaustion, lone working and mental ill-health may impact on workers just as much as the traditional issues such as noise, chemical hazards and falls.

The early 'pioneers' in occupational safety and health tended to come from industrial backgrounds, such as engineering, mining and so on. Their work in the 1950s onwards laid the foundations for the various professional bodies and the qualifications that we have today. I look back with great pride and affection having known many of these (very!) stubborn people who worked so hard to 'make things better' (that seems to have been the unspoken motto for all of them).

A huge thanks needs to go to those who have contributed to the making of this coursebook, especially Dr Luise Vassie, Tony Morriss and Dr Mark Cooper. The authors have worked together with NEBOSH in creating this content to make it as accurate and knowledge-inducing as possible.

I hope this course book plays a part in guiding you towards achieving your qualification.

Paul Randall

AUTHORS

Paul Randall

Paul Randall studied Materials at Imperial College in the 1960s, qualifying with a BSc and a PhD. Aside from time spent working in record shops, he has spent his working life teaching at schools and universities and through his own company. For the last 40 years he has been delighted to be working with occupational health and safety, a subject which has its roots in both human behaviour and engineering.

Dr Luise Vassie

Luise is a consultant and trustee/NED assisting organisations with governance, health and safety risk management, and assurance. Luise has extensive experience in developing the evidence base for health and safety practice and in assessment of health and safety management learning. Luise is a Chartered Fellow of the Institution of Occupational Safety and Health.

Tony Morriss

Tony has devised and presented H&S training courses for more than 25 years, including provision of NEBOSH Certificate and Diploma courses for further education in colleges. He has also provided a H&S consultancy service during this time to private companies and national organisations such as the British Printing Industries Federation. His qualifications are BSc; MSc; CMIOSH(ret): MIIRSM(ret).

Dr Mark Cooper

Mark has been a postgraduate Admissions Tutor at the Health and Safety Unit, Aston University, Birmingham. He also has extensive experience in providing foundation training for health and safety professionals from the UK and overseas.

Caroline Copson

Caroline Copson is Head of Assessment Development at NEBOSH. She has been involved in both qualification and assessment development for over 10 years. She gained the NEBOSH National Diploma in Occupational Health and Safety in 2012 and has been a Chartered member of IOSH since 2015.

Dr David Towlson

David initially trained as a physical chemist (that's applying physics to chemical systems), working in the industrial chemical process sector for large multi-nationals. His current role is to make sure NEBOSH's products (including assessments, qualifications, courses and publications) are designed, developed and maintained to be fit for purpose.

ELEMENT 1

WHY WE SHOULD MANAGE WORKPLACE HEALTH AND SAFETY

1.1: Morals and money

> **Syllabus outline**
> In this section, you will develop an awareness of the following:
> - Moral expectations of good standards of health and safety
> - The financial cost of incidents (insured and uninsured costs)

The reasons for managing health and safety in workplaces fall into two broad categories: moral and financial. Both categories can be used to describe the reasons for, or benefits of, managing health and safety, but equally can be used to explain the costs of getting it wrong.

Moral reasons are concerned with our judgement about what is right and wrong. This reflects society's view that it is wrong for anyone to be injured or made ill by their work and that there should be good standards of health and safety in all workplaces. In most countries, these expectations have led to the enactment of laws that regulate health and safety. This means that moral reasons are often split into societal and legal expectations.

Financial reasons are concerned with the costs of health and safety incidents and also the costs of trying to prevent them. The organisation must invest in health and safety to protect its workforce. This investment will obviously have a financial cost for the organisation but a lack of investment could have a bigger financial impact if things go wrong. After an incident, an organisation could incur both direct and indirect costs. For example, there could be fines from enforcement actions; in such cases the organisation would also be expected to put right the original cause of the accident. While investing in health and safety measures is a cost to the employer, it has the benefit of reducing the costs of failure and protecting the workforce. This means not investing in health and safety could be seen as a 'false economy'.

The individual worker and their family will also face costs, such as loss of income and medical bills. Additionally, wider society also faces costs resulting from, for example, dependency on the State's benefits system for people who are no longer able to work due to their injuries or ill-health.

Now we have seen the reasons for managing health and safety, we need to consider the scale of the issue, which is large.

Why we should manage workplace health and safety

Every day, across the world, people die because of occupational accidents or work-related diseases. Latest estimates are that there are more than 2.78 million deaths per year. Work-related diseases account for 2.4 million (86.3%) of the estimated deaths and fatal occupational accidents account for the remaining 380,000 (13.7%). Additionally, there are some 374 million non-fatal work-related injuries each year that have resulted in each injured person taking more than four days of absence from work. Respiratory diseases, circulatory diseases and cancers contributed more than three-quarters of the total work-related mortality.[1]

In Great Britain, the average annual cost of work-related injury and ill-health is £1.6 billion, with around two-thirds of this cost attributable to ill-health. Most costs fall on individuals, driven by human costs – loss of life and loss of quality of life – while employers and government/taxpayers carry a similar proportion of the remaining costs.[2]

Statistics on occupational accidents and illnesses are often incomplete because under-reporting is common and official reporting requirements frequently do not cover all categories of workers – those in the gig economy, for example. Other indicators, such as compensation data, disability pensions and absenteeism rates, could also be considered, although these too provide incomplete data. Hence the statistics do not truly reflect the scale of poor health and safety management.

In Great Britain, each year there are approximately 130 fatal accidents at work. However, this figure is a small proportion of the true toll of health and safety failure. For example, it does not include those who are killed in work-related traffic accidents (which are not separately recorded in Great Britain). It is estimated that between a quarter and a third of road fatalities involve someone driving for work. Those who die from diseases such as asbestos-related cancer, which claims over 5,000 lives per year in Great Britain, are also not included.

We all know that it is not just the injured person that is affected; it is their families, friends and work colleagues. Multiplying the 2.78 million cases by the number of people affected, we can readily see that the impact of poor health and safety standards goes far beyond the global estimates. From a financial perspective, the economic burden of poor occupational safety and health practices is estimated at 3.94% of global gross domestic product (GDP) each year.

1.1.1 Moral expectations of good standards of health and safety

Understandably, society takes the view that suffering as a result of poor health and safety standards is unacceptable and should be prevented whenever possible. Put simply, society expects that workers should leave work at the end of the day in the same condition as when they arrived to start their work – being injured or becoming ill as a result of your work or workplace is morally unacceptable.

Moral duties can be considered as underpinning legal requirements. The common law duty of care places duties on employers to take reasonable care of their workers. Moral standards are often enshrined in laws, so there is a considerable overlap between morals and legal obligations.

Society will expect to see organisations (or individuals) penalised for breaches of health and safety law – especially when a worker has been injured or killed as a result of obvious breaches.

The Health and Safety at Work etc Act 1974 places a general duty on an employer to ensure, so far as is reasonably practicable, the health, safety and welfare of all employees. This means providing a safe place of work, safe plant and equipment and safe systems of work and ensuring competent workers. There are also individual duties placed on employees to work safely. Any breach of the statutory duties can result in the employer being involved in enforcement action, which could result in criminal proceedings.

Legal obligations will be discussed in 1.2: The force of law – punishment and compensation.

1.1.2 Financial costs of incidents

The social and legal reasons for managing safety and health have already been briefly outlined. At the very least, an employer should have a sufficiently well-developed social conscience not to unnecessarily place workers at risk. However, what might really motivate employers to focus on health and safety is that accidents and ill-health cost an organisation money in terms of financial losses resulting from a failure to manage risk. Therefore, employers are often more likely to provide safe systems if these costs, and ways to avoid incurring them, are made clear.

Direct and indirect costs

There are two broad types of accident cost – 'direct' and 'indirect'. These may also be referred to as 'tangible' and 'intangible' costs.

Direct costs
Direct costs are those that come directly from an accident. They are measurable and quantifiable and include fines paid when convicted of a criminal offence, the cost of first-aid treatment, sick pay, payments for medical treatment, equipment repairs and replacement costs, the value of lost and damaged product, lost production time and overtime for workers who are having to cover the injured worker's job as well as their own.

Indirect costs
Indirect costs are those that arise indirectly because of an accident and cannot really be said to be a direct consequence of it. Nevertheless, they can be traced back to the event. So, if the event never occurred, these costs would not have been incurred. They are hard to quantify, not just because of the less direct relationship with the cause, but also because they may be incurred a considerable time after the event. Examples include workers being taken from their normal duties to take part in an investigation, loss of goodwill, reputational damage, a reduction in workforce morale that could lead to lower production rates, activation of penalty clauses for failing to meet delivery dates, cost of recruitment of replacement labour, increased insurance premiums and loss of experienced, competent workers that will result in recruitment and training costs for replacement workers.

Insured and uninsured costs

Some of the costs, whether direct or indirect, are insured; many are not. It is therefore important to understand the relationship between 'insured' and 'uninsured' costs (see Figure 1).

Insured costs

In some cases, it is possible to take out insurance cover that will pay out in the event of a loss, provided all the relevant terms of the contract of insurance have been met. Examples include coverage for equipment damage and medical costs.

Employers are required to take out compulsory employers' liability insurance, which ensures that there will be money available to meet compensation claims should they arise. If such insurance was not in place, workers might win their claim against the employer but find that the employer has no money to meet the judgment sum or cover legal costs. Personal injury costs are therefore 'insured'.

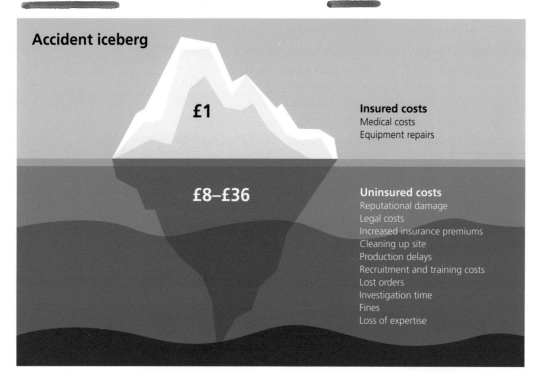

Figure 1: Insured and uninsured costs

Uninsured costs

In most cases, the employer will find that 'uninsured' costs exceed those for which an insurance claim may be made. This means that the organisation will lose money as these types of loss cannot be reclaimed through insurance. Uninsured costs can be more than 10 times greater than the insured ones.

Losses may be uninsured for two main reasons: either it is impossible to get insurance for them (for example the payment of a fine levied by a criminal court) or the employer has not taken out insurance cover for a specific loss. In addition, the small size of an individual loss may mean that the employer does not bother to make a claim, even if there is insurance in place that theoretically covers it. In any of these situations, the costs will have to be met straight from company profits.

Alternatively, the amount being claimed may be less than the policy excess amount, which is set by agreement with the insurer to reduce premiums. In such cases, the employer agrees to meet the cost of the first £x of a claim, beyond which the insurance takes over. This means that, if individual claims are less than the 'excess' amount, the employer will need to pay these claims and the insurer is not involved.

Examples of 'uninsured' costs include:

- time spent investigating accidents, ie lost production time due to workers being taken away from their normal jobs;
- fines from criminal courts;
- cost of overtime;
- recruitment costs;
- costs associated with downturns in morale that affect productivity; and
- costs associated with reputational damage, for example loss of orders if customers switch to buying goods and services from other suppliers with a better reputation.

Preventing workers being killed, injured or made ill by their work is not only morally the right thing to do but also makes good business sense: the direct and indirect costs associated with accidents and incidents are prevented and it can also mean the employer pays lower insurance premiums. However, investing in health and safety has additional benefits, such as raised productivity levels, increased staff motivation and improved reputation.

KEY POINTS

- Nearly 3 million people each year die from occupational accident and diseases costing approximately 4% of GDP. So, there are moral and financial drivers for managing health and safety.
- In Great Britain, the average annual costs of work-related injury and ill-health is £1.6 billion.
- The moral reasons can be split into societal and legal expectations.
- From a financial perspective, accident costs are direct and indirect.
- Insurance may cover some but not all the costs associated with accidents.
- Investing in health and safety is associated with raised productivity, increased worker motivation and improved reputation.

References

[1] P Hämäläinen, J Takala and TB Kiat, *Global Estimates of Occupational Accidents and Work-related Illnesses* (2017, Workplace Safety and Health Institute) (www.icohweb.org)

[2] HSE, 'Costs to Great Britain of workplace injuries and new cases of work-related ill health – 2018–19' (2022) (www.hse.gov.uk)

1.2: The force of law – punishment and compensation

Syllabus outline

In this section, you will develop an awareness of the following:

- Sources of law:
  - statute law: the legal status and relationships between Acts of Parliament, regulations, approved codes of practice, official guidance; absolute and qualified duties (practicable and reasonably practicable)
  - common law: precedents and case law; the importance of common law
  - relevance of statute and common law to criminal and civil law
- Types of law:
  - criminal law: offence against the State; prosecution to establish guilt; burden and onus of proof (see s40 Health and Safety at Work etc Act 1974)
  - civil law: private individual seeking compensation; burden of proof; statute-barred
- Criminal law liabilities:
  - role, functions and powers of: the Health and Safety Executive (HSE)/HSE for Northern Ireland (HSENI), Procurator Fiscal (Scotland) and local authorities
  - why fees for intervention (FFI) are charged (material breach of legislation)
  - powers of inspectors (see s20 of the Health and Safety at Work etc Act 1974)
  - enforcement notices (improvement, prohibition): conditions for serving; effects; procedures; rights and effects of appeal; penalties for failure to comply
  - simple cautions
  - prosecution: summary and indictable (solemn in Scotland) offences and relevant penalties (including the disqualification of directors)
  - the Corporate Manslaughter and Corporate Homicide Act 2007: the offence and available penalties and defences
- Civil law liabilities:
  - civil wrong (tort/delict)
  - tort/delict of negligence
  - duty of care (neighbour principle)
  - tests and defences for tort/delict of negligence: duty owed/duty breached/injury or damage sustained
  - contributory negligence
  - vicarious liability

> - the employer's legal duty to provide a safe place of work, safe plant and equipment, safe systems of work, training and supervision, and competent workers
> - breach of statutory duty in relation to new and expectant mothers.

The general legal system in the UK is a common law system. The early English legal system's judgments were based on custom and practice. Principles from judgments of more senior courts in earlier cases create binding precedents for future cases.

You should bear in mind that the legal system is complex. Here we will focus on the legal system as it applies to health and safety. While there are technically several nuances in aspects/terminology of the law, some have been grouped for simplicity.

The law relating to health and safety is a mix of criminal law set out in statutes and regulations and common law created by previous cases. Legal decisions are based on statute and common law. Legal precedents created by the courts can apply to criminal prosecutions and civil actions.

Legislation sets the boundaries within which organisations and society must operate. Health and safety legislation has evolved over many decades from the 19th century industrial revolution. Initially, laws were piecemeal and prescriptive. In 1974, the Health and Safety at Work etc Act was introduced, providing the principal legal framework for health and safety in UK workplaces.

Enforcement of health and safety is a criminal matter, for which there is punishment if found guilty, whereas the route to redress for injured parties is a separate civil matter, for which compensation may be awarded if the claimant is successful.

We will now expand on the sources and types of law, criminal and civil liabilities in relation to health and safety at work.

> **TIP**
>
> The main legislation relevant to health and safety is:
>
> - Health and Safety at Work etc Act 1974 (HASAWA).
>
> The Health and Safety Executive for Northern Ireland (HSENI) is the equivalent of the Health and Safety Executive (HSE).

1.2.1 Sources of law

There are two sources of law:

- legislation, which is written law in the form of Acts (statute law) and regulations etc; and
- common law, which is based on custom and previous decisions made by judges (known as case law).

Legislation

Legislation takes the form of:

- Acts of Parliament (statutes); and
- regulations (a type of Statutory Instrument).

Where there is a conflict, statute law takes precedence over common law. Statutes are sometimes produced to overcome difficulties that have arisen in common law.

The Occupiers' Liability Act 1957, for example, was partly introduced to overcome problems in common law regarding different duties of care that the occupier owed towards different types of lawful visitor.

The main differences between Acts and regulations concern the way in which they are introduced and made. There is a difference in their presentation in as much as Acts contain numbered sections whereas regulations contain numbered regulations.

Acts

Acts of Parliament are initiated by a Bill being introduced in either House of Parliament (Commons and Lords) with the intention of creating a new law or amending an existing law on a given issue. A lengthy process takes place involving debates and considerations in both Houses before eventually being presented for Royal Assent, at which point they pass into law.

Regulations

Acts often contain enabling clauses within them that delegate authority to government ministers to introduce regulations within the scope of the main Act. From the perspective of health and safety, the HASAWA is an example of such an enabling Act. Section 15 of the Act grants powers to the Secretary of State for Work and Pensions to introduce regulations under the Act that relate to health, safety and welfare at work. Legislation introduced in this way is sometimes termed subordinate, delegated or secondary. Although the power to introduce regulations is given to the Secretary of State, in practice health and safety regulations are usually proposed by the Health and Safety Executive (HSE).

Approved codes of practice

Whereas Acts and regulations are law, approved codes of practice (ACoPs) contain examples of good practice that have been approved by the HSE under the powers contained in s16 of HASAWA. ACoPs are usually published to accompany regulations and contain the technical and practical detail that is required to be followed to achieve compliance with the regulation. For example, the Control of Asbestos Regulations 2012 are accompanied by HSE Approved Code of Practice and Guidance L143.[1]

As ACoPs are not law, you cannot be prosecuted for being in breach of the requirements of an ACoP. You may use alternative methods to those set out in a Code to comply with the law. However, ACoPS have a special legal status (sometimes referred to as quasi-legal). If you are being prosecuted for being in breach of a regulation or a section of HASAWA, and it is proved that you did not follow the relevant provisions of the ACoP, you will need to show that you have complied with the law in some other way, or a court will find you at fault.

Guidance notes

Guidance notes (GNs) are usually produced by the HSE and contain best technical and practical advice on specific health and safety issues. They generally aim to:

- interpret Acts and regulations;
- help people to achieve compliance with legal requirements; and
- offer sound technical advice.

In addition to GNs, other materials containing good practical and technical advice are British and European Standards and codes of practice (often produced by organisations, trade federations and so on). Again, these are not legally binding.

Absolute and qualified duties

Duties can be either 'absolute', meaning they are mandatory, or 'qualified', in which case there is some scope for consideration. 'Practicable' and 'reasonably practicable' are examples of qualified duties. Whether a duty is absolute or qualified generally depends on the risk of harm and how important a control is for reducing the risk.

Absolute duties

An employer is expected to comply fully with an absolute duty whatever the cost or the difficultly. These duties will normally be set in legislation with the words 'must' and 'shall'.

For example, reg 3 of the Management of Health and Safety at Work Regulations 1999 requires that:

> Every employer shall make a suitable and sufficient assessment of:
> (a) the risks to the health and safety of his employees to which they are exposed while they are at work; and
> (b) the risks to the health and safety of persons not in his employment arising out of or in connection with the conduct by him of his undertaking,
> for the purpose of identifying the measures he needs to take to comply with the requirements and prohibitions imposed upon him by or under the relevant statutory provisions…

Qualified duties

Qualified duties, where the word 'shall' or 'must' is qualified by a phrase, follow a hierarchical structure from 'practicable' to 'reasonably practicable' and on to 'reasonable'. We will focus on the first two.

'So far as is practicable'

When a legal requirement introduces this phrase, then the extent to which an employer has to comply with this type of duty is limited by the current state of know-how, irrespective of the cost or difficulty. It is often used for controlling high risks with technical solutions that can evolve over time.

The judgement must be kept under review. What may be physically impossible to perform today may, with developments in technology, become physically possible to perform tomorrow. An example of a duty qualified by the phrase 'so far as is practicable' is reg 11 of the Provision and Use of Work Equipment Regulations 1998, which introduces a hierarchy of measures to prevent access to dangerous parts of machines, with each step in the hierarchy being qualified.

'So far as is reasonably practicable'

This qualification requires the employer to balance the costs of compliance with the duty against the risk. This can include time, effort and money. Risk will depend on the chance and severity of the harm that might occur from non-compliance. Where the risk is low, the costs can be balanced accordingly, and may also be low. For higher risks, the costs may be expected to be higher. If the costs greatly outweigh the risk, alternative lower-cost options should be considered.

In considering the costs involved, no allowance can be made for the size, nature and profitability of the business concerned. It is the risk that determines whether the cost involved is justified. These points were summarised in a leading case law decision in which the judge stated that:

> Reasonably practicable is a narrower term than physically possible and implies that a computation must be made in which the quantum of risk is placed in one scale and the sacrifice, whether in money, time or trouble involved in the measures necessary to avert the risk is placed in the other; and that, if it is shown that there is a gross disproportion between them, the risk being insignificant in relation to the sacrifice, the person upon whom the duty is laid discharges the burden of proving that compliance was not reasonably practicable. This computation should be made at a point of time before the incident complained of.
>
> *Edwards v National Coal Board*[2]

Most, but not all, of the HASAWA is qualified by the term 'so far as is reasonably practicable'.

Common law

Before the Norman Conquest, each area of the realm was ruled over by local lords and barons. These had their own established systems of law. Following the Norman Conquest, a unified system of law, known as the King's Justice, was imposed on the entire realm. Because this law was the same for everyone and for every region, it became known as the common law. It would have been physically impossible for the King to have heard and tried every case of alleged law-breaking. The Curia Regis were therefore appointed to administer the law on the King's behalf. Although the King's law was never written down, these itinerant justices, who dealt with the legal disputes, often met and recorded important decisions to guide future justices faced with similar cases. This helped both to establish a common law for all people and to establish legal continuity without undue rigidity.

This developed over time and, today, when making decisions, judges are guided by decisions made in earlier cases (known as judicial precedents). The doctrine of binding precedent, or *stare decisis*, means that courts must follow the earlier decisions of more senior courts.

Case law has an important role in both establishing and modifying the common law and in interpreting the meaning of terms, phrases and requirements found in statute law but not defined or clarified in the relevant Act or regulation.

1.2.2 Types of law

The law is divided into two branches: criminal and civil. The same wrongful act can lead both to a civil action and a criminal prosecution. For example, if an employer fails to guard a piece of machinery and this failure leads to the injury of an employee, then the employer could be prosecuted in the criminal courts by the enforcement authority and sued for damages in the civil courts by the injured employee. Yet, the wrongful act of failing to guard the machine is the same in both cases. What really establishes the difference between civil and criminal law is the purpose to which the law is put and the intended results.

> **TIP**
>
> It is not the wrongful act that distinguishes between civil and criminal law; rather it is the legal consequences that follow from the wrongful act.

Criminal law

Criminal law characterises certain kinds of wrongdoing as offences against the State, not necessarily violating the rights of individuals, and punishable by the State. Crime is an act of disobedience of this type of law, forbidden under pain of punishment.

Because criminal law involves guilt and punishment, the courts require that the defendant be found guilty beyond reasonable doubt. That is, the evidence must be such that the guilt has been established to a degree that is beyond that which a reasonable person could continue to doubt.

There is some overlap between the two types of law. If, for example, the employer we mentioned earlier was found guilty, beyond reasonable doubt, in the criminal courts for failing to guard a piece of equipment that injured an employee, it would be very difficult for that same employer to defend a civil action, on the balance of probabilities, taken by the injured employee for compensation for the loss suffered.

Usually, the presumption of innocence is a legal principle that underpins the criminal justice system. The prosecution brings the case against a defendant, and it is for the prosecution to prove their case. However, in the case of a prosecution under the HASAWA, a reverse burden of proof applies:

> In any proceedings for an offence under any of the relevant statutory provisions consisting of a failure to comply with a duty or requirement to do something so far as is practicable or so far as is reasonably practicable, or to use the best practicable means to do something, it shall be for the accused to prove (as the case may be) that it was not practicable or not reasonably practicable to do more than was in fact done to satisfy the duty or requirement, or that there was no better practicable means than was in fact used to satisfy the duty or requirement.
>
> HASAWA, s40

Put another way, an employer who has an employee who is injured in the workplace and, as a result, is charged with a criminal offence under the HASAWA, would be expected to prove to a criminal court that they did everything that was reasonably practicable to prevent that accident from happening.

Civil law

Civil law is concerned with the rights and duties of private parties towards each other. In many ways this is a formalisation of the moral duties we have to each other. Civil law includes:

- law of contract with its legal consequences, ie whether a promise is legally enforceable; and
- law of tort, which is a civil wrong independent of contract for which the remedy is an action for damages; examples of torts are:
  - nuisance;
  - negligence;
  - defamation; and
  - trespass.

Civil law involves liability for loss suffered rather than punishment; therefore the onus of proof demanded by the courts is less stringent. In civil law it is enough that the defendant is found liable for the loss on the 'balance of probabilities'.

That is to say that when the evidence from both parties is weighed in the balance, the evidence of the claimant (the person making the claim) is found to be more convincing than that of the defendant.

Although we have indicated earlier that a wrongful act may lead to both criminal and civil action, some statutes specifically disallow a civil action. For example, s47 of the HASAWA explicitly prevents civil action for a breach of the general duties of the Act (ss2–9); civil action is said to be 'statute-barred'. However, it goes on to say that duties imposed by regulations made under the Act will be actionable under civil law unless the regulations specify otherwise. In this regard, actions under the Management of Health and Safety at Work Regulations 1999 and parts of the Construction (Design and Management) Regulations 2015 are statute-barred with respect to 'third parties' (persons other than the employer or their employees).

Table 1 gives a summary of both criminal and civil law.

Criminal law:	Civil law:
• Action brought by the State • Intended result is punishment • Punishment is non-insurable • Action can be taken irrespective of whether loss has occurred • Mainly draws on statute law	• Action brought by a private party • Intended result is to compensate the injured party • Insurance can be obtained to cover liability for damages • Action can be taken only where loss has occurred • Mainly involves common law

Table 1: Overview of criminal and civil law

1.2.3 Criminal law liabilities

Role, functions and powers of enforcement agencies

Enforcement of health and safety at work legislation is governed by a series of measures, the principal of which is the HASAWA. Both the Act and regulations made under the Act can provide for enforcement. Enforcement of health and safety at work legislation is the responsibility of various authorities, including the Secretary of State, HSE in England and Wales, its counterpart in Northern Ireland (HSENI), the Crown Office and Procurator Fiscal Service (COPFS) in Scotland and local authorities.

The HSE, for example, will decide if new legislation will be required and will target industries and activities that HSE needs to concentrate on. In addition, the HSE has the power to produce and issue approved codes of practice. To perform its enforcement role, the HSE appoints specialist and general inspectors to work on its behalf. These inspectors are granted certain rights under the HASAWA; inspectors are issued with warrant cards by the HSE. In addition, local authority environmental health inspectors carry out certain roles of behalf of the HSE. Inspectors inspect work activities and investigate incidents and complaints. In general, HSE inspectors enforce industrial premises and fairgrounds, whereas local authority inspectors enforce commercial premises, retail premises and food premises. This division of enforcement responsibility is provided for in s18 of the Act.

Northern Ireland

In Northern Ireland, the Health and Safety Executive for Northern Ireland (HSENI) performs the same role as HSE in England and Wales.

Scotland

In Scotland, HSE is the regulator but health and safety cases and most fatal accident inquiries relating to work-related deaths, are dealt with by a dedicated team of Procurators Fiscal and support staff, including victim information and advice (VIA) officers, called the Health and Safety Investigation Unit.

Procurators Fiscal are civil servants, who are qualified as solicitors, solicitor-advocates or advocates and are independent prosecutors, constitutionally responsible to the Lord Advocate. They receive and consider reports from the police and other agencies and decide whether to raise criminal proceedings in the public interest. The Procurators Fiscal (and Procurators Fiscal Depute) prosecute all criminal cases in the sheriff courts.

In addition, the COPFS has a responsibility to investigate all sudden, suspicious and unexplained deaths in Scotland, including all deaths resulting from an accident in the course of employment or occupation. This results in a Fatal Accident Inquiry unless the death results from natural causes.

Fees for Intervention (FFI)

Under the Health and Safety (Fees) Regulations 2012, if, when visiting a business, HSE inspectors see 'material breaches' of the law, the business or organisation will have to pay a fee. The fee is based on the amount of time that the inspector has had to spend identifying the material breach, helping the business to put it right, investigating and taking enforcement action. The Regulations were introduced to shift the costs of health and safety failure from the public to the businesses and organisations that breach health and safety laws.

A material breach is when, in the opinion of the HSE inspector, there is or has been a contravention of health and safety law that requires them to issue notice in writing of that opinion to the duty holder.

To decide whether a duty holder is in material breach of the law, HSE inspectors must use the FFI guidance together with the principles of HSE's enforcement decision-making frameworks, the Enforcement Management Model (EMM) and the Enforcement Policy Statement (EPS).[3]

Under FFI, HSE will only recover the costs of its regulatory work from duty holders who are found to be in material breach of health and safety law. Duty holders who are compliant with the law, or where a breach is not material, will not be charged FFI for any work that HSE does with them.[4]

Powers of inspectors

The appointment of inspectors is governed by s19 of the HASAWA. The powers of inspectors are contained in s20; these can briefly be outlined as follows:

- to enter premises at any time they deem to be reasonable;
- to take along a police constable if they believe they will be obstructed;
- to take along another person and any equipment (perhaps a specialist inspector);
- to examine and investigate;
- to direct that premises and anything within them be left undisturbed;
- to take measurements and photographs and make recordings;
- to take samples;
- to order the dismantling or testing of any article or substance that appears to have caused danger (only damaging or destroying it where necessary);

- to take possession of and detain an article or substance (for examination, to ensure it is not tampered with and to ensure it is available in evidence for any proceedings);
- to require any person to answer questions and to sign a declaration of the truth of the answers given;
- to require the production of books and documents for the purpose of inspection and copying;
- to require any necessary facilities and assistance; and
- any other powers that are necessary.

Under s25 of the Act, an inspector has the power to render harmless an article or substance that is a cause of imminent danger of serious personal injury.

Enforcement notices

There are two types of enforcement notice that can be issued: improvement notices and prohibition notices. Each notice has strict criteria related to its issue that the inspector must conform to; the reason for the issue of the notice is completed on the notice itself.

Improvement notices (s21 of the HASAWA):

- These may be issued for contravention of a relevant statutory provision at the time of issue.
- They may also be issued for a contravention in the past that is likely to be continued or repeated.
- They take effect on a specified date but no earlier than 21 days after issue.
- Organisations have the right to make an appeal against the issue of the notice; this must be made to an employment tribunal within 21 days of the issue of the notice.
- The notice is suspended while the appeal is being heard.

Prohibition notices (s22 of the HASAWA):

- These are issued if there is a risk of serious personal injury or illness.
- They take effect immediately; for example they can result in cessation of working in an area or removal of a machine or equipment from operation.
- Organisations have the right to make an appeal against the issue of the notice; this must be made to an employment tribunal within 21 days of the issue of the notice.
- The notice stays in effect during the appeal.

The maximum sentences for failure to comply with either of these notices are a fine and/or 12 months' imprisonment in a magistrate's court[5] or an unlimited fine and/or two years' imprisonment in the Crown Court.

Simple cautions

HSE inspectors may, in certain circumstances, issue a simple caution to a duty holder. It is a formal warning that may be given to persons aged 18 or over who admit to committing an offence. The simple caution scheme is designed to provide a means of dealing with offending without a prosecution when there is evidence of an offence but the public interest does not require a prosecution. However, it is unusual for an HSE inspector to take this course of action and there will need to be exceptional circumstances, specifically related to the proposed defendant, that outweigh the general public interest factors.

A simple caution is defined in the HSE Enforcement Policy Statement as:

> a statement by an Inspector, that is accepted in writing by the duty holder, that the duty holder has committed an offence for which there is a realistic prospect of conviction. A simple caution may only be used where a prosecution could be properly brought.

The issue of a simple caution depends on the alleged offender agreeing to accept the caution. If an offender refuses to accept the caution, a prosecution should normally follow. Also, where there is repetition of a breach that was the subject of a simple caution, HSE will normally treat this in the same way as a failure to comply with an enforcement notice: with criminal proceedings.

Prosecution

In England and Wales, most health and safety offences are known as 'hybrid offences' and are 'triable either way', which means summarily before a magistrate (or magistrates) or on indictment before a judge and jury in the Crown Court. The maximum penalties for nearly all offences are, in a magistrate's court, an unlimited fine and 12 months' imprisonment, and in the Crown Court, an unlimited fine and/or up to two years' imprisonment.

TIP

The Health and Safety (Offences) Act 2008 increased maximum penalties for some offences committed after 16 January 2009 (by increasing the maximum fine and introducing imprisonment for certain offences) and s85 of the Legal Aid, Sentencing and Punishment of Offenders Act 2012 had the effect of increasing the level of most fines available for magistrates' courts to an unlimited fine for offences committed after 12 March 2015. For offences committed after 2 May 2022 the maximum term of imprisonment that can be imposed in the magistrates' courts rose to 12 months.

The offences of obstructing an inspector or impersonating an inspector are summary offences only (ie they can be heard only by a magistrate's court) and the penalties are unlimited fines.

Further details on penalties can be found on the HSE website[6] and in Schedule 3A of the HASAWA.[7] While the tables on the HSE website are more accessible, there may be a delay before changes to legislation are included. Always check the official legislation website for updates.

Scotland

In Scotland, offences are triable through two routes – summarily or on indictment. The latter is referred to as 'Solemn'. For offences committed after 16 January 2009, the penalties are set out in the Health and Safety (Offences) Act 2008. The maximum penalties for nearly all offences are, for summary conviction, £20,000 and/or 12 months' imprisonment and, for Solemn, an unlimited fine and/or two years' imprisonment. The offences of obstructing an inspector and impersonating an inspector, are summary offences only and the maximum penalties are £5,000 and £5,000 and/or 12 months' imprisonment respectively.

Disqualification of directors

A disqualification order may be an appropriate penalty in the case of directors or managers convicted of an offence under s7 or s37 of the HASAWA when the offence relates to the management of a company. Such offences are indictable and may call into question that person's suitability to fulfil the role of director.

Disqualified persons cannot be a director, liquidator or administrator of a company, or the manager of company property, or in any way, directly or indirectly, be involved in the promotion of, setting up or running a company for a specified period. The maximum period of disqualification is five years in a magistrate's court and 15 years in the Crown Court.

APPLICATION 1

> An organisation's worker is injured when their arm is 'pulled into' moving parts of a machine; the doctors advise that the arm cannot be saved and has to be amputated. The machine should have been guarded but the supervisor was instructed by the organisation's management to remove the guard to speed up production. HSE visits the premises to investigate the accident; the organisation's management try to stop the visit.
>
> What actions can the HSE inspector take before and during their visit?
>
> What are the outcomes of the HSE's visit likely to be?
>
> A suggested answer is provided at the end of 1.2: The force of law – punishment and compensation

The Corporate Manslaughter and Corporate Homicide Act 2007

Under the Corporate Manslaughter and Corporate Homicide Act 2007 an organisation can be convicted of an offence when a gross failure in the way activities were managed or organised results in a person's death.

In England, Wales and Northern Ireland the offence is called corporate manslaughter. It is called corporate homicide in Scotland. Courts look at management systems and practices across the organisation, providing a more effective means for prosecuting the worst corporate failures to manage health and safety properly.

The Act applies to all companies and other corporate bodies operating in the UK in the private, public and third sectors. It also applies to partnerships (and to trade unions and employers' associations) if they are an employer, as well as to government departments and police forces. Specifically, the Act applies to public bodies incorporated by statute such as local authorities, NHS bodies and a wide range of non-departmental public bodies.

The offence

An organisation can be found guilty of the offence if the way in which its activities are managed or organised causes a death and amounts to a gross breach of a duty of care to the deceased.

Juries consider how the fatal activity was managed or organised throughout the organisation, including any systems and processes for managing safety and how these were operated in practice. Juries also consider the attitudes towards safety that exist in the organisation, so issues of culture can be a legitimate consideration.

A substantial part of the failure within the organisation must have been at a senior level. Senior level means the people who make significant decisions about the organisation or substantial parts of it. This includes both those in centralised, headquarters functions and those in senior operational management roles.

1.2 The force of law – punishment and compensation

> **CASE LAW**
>
> In the case of *R v Lion Steel Equipment Ltd*[8], a general maintenance worker tragically died when they stepped on a fragile roof light and fell 13 metres to the floor below. They had gone onto the roof to locate the source of a leak at their employer's premises in Greater Manchester. Following a lengthy investigation, the company was charged with corporate manslaughter.
>
> A quote had been obtained for the repair of the roof that was considered by the management board to be unacceptably expensive and so they agreed to do the work in-house. The maintenance worker was asked to do the job by the works manager (also a director of the company). They went onto the roof, which was wet, without any safety mechanism and, while on the roof, fell through a fibreglass roof light to the floor below, suffering fatal injuries.
>
> Lion Steel was fined £480,000 which was to be paid over four years.

'Gross breach' and 'duty of care'

To establish a 'gross breach' the organisation's conduct must have fallen far below what could have been reasonably expected. Juries consider any health and safety breaches by the organisation and how serious and dangerous those failures were. A 'duty of care' exists, for example, in respect of the systems of work and equipment used by employees, the condition of worksites and other premises occupied by an organisation and in relation to products or services supplied to customers. The Act did not create new duties of care – these were already owed in the civil law of negligence and the offence is based on these.

> **APPLICATION 2**
>
> Two years after the amputation incident outlined in Application 1, the machine is still in use. The organisation has also undergone a change of management. The management asked if the guard was necessary. The supervisor explained about the previous incident. The management were not concerned about this and wanted production rates increased. The machine guard was again removed. Workers started reporting minor injuries caused by the moving parts but were told to 'get on with the job'. Six months later, a worker is killed when their arm is drawn into the machine. By the time they are released from the machine the worker had lost a lot of blood and surgeons are unable to save the worker's life.
>
> Do you think that there has been a gross breach of duty of care by the organisation?
>
> Do you think that a charge of 'corporate homicide' could be brought against the organisation?
>
> In both cases explain your reasoning.
>
> A suggested answer is provided a the end of 1.2: The force of law – punishment and compensation

Penalties

An organisation guilty of the offence will be liable to an unlimited fine. The Act also provides for courts to impose a publicity order, requiring the organisation to publicise details of its conviction and fine. Courts may also require an organisation to take steps to address the failures behind the death (a remedial order).

Investigation and enforcement

The police lead an investigation if a criminal offence (other than under health and safety law) is suspected. They work in partnership with the HSE, local authority or other regulatory authority.

Cases of corporate manslaughter/homicide following a death at work are much less common than other health and safety offences, as the offence covers only the worst instances of failure across an organisation to manage health and safety properly.

Cases of corporate manslaughter are prosecuted by the Crown Prosecution Service in England and Wales and corporate homicide cases will be prosecuted by the Crown Office and Procurator Fiscal Service in Scotland.

Health and safety charges may be brought at the same time as a prosecution for this offence, as well as in cases where it is not prosecuted.

Individual liability

Under s18 of the Act there is no secondary liability, meaning individuals cannot be found guilty of aiding, abetting, counselling or procuring the commission of the offence of corporate manslaughter. However, the Act does not preclude individuals being prosecuted for primary liability offences, including gross negligence manslaughter (which remains unaffected by the Act so far as it concerns individuals) and/or breaching s7 or s37 of the HASAWA. It is possible that, for the prosecution to prove failure at senior management level, individual directors and managers can find themselves more of a target for personal prosecution.

1.2.4 Civil law liabilities

A tort (delict in Scotland) is a civil wrong that causes a claimant to suffer loss or harm, resulting in legal liability for the person who commits the tortious act. In the case of harm resulting from occupational injury or ill-health, the claimant may pursue a civil action under the *tort of* negligence. For example, if a person suffers an injury from slipping and falling on a cooking oil spillage in a supermarket, they may pursue civil action claiming that the supermarket has acted negligently in not managing the risk and is legally liable.

Negligence means failing to act in a way a reasonable person would have acted under the circumstances or acting in a way a reasonable person would not have acted in the circumstances. In other words, it involves both acts and omissions.

Duty of care (neighbour principle)

This sets out an obligation to take reasonable care to avoid acts or omissions that foreseeably could injure your neighbour. It was developed by Lord Atkin in the case of *Donoghue v Stevenson*:[9]

> You must take reasonable care to avoid acts or omissions which you can reasonably foresee would be likely to injure your neighbour. Who, then, in law, is my neighbour? The answer seems to be persons who are so closely and directly affected by my act that I ought reasonably to have them in contemplation as being affected when I am directing my mind to the acts or omissions which are called in question.

The neighbour principle allows for claims in negligence from injured parties by identifying the class of people to whom a duty may be owed in any given scenario. That class of people includes those who are close enough to be directly affected by the allegedly negligent act and close enough that the alleged 'tortfeasor' (person who commits a tort) should have had their interests in contemplation when acting as they did. A tortfeasor will not be held to owe a duty of care to those who are not close enough to be in their contemplation when the tortious act or omission occurred.

Tests

To be successful in an action under the tort of negligence, the claimant must establish three things:

- that the defendant owed the claimant a duty of care;
- that the duty of care was breached through the negligent acts of the defendant; and
- that loss resulted from the breach of the duty of care.

Once these things have been established, the claimant must also demonstrate that the damage suffered was causally connected with the breach and not too remote from it. This means that the type of damage had to be foreseeable. For example, a premises carries out maintenance on a tree-lined boundary, causing tree debris to fall into the yard of the neighbouring premises. During the hours of darkness, the neighbour goes into the yard and trips over the fallen debris, hitting their head on some stone steps. It could be argued that the action of cutting the trees was not too remote (ie proximate) to the injury.

Under the Limitation Act 1980, action must be initiated within three years from either the date on which the cause of action accrued or the date of knowledge (if later) of the person injured.

Defences

Defences against negligence rely on disproving one of the three elements:

- the defendant did not owe the claimant a duty of care;
- the duty of care was not breached (the defendant had taken reasonable care/the loss was not foreseeable/it was an 'act of God'); or
- the breach of the duty of care did not give rise to the injury or loss.

The defendant may also argue that the proceedings were not brought within the specified time limit.

Consent

A final defence that can be used involves the voluntary assumption of risk of injury (*volenti non fit injuria*). Roughly translated this means that those who volunteer to place themselves at risk cannot then claim compensation if they are injured. The courts are extremely reluctant to accept this defence and, even where this defence is accepted, the person held to have volunteered must have done so, without coercion, in respect of a specific risk of which they had full knowledge. If this defence is accepted, it provides a full defence to the claim of negligence and no damages will be paid.

Contributory negligence

The Law Reform (Contributory Negligence) Act 1945 provides that when injury is partly the fault of the injured person, the court must decide how much the injured person is to blame. The payment of damages is then reduced by a percentage to reflect this. One of the persons involved may be held to be the worker who is taking an action against their employer following an accident or case of ill-health at work.

Why we should manage workplace health and safety

The worker may be apportioned some blame if, for example, they disobeyed a safety notice or did not use some personal protective equipment (PPE) that had been supplied.

Contributory negligence is not strictly a defence against a civil action, for, if a defence succeeds, the claim is lost completely. Rather, it is used when the defendant is held liable, with the intention of shifting some, but not all, of the liability onto the claimant to reduce the level of damages awarded. Hence it is often regarded as a 'partial defence'.

> **CASE LAW**
>
> In the case of *Jones v Livox Quarries Ltd*[10], the claimant was an employee of the defendants. They were going from work to the canteen for lunch. Contrary to instructions, they jumped on the back of one of the defendants' pieces of plant equipment to get a ride to the canteen. The driver was unaware of the fact that the claimant was standing on the tow bar of their vehicle and, when they made a sharp turn, the dumper behind them ran into the plant. As a result, the claimant was injured. Consequently, the claimant claimed damages for personal injuries. The defendants alleged there was contributory negligence by the claimant.
>
> The Court of Appeal dismissed both a cross-appeal by the defendants that the dumper driver was responsible and an appeal by the claimant. The Court held that "if a man carelessly rides on a vehicle in a dangerous position, and subsequently there is a collision in which his injuries are made worse by reason of his position than they would otherwise have been, then his damage is partly the result of his own fault, and the damages recoverable by him fall to be reduced accordingly".
>
> If a person exposes themselves to an unreasonable risks, they cannot then claim that their act is not an actionable cause to produce the damage.

Vicarious liability

Vicarious liability means that one party is liable for the acts of another party where loss or injury has occurred. You might think that this is an unusual situation but, in fact, employers usually take on the liability for losses caused by their employees even when those employees are acting in a negligent way. A key point is whether the employee's negligent acts or omissions were in the course of their employment. The following two examples illustrate this point.

> **CASE LAW**
>
> In the case of *Rose v Plenty*[11], a boy was injured on a milk float and the dairy company was held vicariously liable for the injuries suffered by the child (who was helping to deliver the milk) due to the negligence of the milkman. Even though the company had expressly forbidden the driver, their employee, to 'employ' children to deliver milk and collect empties, it was held that the dairy had received benefit – more milk sold – and therefore the milkman was acting 'in the course of employment' when the accident occurred.

> **CASE LAW**
>
> The case of *Wm Morrisons Supermarket plc v Various Claimants*[12] is not a health and safety case, but it sets out when an employee can be said to be acting in the course of their employment.

> A copy of the payroll data was given to an internal auditor at Morrisons to pass on to external auditors. The internal auditor also made the payroll data publicly available to cause harm to Morrisons as a vendetta against disciplinary proceedings that had been brought against them months prior.
>
> Morrisons' employees whose data had been disclosed brought claims against Morrisons based on vicarious liability for the internal auditor's acts. These claims were initially successful but the Supreme Court allowed Morrisons' appeal on the basis that the public disclosure could not be regarded as being so closely connected with the task of passing the data to the external auditors that it could be considered to be in the ordinary course of the internal auditor's employment.

The key points here are that:

- The employee's motive is relevant to deciding whether their wrongdoing was for the employer's business.
- A wrongdoing motivated by a desire to harm an employer is not within the scope of employment.

Insurers usually take on the liability of employers in such situations. Although civil liability can be passed on in this way, criminal liability, in the form of guilt, stays with the guilty party. Therefore, vicarious liability has become the principal ground for actions by employees against employers since injury-causing accidents rarely arise through the personal negligence of employers but rather through the negligence of others.

Employer's legal duty of care

Duties of care are established wherever a legal relationship exists between the parties involved. It is the legal relationship that establishes the relevant duty of care.

For example:

- the master–servant (or employer–employee) relationship;
- the manufacturer/supplier–customer relationship; or
- the occupier–visitor relationship.

Wilsons and Clyde Coal Co Ltd v English[13] identified the four main elements of the duty of care an employer owes their employees:

- the provision and maintenance of a safe place of work;
- the provision and maintenance of a safe system of work;
- the provision and maintenance of safe plant and appliances; and
- the provision of competent fellow employees.

The duty of care extends to the individual (a person's unique characteristics) and not towards what we might call 'the average person'. If an employer knows, or if a reasonable employer could have reasonably foreseen, that an individual employee is at greater risk than an average employee, then the employer's duty of care towards that individual employee is correspondingly greater. Employees at greater risk include those who:

- have disabilities or physical or mental health conditions;
- are inexperienced;
- are young;
- are, or might become, pregnant; or
- are experiencing stress.

Was the injury foreseeable?

It should always be kept firmly in mind that the existence of a duty of care does not automatically mean that that person is always liable for any losses that may occur. All that is required is that the person owing the duty of care does all that a reasonable person would do in the circumstances to satisfy that duty of care. A key test is: 'Was the injury suffered reasonably foreseeable?' Where the answer is 'yes', the next question would be: 'What would a reasonable person have done to prevent this injury?'

> **CASE LAW**
>
> In the case of *Paris v Stepney Borough Council*[14], the claimant had suffered damage to one of his eyes in war. He was employed in a garage, but was not provided with safety goggles while working with dangerous equipment. As a result, he was blinded when a piece of metal hit him in his undamaged eye. He argued that the defendant breached their duty of care to him as they failed to act as a reasonable person would in their position.
>
> The defendant argued that it was not normal practice to provide (normally sighted) employees with safety goggles, and so they were under no obligation to provide them to the claimant.
>
> The court noted that the duty the defendant owed was to the particular employee (with all his known characteristics), not to a hypothetical 'reasonable' employee.
>
> Because the claimant had sight only in one eye, there was a high likelihood that the harm would be significant; more so than would be inflicted on a normally sighted person. This meant that a reasonable person would take greater steps than usual to protect him.

Within the course of employment

Regarding the master–servant relationship, the employer's duty of care only extends to cover employees acting within the course of their employment and as a reasonable employee would act. Travelling to and from work is not normally held to be within the course of employment (although it may be if you are paid travelling expenses from home rather than from your workplace), whereas travelling between sites on company business will normally be held to be part of the course of your employment. Using the company's equipment to repair parts for your car during the lunch break will not normally be regarded as part of the course of your employment, unless your employer is aware that you are doing it and has condoned your actions by not taking precautions to stop them.

The test of foreseeability is crucial. The court will not judge whether the employer did foresee the likelihood of injury, but rather whether a reasonable employer would have foreseen such a likelihood and, if so, what actions such a reasonable employer would have taken to prevent the likelihood from becoming a reality.

Responsibilities of employees

The duty of care created by the master–servant relationship operates in both directions. As well as placing a duty of care on employers to act as a reasonable employer regarding the health and safety of their employees, it also places a duty of care on employees to act as a reasonable employee would act and not to cause their employer loss.

> **CASE LAW**
>
> This was established in the case of *Lister v Romford Ice and Cold Storage Co Ltd*[15], where a father and son were working together. The son was a lorry driver employed by the company to drive their lorry to a slaughterhouse to collect waste. During the journey the son reversed the lorry into his father, injuring him. The father successfully made a claim against the employer for the injuries caused by his son's driving. The employer's insurers then made a successful claim against the son, holding him responsible for their losses, arguing he breached an implied term in his service contract that he would use reasonable skill and care during the course of driving.

Often statutory obligations are imposed in addition to, or instead of, the common law duty of care. The common law duty of care is also captured in the statutory duties in s2(2) of the HASAWA, which requires the employer, so far as is reasonably practicable, to:

- provide and maintain safe plant and safe systems of work;
- make arrangements for ensuring safe means of handling, use, storage and transport of articles and substances;
- provide information, instruction, training and supervision (competent employees);
- provide a safe place of work;
- provide and maintain safe access to and egress from that workplace; and
- provide and maintain a safe working environment and adequate welfare facilities.

We will consider the main section of the HASAWA 1974 in 1.3: The most important legal duties for employers and workers.

Breach of statutory duty in relation to new and expectant mothers

New or expectant mothers can bring a claim for damages for injuries caused by breaches of regs 16 or 17 of the Management of Health and Safety at Work Regulations 1999.

Regulation 16 requires the risk assessment completed by an employer to include any risks to workers of childbearing age who could become pregnant, and any risks to new and expectant mothers. Regulation 17 states that, where a new or expectant mother works nights, but provides a medical certificate from her GP or midwife saying night shifts will affect their health, their employer must suspend them from work on full pay for as long as necessary.

Under the Employment Rights Act 1996, where appropriate, suitable alternative work should be offered on the same terms and conditions before any suspension from work is considered.

A claim brought in this way uses an alternative route to the tort of negligence and is referred to as a tort of breach of statutory duty. This route is only available if the regulations allow for it (see earlier section on civil law in 1.2.2: Types of law). As the statutory duty set out in regs 16 and 17 is very specific, it may provide a more straightforward course of action for the people concerned than the general duty of care.

Why we should manage workplace health and safety

A breach of a statute is not sufficient on its own to prove the tort/delict of breach of statutory duty in a civil claim. The following conditions must be fulfilled:

- the statute must apply to the claimant, ie an employee can only sue if the statute applies to employees;
- the statute must have been designed to prevent the type of injury incurred by the claimant;
- the statutory duty must have been breached; and
- the injury must have been caused by the breach of statutory duty.

KEY POINTS

Sources of law

- There are two sources of law: legislation and common law.
- Legislation consists of written law in the form of Acts of Parliament and regulations.
- Duties can be either 'absolute', in which case they are mandatory, or 'qualified', in which case there is some scope for consideration.
- Common law is based on custom and decisions made in previous decisions of the courts, also known as case law.
- Legislation and common law provide a body of law that can be used in both criminal and civil cases.

Types of law

- Criminal law is concerned with offences against the State.
- Criminal law involves guilt and punishment; the courts require that the defendant be found guilty beyond reasonable doubt.
- Civil law is concerned with the rights and duties of private parties towards each other.
- Civil law involves liability for loss suffered; therefore the onus of proof demanded by the courts is that the defendant is found liable for the loss on the balance of probabilities.

Criminal law liabilities

- The Health and Safety at Work etc Act 1974 (HASAWA) and regulations made under the Act provide for enforcement of health and safety.
- Various authorities including the Secretary of State for Work and Pensions, the Health and Safety Executive (HSE) in England and Wales, its counterpart in Northern Ireland (HSENI), the Crown Office and Procurator Fiscal Service (COPFS) in Scotland and local authorities are responsible for enforcing health and safety.
- Under the Health and Safety (Fees) Regulation 2012, when an HSE inspector sees material breaches of the law, a fee can be charged based on the amount of time that the inspector has had to spend identifying the material breach, helping businesses to put it right, investigating and taking enforcement action.
- HSE inspectors have a range of powers available to them, which are detailed under s20 of the HASAWA.

- HSE inspectors may issue two types of enforcement notice: improvement and prohibition. Both can be appealed within 21 days of issue by an appeal to an employment tribunal. Prohibition notices stay in force during the appeal period but improvement notices are suspended.
- Improvement notices are issued where there is a contravention of a statutory requirement.
- Prohibition notices are issued where there is serious risk of personal injury or illness.
- An inspector may issue a simple caution where there is evidence of an offence but the public interest does not require a prosecution.
- The maximum penalties for nearly all HASAWA offences are: in a magistrate's court, an unlimited fine and 12 months' imprisonment (or six months for offences committed before 2 May 2022); and, in the Crown Court, an unlimited fine and/or up to two years' imprisonment.
- In Scotland offences are triable through two routes – summarily or on indictment (Solemn). The maximum penalties for nearly all offences are: for summary conviction, £20,000 and/or 12 months' imprisonment; and, for Solemn, an unlimited fine and/or two years' imprisonment.
- Disqualification orders may also be issued to directors. The maximum period of disqualification is five years in a magistrate's court and 15 years in the Crown Court.
- The Corporate Manslaughter and Corporate Homicide Act 2007 sets out an offence for convicting an organisation where a gross failure in the way activities were managed or organised resulted in a person's death. A substantial part of the failure within the organisation must have been at a senior level.
- The penalties are unlimited fines and, potentially, publicity orders and remedial orders.

Civil law liabilities
- A tort (delict in Scotland) is a civil wrong independent of contract that causes a claimant to suffer loss or harm, resulting in legal liability for the person who commits the tortious act.
- A claimant may pursue a civil action under the tort of negligence when harm results from occupational injury or ill-health.
- The neighbour principle allows for claims in negligence for injured parties by identifying the class of people to whom a duty may be owed in any given scenario.
- To be successful in an action under the tort of negligence, the following elements must be satisfied:
  - that the defendant owed the claimant a duty of care;
  - that the duty of care was breached through the negligent acts of the defendant; and
  - that loss resulted from the breach of the duty of care.
- Action must be initiated within three years from either the date on which the cause of action accrued, or the date of knowledge (if later) of the person injured. The damage must also be foreseeable.
- Defences against negligence rely on disproving the three elements.

- Contributory negligence is used when a defendant is held liable, with the intention of shifting some, but not all, of the liability onto the claimant to reduce the level of damages awarded.
- Vicarious liability means that one party is liable for the acts of another party when loss or injury has occurred.
- Under the duty of care, an employer owes their employees:
  - provision and maintenance of a safe place of work;
  - provision and maintenance of a safe system of work;
  - provision and maintenance of safe plant and appliances; and
  - provision of competent fellow employees.
- The person owing the duty of care must do all that a reasonable person would do in the circumstances to satisfy the duty.
- Foreseeability is important – a court would consider whether a reasonable employer would have foreseen likelihood of injury and, if so, what actions such a reasonable employer would have taken to prevent the likelihood from becoming a reality.
- The duty of care operates in both directions. It also places a duty of care on employees to act as a reasonable employee would act and not to cause their employer loss.
- New or expectant mothers can bring a claim for damages caused by breaches of regs 16 or 17 of the Management of Health and Safety at Work Regulations 1999.

APPLICATION 1: ANSWER

The inspector could arrange for a police officer to accompany them if they feel they are going to be obstructed by the company. They could also arrange for an inspector with specialist knowledge about the machine in question to join them. The inspector is likely to request that the machine and surrounding area are left undisturbed. The inspector will take photos and measurements, request documents, such as the risk assessment for the machine and training records of the worker involved, and ask questions of those involved, which they will record.

The inspector will consider the Enforcement Management Model (EMM) to help them decide what action to take and whether to serve an enforcement notice.

APPLICATION 2: ANSWER

There appears to have been a gross breach of duty of care by the organisation. There has been wilful disregard by the senior management for the health and safety measures in place. They have instructed workers to remove a safety device designed to protect workers to speed up production.

A charge of 'corporate homicide' could be brought against the organisation because a gross failure in the way the company managed health and safety has resulted in the death of a worker.

References

[1] HSE, *Managing and working with asbestos. Control of Asbestos Regulations 2012. Approved Code of Practice and guidance* (L143, 2nd edition, 2013) (www.hse.gov.uk)

[2] *Edwards v National Coal Board* [1949] 1 All ER 743 (CA)

[3] HSE, 'Enforcement Policy Statement' (2015) (www.hse.gov.uk); HSE, 'The Enforcement Management Model' (EMM) (2013) (www.hse.gov.uk)

[4] HSE, *Guidance on the application of Fee for Intervention (FFI)* (HSE 47, 2012, amended 2022) (www.gov.uk)

[5] The maximum fine that can be imposed differs between England and Wales and Scotland. For further information, see HASAWA, Schedule 3A and HSE, 'Enforcement guide (England and Wales): Court Stage. Model examples' (www.hse.gov.uk); HSE, 'Enforcement guide (Scotland): Penalties' (www.hse.gov.uk)

[6] HSE, 'Enforcement guide (England and Wales)' and 'Enforcement Guide (Scotland)'. See note 5

[7] Health and Safety at Work etc Act 1974, Schedule 3A (www.legislation.gov.uk)

[8] *R v Lion Steel Equipment Ltd*, sentencing remarks of Judge Gilbart QC, 20 July 2012 (www.judiciary.uk/judgments/r-v-steel-equip-ltd-sentencing-remarks/)

[9] *Donoghue v Stevenson* [1932] AC 562 (HL Sc)

[10] *Jones v Livox Quarries Ltd* [1952] 2 QB 608

[11] *Rose v Plenty* [1976] 1 WLR 141

[12] *Wm Morrisons Supermarket plc v Various Claimants* [2020] UKSC 12

[13] *Wilsons and Clyde Coal Co Ltd v English* [1938] AC 57

[14] *Paris v Stepney Borough Council* [1951] AC 367

[15] *Lister v Romford Ice and Cold Storage Co Ltd* [1957] AC 555

1.3: The most important legal duties for employers and workers

> **Syllabus outline**
>
> In this section, you will develop an awareness of the following:
> - Health and Safety at Work etc Act 1974: ss2–4, 6–9, 36 and 37
> - Management of Health and Safety at Work Regulations 1999: regs 3–5, 7, 8, 10 and 13–14

Having introduced the UK legal system and its provisions in 1.2: The force of law – punishment and compensation, we now look in more detail at two key pieces of statute law:

- the Health and Safety at Work etc Act 1974; and
- the Management of Health and Safety at Work Regulations 1999.

The Health and Safety at Work etc Act 1974 (HASAWA) provides the principal legal framework for health and safety in UK workplaces. Introduced following a review of health and safety by the government-appointed Committee on Health and Safety at Work, chaired by Lord Robens, in 1970–1972, it represented a key change from piecemeal, prescriptive legislation to enabling and goal setting legislation. The enabling nature of HASAWA meant that it allowed other regulations to be introduced to support it and greater involvement of the workforce in health and safety through representation and consultation. It also established the Health and Safety Commission, replaced in 2008 by the Health and Safety Executive (HSE).

The Management of Health and Safety at Work Regulations 1999 (MHSWR) is one such set of supporting regulations. They build on the HASAWA and importantly set out the requirement for employers to assess the risks presented by all the hazards in their workplaces.

> **TIP**
>
> In Northern Ireland the Management of Health and Safety at Work Regulations (Northern Ireland) 2000 are equivalent to the 1999 regulations applicable in Great Britain.
>
> In relation to the self-employed, the following regulations are relevant:
> - The Health and Safety at Work etc Act 1974 (General Duties of Self-Employed Persons) (Prescribed Undertakings) Regulations 2015.

1.3.1 Health and Safety at Work etc Act 1974: sections 2–4, 6–9, 36 and 37

The HASAWA imposes general duties on everybody connected with work. The Act sets out to (s1):

- maintain or improve standards of health and safety at work;
- protect other people against risks arising from work activities; and
- control the storage and use of dangerous substances.

1.3 The most important legal duties for employers and workers

Sections 2–6 of the HASAWA impose general duties on employers, self-employed persons, persons otherwise in control of premises and designers, manufacturers, importers or suppliers of articles for use at work. These duties are owed to employees and any other persons who may be affected by the work activities and are mainly qualified by "so far as is reasonably practicable".

Section 2 duties are the main basis for criminal prosecution and administrative enforcement by the HSE.

Section 2(1) sets out the general duty of employers to ensure, so far as is reasonably practicable, the health, safety and welfare at work of all their employees.

Section 2(2)(a)–(e) amplifies this general requirement by explicitly requiring the employer, so far as is reasonably practicable, to:

(a) provide and maintain safe plant and safe systems of work;
(b) make arrangements for ensuring safe means of handling, use, storage and transport of articles and substances;
(c) provide information, instruction, training and supervision (competent employees);
(d) provide a safe place of work and provide and maintain safe access to and egress from that workplace; and
(e) provide and maintain a safe working environment and adequate welfare facilities.

The requirements of s2(2) are also the expectation of employers under common law (see 1.2: The force of law – punishment and compensation). However, duties in s2(2) may not be used to support civil liability cases.

Section 2(3) requires every employer to:

- Prepare (and keep up to date) a written statement (the safety policy only needs to be in writing where five or more are employed) of their general policy with respect to:
  - the health and safety at work of their employees; and
  - the organisation and arrangements in force for the time being for carrying out that policy.
- Bring the statement and any revision of it to the notice of all their employees.
- Include in the health and safety policy:
  - general statement of intent, dated and signed by the owner or chief executive;
  - details of the organisation and who is responsible for doing what regarding health and safety;
  - arrangements for implementing the policy; in practice these are often divided into:
    - general arrangements, for example, fire and emergency procedures; and
    - specific arrangements, such as how to use certain pieces of equipment.

CASE LAW: SECTION 2(1)

In the case of *R v Gateway Foodmarkets Ltd*[1], the company was prosecuted for failing to discharge their duty under s2(1) of the HASAWA to ensure, so far as is reasonably practicable, the health, safety and welfare of their employees. The case centred on the failure of the management of one of the company's stores to adopt the company maintenance procedure for lifts in its stores. The store management had, without the knowledge of the head office, developed their own procedures for rectifying a recurrent defect to the store's lifts. The store manager, adopting this procedure, suffered a fatal fall down the lift shaft.

Why we should manage workplace health and safety

Section 2(4) concerns the appointment of safety representatives from among the employees by trade unions that are recognised by the employer. Those representatives shall represent the employees in consultations with the employers under s2(6) and shall have such other functions as may be prescribed.

Section 2(6) requires employers to consult any such representatives with a view to the making and maintenance of arrangements that will enable them and their employees to co-operate effectively in promoting and developing measures to ensure the health and safety at work of the employees, and in checking the effectiveness of such measures.

Section 2(7) requires employers, if requested to do so by these safety representatives, to establish a safety committee that has the function of keeping under review the measures taken to ensure the health and safety at work of their employees and such other functions as may be prescribed. The functions and rights of safety representatives will be discussed further in 3.2: Improving health and safety culture.

Section 3 deals with duties of employers and self-employed persons to those other than their employees, such as contractors and members of the public, and requires employers and the self-employed to:

- ensure that their activities do not endanger anybody (with the self-employed, 'anybody' includes themselves); and
- provide information, in certain circumstances, to the public about any potential hazards.

An exception to these requirements applies when the activities carried out by self-employed persons are not on the list of prescribed activities listed in the schedule of the Health and Safety at Work etc Act 1974 (General Duties of Self-Employed Persons) (Prescribed Undertakings) Regulations 2015 and do not pose a risk to the health and safety of others. In this case, self-employed persons are exempted from the scope of s3(2) HASAWA and are not subject to any duty to conduct their work in a way that minimises health and safety risks.

Self-employed persons who are themselves employers are still subject to health and safety duties, not only to their employees (by virtue of s2 HASAWA) but also to persons other than their employees (by virtue of s3(2) HASAWA).

The case of *R v Swan Hunter Shipbuilders Ltd*[2], highlighted in 1.4: Managing contractors effectively, is an illustration of a failure to discharge s3 duties to non-employees.

> **CASE LAW: SECTION 3**
>
> In the case of *R v Board of Trustees of the Science Museum*[3], the organisation was charged with failure to maintain the air-conditioning system of the museum buildings with the result that bacteria, which causes legionnaires disease, existed in the water in the cooling system. This meant that non-employees, such as members of the public, were also potentially exposed to the bacteria.

Section 4 places a duty on those in control of premises that are non-domestic and used as a place of work to ensure they do not endanger those who work within them. This extends to plant, substances and means of access and egress as well as to the premises themselves.

Section 6 places duties on manufacturers, suppliers, designers, importers etc in relation to articles and substances used at work. Basically, they must research and test them and supply information to users.

Section 7 requires all employees while at work to:

- take reasonable care for their own health and safety and that of others who may be affected by what they do or by their failure to do what they should; and
- co-operate with their employer or any other person, so far as is necessary, to enable the employer or other person to perform or comply with any requirement or duty imposed under a relevant statutory provision.

Section 8 places a duty on everyone not to interfere with or misuse (whether intentionally or recklessly) anything provided in the interests of health, safety and welfare.

Section 9 provides that an employer may not charge their employees for anything done or equipment provided for health and safety purposes under a relevant statutory provision.

Section 36 is concerned with offences due to the fault of another person. It provides for an individual to be charged with an offence if, due to their fault, another person committed an offence. This is irrespective of whether proceedings are taken against the person who actually committed the offence. An example of this could be a furniture-making company that relies on an occupational hygienist consultant to assess and advise on exposure of its employees to wood dust. If the company is regularly exposing its employees to levels of wood dust in excess of legal limits, the company will be in breach of regulations to minimise the exposure of workers to harmful substances. If this is due to poor advice from the hygienist (such as an inadequate assessment and wrongly concluding that controls are adequate), the breach by the company is (at least in part) the fault of the occupational hygienist.

Section 37 is concerned with offences by corporate bodies, for example, companies and organisations, and permits the prosecution of an individual senior manager or officer, if the offence was committed with the consent, connivance or negligence of a director, manager, secretary or other officer.

The individual has to be the 'directing mind' of the company. Clearly, not every manager in a company could be prosecuted as they could not be considered to have the 'directing mind'.

The Corporate Manslaughter and Corporate Homicide Act 2007, discussed in 1.2: The force of law – punishment and compensation, provides that an organisation may be convicted of an offence where a gross failure in the way activities were managed or organised resulted in a person's death.

> **CASE LAW: SECTION 37**
>
> In the case of *Tesco Supermarkets v Natrass*[4], Tesco had a discount on washing powder advertised on posters in the store. When the discounted product ran out it was replaced with the regular priced product. The store manager did not take down the promotional price posters and a customer got charged the higher price. Tesco was charged for falsely advertising the price (under the Trade Descriptions Act 1968). The company argued that the conduct of the manager could not attach liability to the organisation.

It was held that the store manager did not act **as the company** (ie as the directing mind) but **for the company**. When someone acts **as the company**, their mind, which directs their acts, **is the mind of the company**. So, if their mind is guilty of some act or omission then their guilt is the guilt of the company. The guilty mind of the company could only be derived from the actions of those who represented the company's directing mind.

In the case of *Armour v Skeen*[5], a Scottish local authority and its Director of Roads were both prosecuted under s37 following the death of a worker while carrying out road bridge repairs. The breach concerned safety regulations, lack of a safe system of work and a failure to notify an inspector that notifiable work was being carried out. The director was deemed negligent for failing to have a safety policy and providing information and training to subordinates in safe working practices. Although s2 requires an employer to provide a safe system of work, there was also a duty on the part of the director to carry out that duty. The offence was committed by the company through the negligence of the director.

1.3.2 Management of Health and Safety at Work Regulations 1999

The Management of Health and Safety at Work Regulations 1999 (MHSWR) set out broad general duties that apply to almost all work activities in the UK and offshore. They are aimed mainly at improving health and safety management and can be seen as a way of making more explicit what is required of employers and employees under the HASAWA.

Table 1 details the requirements of regs 3–5, 7, 8, 10 and 13–14.

MHSWR reg 3: Risk assessment	Requires employers to assess the risks to the health and safety of their employees and to anyone else who may be affected by their work activity. This is so that the necessary preventive and protective measures can be identified. Employers with five or more employees must record the significant findings of the assessment. Risk assessment is discussed further in 3.4: Assessing risk.
MHSWR reg 4: Principles of prevention	Requires employers and the self-employed to introduce preventive and protective measures to control the risks identified by the risk assessment. The following principles of prevention, set out in Schedule 1 of the Regulations, should be followed in doing this: • avoid risks; • evaluate risks that cannot be avoided; • combat the risks at source; • adapt the work to the individual; • adapt to technical progress;

1.3 The most important legal duties for employers and workers

	• replace the dangerous with the non-dangerous or the less dangerous; • develop a coherent overall prevention policy; • give collective protective measures priority over individual protective measures; and • provide appropriate instructions to workers.
MHSWR reg 5: Health and safety arrangements	Requires employers to make arrangements for putting into practice the health and safety measures that follow from the risk assessment. These need to cover planning, organisation, control, monitoring and review of the protective and preventive measures. Again, employers with five or more employees must record their arrangements.
MHSWR reg 7: Health and safety assistance	Requires employers to appoint one or more competent people (either from inside the organisation or from outside) to help them in undertaking the measures they need to take to comply with the requirements imposed on them by or under the relevant statutory provisions. Where more than one competent person is appointed, the employer needs to ensure adequate co-operation between them. The competent person(s) must have time and resources available to fulfil their functions and be provided with the necessary information.
MHSWR reg 8: Procedures for serious and imminent danger and danger areas	Requires employers to: • identify danger areas; • have procedures in place to deal with serious and imminent danger; and • appoint competent persons to take charge and implement evacuation procedures.
MHSWR reg 10: Information for employees	Requires employers to provide employees with comprehensible information about specific health and safety matters.
MHSWR reg 13: Capabilities and training	Employers must ensure that employees have adequate health and safety training and are capable enough at their jobs to avoid risks. Training is required: • on induction; • when job responsibilities change; and • upon the introduction of new technology or new systems of work.
	MHSWR reg 13 also requires refresher training to be provided and states that all training must be within normal working hours.

Managing Health and Safety in the UK

Why we should manage workplace health and safety

MHSWR reg 14: Employees' duties	Extends and clarifies the HASAWA duties of employees to: • report serious and imminent danger; and • point out any shortcomings in their employer's arrangements for health and safety.
	These duties are limited by the training and instruction that the employee has received.

Table 1: Requirements of key MHSWR regulations

APPLICATION

Check how your organisation meets the requirements of the Management of Health and Safety at Work Regulations 1999 we have described.

CASE STUDY

The UK's largest reinforcing steel producer, Celsa Manufacturing (UK) Ltd, pleaded guilty to breaching reg 3 of the Management of Health and Safety at Work Regulations 1999 and was fined £1.8m in October 2019 after an explosion at the company's Cardiff Rod and Bar Mill killed two workers and seriously injured another in November 2015. The men were working on an accumulator vessel when it exploded. An HSE investigation found that a flammable atmosphere developed within the accumulator as hydraulic lubrication oil was being drained from it. The flammable atmosphere was ignited by an electric heater within the accumulator. The investigation also found that the company failed to assess the risks to which its employees were exposed when manually draining lubrication oil from the accumulator, and that the locally developed 'procedure' was not fully understood or consistently carried out by the company's employees.[6]

KEY POINTS

- The Health and Safety at Work etc Act 1974 is an enabling Act and the principal piece of health and safety legislation in the UK.
- It sets out duties for employers, the self-employed, persons in control of premises and designers, manufacturers, importers or suppliers of articles for use at work.
- Employees are required to take care of themselves, co-operate with their employer and not to intentionally or recklessly interfere with anything provided in the interests of safety.
- Individuals and corporate bodies may be charged with offences under the Act.
- The Management of Health and Safety at Work Regulations 1999 build on the Health and Safety at Work etc Act 1974.
- These Regulations set out requirements for employers to carry out risk assessment and to put in place preventative and protective measures on the basis of their assessment of risks in the workplace.

- Other requirements for employers relate to accessing competent advice to support them in managing health and safety risks, providing information and training to employees and having procedures to deal with serious and imminent danger.
- Employees must follow instructions and point out issues of concern.

References

[1] *R v Gateway Foodmarkets Ltd* [1997] 3 All ER 78

[2] *R v Swan Hunter Shipbuilders Ltd* [1982] 1 All ER 264

[3] *R v Board of Trustees of the Science Museum* [1993] 1 WLR 1171

[4] *Tesco Supermarkets v Natrass* [1972] AC 153

[5] *Armour v Skeen* [1977] IRLR 310

[6] HSE Case list, Case number 4547496, Celsa Manufacturing (UK) Ltd, 18 November 2015 (www.hse.gov.uk)

1.4: Managing contractors effectively

> **Syllabus outline**
>
> In this section, you will develop an awareness of the following:
> - Planning and co-ordination of contracted work
> - Pre-selection and ongoing management of contractors
> - Roles and duties under the Construction (Design and Management) Regulations 2015 of the client, principal designer, principal contractor, contractors, workers and domestic clients:
>   - HSE notification, pre-construction information, construction phase plan, health and safety file

A contractor is a person or an organisation that has been brought in to perform a task on behalf of the client, but not under the client's direct supervision or control. For example, contractors have always been used on construction sites, but many other places of work increasingly make use of contractors, such as:

- facilities managers (ie external contractors who are responsible for much of the management and maintenance of a commercial, educational or industrial site); and
- all sorts of 'bought-in'/'contracted-out' services, such as occupational health services, cleaning, catering security and IT.

Using a contractor enables the client to access expertise and resources in situations where a work activity is done infrequently or specialist skills might be needed as a one-off. In both cases, the costs of employing workers to do that work would not be financially viable. Sometimes contractors are engaged on long-term contracts when an employer has an activity they prefer not to do, such as cleaning or security, so that they can focus on core activities.

The client's organisation of the contract and the contractor's execution of the contract can have a big impact on health and safety. Clients can place unreasonable demands on contractors or the way the contractor carries out the contract could generate added risks. If there are conflicts in the way both parties operate, people can be put at risk of harm.

In relation to construction, the Construction (Design and Management) Regulations 2015 (CDM) set out requirements for the planning and management throughout construction projects, from design concept onwards. This means that health and safety considerations are treated as an essential but normal part of a project's development. They set responsibilities for a wide range of duty holders.

> **TIP**
>
> Legislation relating to the control of contractors working in construction can be found in the:
> - Construction (Design and Management) Regulations 2015 (CDM).

> Guidance relating to the control of contractors in construction and more widely can be found in the following Health and Safety Executive (HSE) publications:
>
> - *Need building work done? A short guide for clients on the Construction (Design and Management) Regulations 2015* (INDG411);[1]
> - *Managing health and safety in construction. Construction (Design and Management) Regulations 2015. Guidance on Regulations* (L153);[2] and
> - *Managing contractors: A guide for employers* (HSG159).[3]

1.4.1 Planning and co-ordination of contracted work

Contractor management can pose health and safety issues, which may include the following:

- A contractor's workforce may not be familiar with the client's processes and procedures.
- Contractors may have a different 'world view' to that of the employing organisation, meaning they will have different attributes that make them behave differently from the employing organisation's workforce.
- Necessarily, contract work may involve activities and relationships outside the normal activities of the organisation, many of which are likely to present higher levels of risk, such as demolition, repair and commissioning new equipment.
- The contractor's workforce may change from day-to-day, particularly if subcontractors are used.
- Contractors may be tempted to cut corners to cut costs, save time or finish the work so that they can begin another project, especially if there are bonuses for finishing early or penalty clauses for finishing late.

However, it is in the interests of both parties to have a successful and safe contract. Both parties share the responsibility to collaborate to ensure this happens and that health and safety are built into the working methods.

Health and safety information needs to be exchanged between the client and the contractor to allow the work to be planned effectively. Usually, the client would provide information about known hazards and risks, while the contractor would provide information such as method statements. Collectively the information allows for the planning and co-ordination of the work and foreseeing how activities and people interact with each other.

Collaboration is not only needed prior to contract agreement and at commencement of the work but throughout the duration of the contract. There is also a need to monitor the health and safety performance of the contract throughout its duration. As the client retains the overall responsibility for the contracted-out work, if they have concerns about the performance of the contractor, they should take action to address the shortfalls and make sure the contractor adheres to the health and safety requirements.

So far as risks generated by their own work are concerned, contractors will normally be responsible for their own health and safety and that of others who may be affected by their acts or omissions. However, it would be wrong to say that they are

Why we should manage workplace health and safety

solely responsible for their own health and safety, especially when they are working on a client's premises. That being the case, the client (eg the site owner) will still be responsible for the contractor's safety insofar as the contractor may be exposed to risks that are within the client's control.

Both clients and contractors are employers (usually) and so both have obligations, under the Health and Safety at Work etc Act 1974 (HASAWA), for the health and safety of their own and other workers and the public.

The following HASAWA duties are relevant here:

- Duty of every employer to ensure the health, safety and welfare at work of all their employees: s2(1).
- Duty to provide and maintain plant and systems of work that are safe and without risks to health: s2(1)(a).
- Duty to provide such information, instruction, training and supervision as is necessary to enable the health and safety at work of employees: s2(1)(c).
- Duties of employers and the self-employed to ensure their activities do not endanger anybody not in their employment (including contractors, members of the public and trespassers): s3.
- Duty of those in control of work premises to ensure they do not endanger those who work within them: s4.

In relation to contractors, two cases provide definitive interpretation of s2(2)(c) and s3.

CASE LAW

R v Swan Hunter Shipbuilders and Telemeter Installations Ltd[4]

Telemeter Installations were subcontracted to Swan Hunter to work on building a warship. One of their employees, who was welding alongside Swan Hunter's own employees, started an intense fire in an oxygen-rich part of the ship in which eight men died. Swan Hunter had informed their own employees of the dangers of fire in such an environment but were prosecuted for failing to inform their contractors' employees of the risks. They were found guilty and thus the duty under s2(2)(c) of the HASAWA was interpreted as making clear that a company must pass on information to non-employees if the safety of their own employees would be placed at risk.

The outcome is that you must pass information on to others if their actions could endanger themselves or your own employees.

CASE LAW

R v Associated Octel Co Ltd[5]

A contractor was badly burned when they started a fire while repairing the lining of a tank at the Ellesmere Port site of Associated Octel. The injured person was employed by Resin Glass Products, who were contracted by Associated Octel to undertake repairs during Octel's annual maintenance shutdown.

Associated Octel argued that it had no duty to control the way independent contractors did their work. However, the House of Lords said that repair and maintenance was as much a part of Octel's undertaking as the manufacture of the chemicals and thus the employer had to stipulate to the contractor(s) the conditions that were required to avoid risks to health and safety.

1.4 Managing contractors effectively

> Being part of the 'undertaking' means that, under s3 HASAWA, Octel had to ensure that the work was carried out in a way that did not endanger non-Octel employees, including the contractor carrying out the work. Additionally, because the work was carried out under Octel's permit system and because, prior to work starting, an Octel engineer had checked the contractor's risk assessment and method statement, Octel was held to be in control of the work.

Use of equipment

If the contractor brings their own equipment then they are responsible for it. If the contractor makes use of the employing party's tools then the responsibility lies with the employing party. For that reason, it is invariably not a good idea to 'loan' tools and equipment to contractors – besides, if they do not already have the right tools and equipment, then a question mark has to be raised about their ability to do the job.

Serious accidents have been caused (see case law examples) when contractors were using their own equipment and the employing parties were negligent in not informing the contractors of the hazards. Section 2 of the HASAWA requires the provision of a safe system of work in the workplace whoever is in control of the work being undertaken. The responsibility for providing this safe system of work could lie with the employing party or with a fully briefed contractor.

1.4.2 Pre-selection and ongoing management of contractors

As clients retain the overall responsibility for the health and safety of the contracted activities, they must set up contracts that can be achieved safely and to do this they need to select contractors carefully and ensure that the contractor can meet their obligations.

Selecting contractors usually involves balancing a number of issues such as cost, time to carry out the contract and performance of the contractor, including quality and health and safety. Contractor selection based on cost alone may not meet the health and safety responsibilities of the client. The client must take reasonable care to ensure the quality of the contractor they have chosen.

Considering health and safety prior to the price being agreed is preferable. This can be done using the following key steps:

- Identify the most likely hazards and risks associated with the work being contracted out.
- Identify suitable contractors from the pool of those contractors potentially interested in bidding. This involves assessing the contractors against a range of health and safety criteria to establish a 'track record' for the contractor. Criteria could include:
    - existence and adequacy of their health and safety policy;
    - their health and safety competence – this might involve considering their competent person, their qualifications and experience and memberships of professional or trade bodies;
    - accident/incident history;
    - details of any enforcement agency action;
    - details of any prosecutions or civil claims;
    - competence of workers;

- presence of a health and safety management system and whether this is certified by a third party;
- insurances held – level and type; and
- references from previous clients.
* Assess the health and safety aspects of the applications submitted. This could be done using a questionnaire.
* Select contractors to proceed to the final selection based on a range of aspects such as quality, environmental and financial as well as health and safety.

The extent to which each of the steps is relevant depends on the nature of work and the risk involved.

Once the contractor has been selected and appointed, they need to be managed and monitored throughout the contract. Once the contract has been completed, their health and safety performance should be reviewed and noted on the list of potential contractors, so that should further opportunities arise for contract work of this type there is up to date information on file. HSG159 provides useful information on contractor management. Although focused on the chemical industry, it should be straightforward to relate the contents to your own work situation.

APPLICATION

Find out how contractors are selected in your own organisation. How is health and safety assessed in the selection process?

1.4.3 Roles and duties under the Construction (Design and Management) Regulations 2015

The Construction (Design and Management) Regulations 2015 (CDM) focus attention on planning and management throughout construction projects, from design concept onwards. The aim is for health and safety considerations to be treated as an essential but normal part of a project's development, not an afterthought or bolt-on extra.

Virtually everyone involved in a construction project has legal duties under CDM. These 'duty holders' are defined as follows:

- client:
  - commercial clients; and
  - domestic clients;
- designer;
- principal designer;
- principal contractor;
- contractor; and
- worker.

Client

This is the person/organisation that has construction work carried out for them. The main duty for clients is to make sure the project is suitably managed, ensuring the health and safety of all who might be affected by the work, including members of the public.

CDM recognises two types of client:

- Commercial clients have construction work carried out as part of their business.
- Domestic clients have construction work carried out for themselves, but not in connection with any business. The work being undertaken is on the client's home or on the home of a family member. The only responsibility a domestic client has is to appoint a principal designer and principal contractor when there is more than one contractor. However, often this does not happen, in which case the domestic client's duties are automatically transferred to the contractor or principal contractor. If there is a relationship between the client and designer before work starts, the designer can take on the client duties subject to written agreement.

For some individuals and organisations, being a client in terms of CDM may be a rare or infrequent event. It is important that everyone commissioning construction works has an appreciation of the Regulations. More information for clients can be found in INDG411.

Designer

This is an organisation or individual whose work involves preparing or modifying designs, drawings, specifications, design calculations etc. Designers can be architects, consulting engineers, quantity surveyors or anyone who works with designs. The main duty for any designer is to eliminate, reduce or control foreseeable risks that may arise during construction work, or in the use and maintenance of the building once built. For example, if access is required for cleaning, maintenance or repair of windows in a building roof using temporary access equipment, then a suitable means of fixing it to the structure to permit safe access should be incorporated in the design. In situations where there is more than one contractor, designers will work under the control of a principal designer.

Principal designer

They will be appointed by the client to control the pre-construction phase on projects with more than one contractor. Their main duty is to plan, manage, monitor and co-ordinate health and safety during this phase, when most design work is carried out.

Principal contractor

They will be appointed by the client to manage the construction phase on projects with more than one contractor. The principal contractor's main duty is to plan, manage, monitor and co-ordinate health and safety during this phase, when all construction work takes place.

Contractor

This is an individual or business in charge of carrying out construction work (such as building, altering, maintaining or demolishing). Their main duty is to plan, manage and monitor the work under their control in a way that ensures the health and safety of anyone it might affect (including members of the public). In projects with more than one contractor, contractors work under the control of the principal contractor.

Worker

This is the individual who carries out the work involved in building, altering etc. Workers include: plumbers, electricians, scaffolders, painters, decorators, steel erectors and labourers, as well as supervisors like foremen and chargehands. Duties include co-operating with their employer and other duty holders and reporting anything that might endanger the health and safety of themselves or others. Workers must be consulted on matters affecting their health, safety and welfare.

Notification under CDM

A construction project is 'notifiable' if the construction work on a construction site is scheduled to:

(a) last longer than 30 working days and have more than 20 workers working simultaneously at any point in the project; or

(b) exceed 500 person days.

When a project is notifiable, the client must give notice in writing to the HSE as soon as is practicable before the construction phase begins. The notice must contain the particulars that are specified in Schedule 1 of the CDM Regulations. These include the address of the site, local authority, brief description of the project, details of principal designer etc. The notice should be clearly displayed in the construction site office where it can be freely viewed by any worker engaged in the construction work.

The form can be completed online through the HSE website.[6] HSE has provided a guide to completing the online form.[7] To see how the form works in practice, you will have to invent details of your construction project before you can progress through the form.

Pre-construction information

This information is in the client's possession before the work starts. It can include surveys, drawings and health and safety information pertaining to prior construction works. The client has the main duty for providing the pre-construction information to designers and contractors. Where there are multiple contractors in a project, the principal designer is expected to co-ordinate the distribution of information to the other designers and contractors.

Construction phase health and safety plan

The construction phase health and safety plan sets out the arrangements for managing health and safety during the construction phase, such as site rules. It must be drawn up by the contractor before the construction phase starts. Where multiple contractors are involved, the responsibility for the plan rests with the principal contractor. The plan must be sufficient to ensure that the construction phase is planned, managed and monitored in a way that enables the construction work to be started so far as is reasonably practicable without risk to health or safety. Risks arising from the work and mitigation measures, such as site rules, should be included. Adequate attention should be given to the information provided by the designer and the pre-construction information provided.

Throughout the project the plan must be updated, reviewed, revised and refined as necessary so that it continues to be sufficient and fulfil its purpose.

Health and safety file

A health and safety file is only required for projects involving more than one contractor. The file should contain information relating to the project that is likely to be needed during any subsequent construction work to ensure the health and safety of any person involved. Such information includes a description of the work carried out, residual hazards, structural principles used in the construction, known underground services, surveys and reports and drawings such as of access routes to voids and fire doors. The principal designer must prepare the file and keep it up to date as the project progresses and hand it over to the client on completion of the project. If their involvement ceases before the end of the project, they should pass it over to the principal contractor, who takes on the responsibility of maintaining it and handing it over to the client.

CASE STUDY
Commercial redevelopment[8]

Here is an example of a commercial project. It concerned the redevelopment of a London site on a main high street near a London Underground station. The project value was over £15 million. The client/developer was an experienced developer in London. The works consisted of:

- stripping out the existing fittings;
- internal structural demolitions for a new core and lifts;
- the basic finishing of the interior space, installation of mechanical and electrical services, finishing internal walls, reception areas and lift lobbies for an existing five-storey building; and
- refurbishment of the existing office building.

Key duty holder roles were:

- project manager;
- architectural design;
- structural design;
- mechanical, electrical and plumbing design;
- principal contractor; and
- principal designer.

Management arrangements

Project definition and strategy: The architectural designer took on the duties associated with the principal designer function. During the early stages of the project, the project manager set up a workshop on behalf of the client and designers to map out the strategy for the project. A CDM strategy brief was produced and included in the project brief. The architectural designer and principal designer prepared an initial set of pre-construction information that included, for example, survey information, photographs and drawings and significant risks.

Design development: The architectural designer produced a CDM analysis and options report that was reviewed with the client and design team at each work stage and at design team meetings. The client appointed their preferred contractor under a pre-contract service agreement to assist with the design risk management and cost finalisation.

Detailed design and construction: Once agreement was reached on the terms of the contract, the principal contractor worked with the principal designer and collectively co-ordinated the technical design together with their specialist sub-contractors, including some temporary works design.

Health and Safety File: An electronic file (as set out in Appendix 4 of L153) was compiled. Any residual risks that remained at the end of the construction phase were recorded in the health and safety file.

Why we should manage workplace health and safety

> **CASE STUDY**
>
> **Domestic project**[9]
>
> Here is an example of a medium-sized domestic project to refurbish a row of two-storey 18th century mews houses in London. The client was from overseas and aiming to convert the existing mews property into a family residence.
>
> The project involved remodelling of the existing elevation with new entrances and windows, gutting the building and ensuring stability when basement excavation, roof extension and floor removal works were being carried out. A significant amount of demolition and fit-out work was included. Access to the properties for construction vehicles and plant was height and width restricted.
>
> Key duty holder roles were:
>
> - Principal designer: an architectural practice was employed by the property owner, which included a structural engineer and building services consultant.
> - Principal contractor: an experienced contractor was appointed for the construction phase.
>
> **Management arrangements**
>
> **Project definition and strategy:** The architectural designer explained the requirements of UK legislation to the client, who appointed the architectural designer as principal designer and to take on their client duties. They jointly developed a CDM Strategy Brief to set out their joint understanding of the management arrangements for the project. The brief highlighted the structural stability and temporary works, which the designers and contractors would have to take account of in their work. The project was notifiable to the HSE using form F10, submitted by the architectural designer as a client duty.
>
> **Design development:** The design team produced a detailed set of design drawings with significant risk issues highlighted, especially the requirement to manage the considerable amount of temporary works. Experienced contractors were invited to tender against this.
>
> **Detailed design and construction:** The successful contractor provided details of their temporary works procedures in their tender submission and subsequently set out the temporary works requirements in their construction phase plan. The architectural designer continued to project manage the scheme on behalf of the client and monitored the principal contractor's temporary works design performance.
>
> **Health and safety file:** A simple file was issued to the client based on the pre-construction information, highlighting any residual risks, and referenced the contractor's operation and maintenance manual.

KEY POINTS

- In addition to construction work, many other places of work make use of contractors, including facilities managers (ie external contractors who are responsible for much of the management and maintenance of a commercial, educational or industrial site) and 'contracted-out' services, such as occupational health services, cleaning, catering, security and IT.

1.4 Managing contractors effectively

- Contractor management can pose health and safety issues; both parties share the responsibility to collaborate to ensure that health and safety are built into the working methods.
- Section 2 of the HASAWA requires the provision of a safe system of work in the workplace, whoever is in control of the work being undertaken.
- Selecting contractors usually involves balancing issues such as cost, time to carry out the contract and performance of the contractor, including quality and health and safety.
- Once the contractor has been selected and appointed, they need to be managed and monitored throughout the contract.
- The Construction (Design and Management) Regulations 2015 set out legal duties for the following duty holders: client (domestic and commercial), principal designer, principal contractor, contractors and workers.
- Where a construction project is to last longer than 30 working days and have more than 20 workers working simultaneously at any point in the project, or to exceed 500 person days, it must be notified in writing to HSE ahead of the work starting.
- Pre-construction information is held by the client and includes surveys, drawings and information from prior construction projects.
- Before the construction phase starts, the principal contractor must prepare a construction phase plan.
- The health and safety file, containing information relating to the project that is likely to be needed during any subsequent construction work to ensure the health and safety of any person, must be given to the client on completion of the project.

References

[1] HSE, *Need building work done? A short guide for clients on the Construction (Design and Management) Regulations 2015* (INDG411, 2015) (www.hse.gov.uk)

[2] HSE, *Managing health and safety in construction. Construction (Design and Management) Regulations 2015. Guidance on Regulations* (L153, 2015) (www.hse.gov.uk)

[3] HSE, *Managing contractors: A guide for employers* (HSG159, 2011) (www.hse.gov.uk)

[4] *R v Swan Hunter Shipbuilders and Telemeter Installations Ltd* [1982] 1 All ER 264

[5] *R v Associated Octel Co Ltd* [1996] 4 All ER 1051 (HL)

[6] HSE, 'F10 – Notification of construction project' (2022) (www.hse.gov.uk)

[7] HSE, 'How to fill in the F10 e-Form' (2022) (www.hse.gov.uk)

[8] Construction Industry Advisory Committee (CONIAC), 'CDM 20-20 Vision - changing the culture: Applying CDM 2015 strategically to diverse projects' (2021) (www.cic.org.uk)

[9] See note 8

ELEMENT 2

HOW HEALTH AND SAFETY MANAGEMENT SYSTEMS WORK AND WHAT THEY LOOK LIKE

2.1: What they are and the benefits they bring

Syllabus outline

In this section, you will develop an awareness of the following:
- The basics of a health and safety management system: the 'Plan, Do, Check, Act' model (see ISO 45001:2018 and HSG65)
- The benefits of having a formal/certified health and safety management system

A health and safety management system is a set of interrelated components/elements that allow an organisation to manage health and safety in a structured way to achieve its objectives. Formal systems are built on the elements of Plan, Do, Check, Act (PDCA), capturing the principle of continuous improvement.

The main components of the system are a *policy*, which sets out a mission statement for health and safety and mechanisms for management control and accountability, and *arrangements* for implementing, monitoring, auditing and continuously improving. A formal system develops consistency and supports a culture that can involve everyone.

Organisations need to:

- work out the issues to be addressed;
- set the direction;
- plan what needs to be done and organise who will do it;
- set them up to do so;
- carry out the plan;
- check completion and efficacy; and
- take on board any learning so that they can continually improve.

Organisations are being encouraged to adopt management systems through their supply chains. There are generic and sector-specific approaches as well as approaches for which independent third-party certification can be obtained. Organisations have freedom to choose the approach they want to follow and can decide whether they want to work towards a certifiable standard.

2.1.1 The basics of a health and safety management system: the 'Plan, Do, Check, Act' model (see ISO 45001:2018 and HSG65)

ISO 45001:2018[1]

ISO 45001:2018 is the first truly international certifiable occupational health and safety management system standard. The development of the standard has drawn on experience gained with OSHSAS 18001 (replaced by ISO 45001 in 2018) and other national approaches. As a result, the new standard is enhanced and more comprehensive, reflecting the approaches of organisations that strive for and succeed at health and safety management. Figure 1 shows elements of the standard.

The management system uses the Plan, Do, Check, Act (PDCA) cycle.

- **Plan** – assess occupational safety and health (OSH) risks and opportunities, taking into account the organisation's operating environment, and set out OSH objectives and delivery plans in line with the organisation's policy.
- **Do** – implementation of the processes.
- **Check** – monitor and measure OSH processes and report results.
- **Act** – take action to continually improve OSH performance and achieve intended outcomes of the system.

> **APPLICATION**
>
> Think about your organisation. Can you identify examples of how the PDCA approach is used in practice?

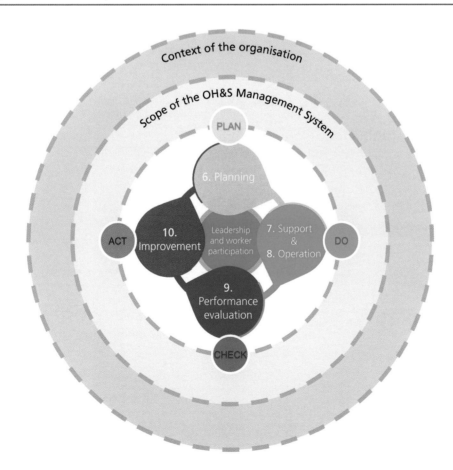

Figure 1: Elements of ISO 45001:2018[2]

Let us look at the key clauses in the ISO 45001:2018 standard and what they mean for organisations.

Context of the organisation

Clause 4 deals with establishing the organisation's context as a major building block that underpins the rest of the standard. This is about identifying and understanding the internal and external environments in which the organisation operates and the influence they exert. Influences may be positive or negative. The scope of the management system must be set out, taking account of these. Setting the scope will determine the boundaries of the system; this is particularly important if the organisation is part of a larger organisation.

External aspects could include: cultural, social, political, legal, financial, technological, economic and natural surroundings; and market conditions and key drivers and trends relevant to the industry or sector. Internal considerations could include: the organisational structure, roles and accountabilities and culture; policies, objectives and strategies; and capabilities and decision-making processes.

There are various tools that can be used to gain an understanding of these external elements. One of these is STEEPLE (see Figure 2). Other models include PEST (Political, Economic, Social, Technological) or PESTEL (Political, Economic, Social, Technological, Environment and Legal).

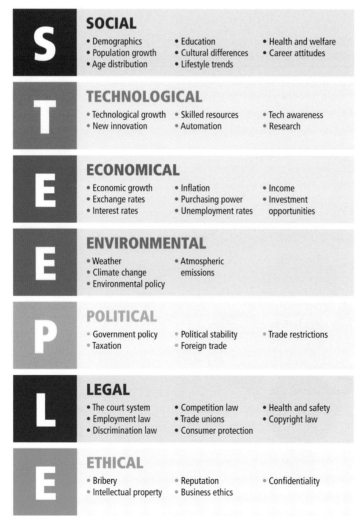

Figure 2: The STEEPLE model

There is also a requirement to consider relationships with external interested parties, such as shareholders, customers, suppliers and visitors.

Leadership and worker involvement

Clause 5 sets out a requirement for responsibility and accountability of top management to be defined and visibly demonstrated. They must develop, lead and promote a culture that supports the management system. This means ensuring that the management system requirements are integrated into business processes and the health and safety policy and that objectives align with the strategic direction of the organisation. That way, health and safety becomes integrated. They must also help to ensure active worker participation when developing and maintaining the system; this includes processes for consultation and communication. Combining leadership and worker involvement in this clause recognises that effective health and safety management is about teamwork.

Planning

Planning will be a familiar process for those who have used other management systems and is covered in clause 6. This includes identifying hazards and risks from the conceptual design stage of workplaces, facilities, products or the way work is organised, taking account of routine and non-routine situations and the people involved. Planning also needs to be an ongoing process. Particular attention should be given to:

- consideration of the wider human and social aspects, such as workload, bullying and stress;
- a requirement to identify and maintain a means of keeping up to date with legislative and other requirements, such as having a legal register;
- plans to address the risks and opportunities for the management system and plans for emergency response; and
- establishing measurable health and safety objectives at different levels of the organisation, which can be strategic, tactical or operational, and plans to achieve them. Objectives set by the organisation should achieve specific results consistent with the health and safety policy.

Support

Clause 7 deals with the support required to realise the objectives. Organisations need to determine the level of resources to implement, maintain and continually improve the management system. This should consider human, infrastructure and financial requirements. Other support requirements include competence, awareness and communication.

Organisations need to identify the levels of competence workers need to deliver their OSH performance and ensure they receive the appropriate education and training to meet this need. Put simply, competence is about being able to apply your knowledge and skills to achieve the results you want. In the case of health and safety, being competent allows you to achieve those results safely. Keeping records to provide evidence of competence is required. Everyone in the organisation needs to be made aware of the policy, objectives, hazards and risks and how they contribute to the overall system. To this end the organisation needs to have a process of internal and external communication. Requirements for documented information relating to the planning and evaluation aspects of the management system include creation, update and control of the information.

Operation

Plans need to be implemented with the necessary support, so clause 8 is about implementation. Control processes need to be put in place to support health and safety through eliminating hazards and reducing risk. This can include procedures, methods statements, system of work etc. In deciding on control processes, a hierarchy of control approach is required. This is a systematic method for deciding what control measures are needed (see Figure 3). It starts with elimination and then looks at substitution, engineering controls and administrative controls, down to personal protective equipment (PPE) as a last resort. This method is discussed in more detail in 3.4: Assessing risk.

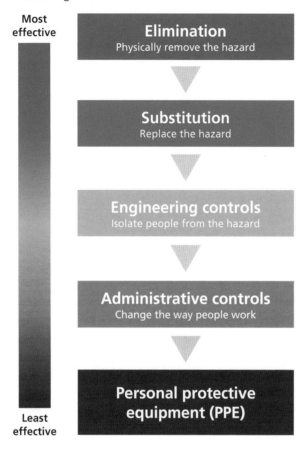

Figure 3: A systematic method for deciding what control measures are needed

Requirements for the management of change are also set out so that change is dealt with in a systematic way to prevent the introduction of new hazards and risks. When changes are proposed, organisations need to identify opportunities to reduce the level of risk or improve the management system.

To further enhance risk control, there are requirements for processes to control the procurement and outsourcing of goods/services. The activities of contractors must be considered, and the requirements of the management system should extend to the contractor's workforce.

Emergency preparedness and response are also covered in this clause. The standard requires organisations to have documented information on emergency plans and to periodically test these plans.

Performance evaluation

Clause 9 covers the requirement for organisations to set up a range of performance evaluation processes: monitoring, measurement and analysis; and evaluation of compliance, internal audit and management review. They will need to identify what information they need to evaluate OSH performance, which in turn will identify what needs to be measured and who by, and how and when it will be done.

The management review needs to consider the adequacy of the resources for maintaining the system. The frequency of monitoring and measuring needs to be appropriate to the risk, the level of OSH performance and the size and nature of the organisation.

Improvement

Clause 10 sets out a general requirement to proactively look for opportunities to effect improvement in health and safety and the performance of the system. Added to this, there is a requirement to address incidents and non-conformities if they arise. Consideration is also needed of whether similar incidents or non-conformities exist or could occur. Where the potential for these is identified, existing risk assessments may need to be revisited and could lead to corrective actions across the organisation.

Continual improvement requirements include improvements to the suitability, adequacy and effectiveness of the system.

HSG65 – *Managing for health and safety*[3]

The Health and Safety Executive (HSE) developed its HSG65 *Managing for health and safety* to assist leaders, owners, trustees and line managers to put in place or oversee their organisation's health and safety arrangements.[4] The document provides a framework that can be tailored to individual organisation's circumstances. The intention is for health and safety to be managed as an integral part of good management rather than a standalone system.

Unlike ISO 45001:2018, HSG65 is not a certifiable health and safety management system. Adopting the framework it sets out, and following the guidance, will help organisations achieve compliance with legal requirements.

The guide is structured as follows:

Part 1: Core elements of managing health and safety and their fit with general business operation
Part 2: Considerations to help decide if you are doing what is needed
Part 3: Advice on delivering effective arrangements
Part 4: Useful resources and information

The guidance allows organisations to take a formal approach to health and safety management but does not allow for certification.

HSE considers the core elements for effective health and safety management to be:

- leadership and management;
- a competent workforce; and
- an environment that fosters trust and involvement.

These elements should be underpinned by an understanding of the organisation's risk profile and the compliance requirements.

Figure 4 shows the how the different parts of the approach fit together. The circular shape reinforces that the process of managing health and safety is continuous and iterative. This is because organisations seldom stand still for long; they grow and develop, constantly evolving to meet the needs of the changing world around them. The cyclical process may need to be repeated more than once; for example, when first introducing the approach, implementing change or developing a new process, product or service.

How health and safety management systems work and what they look like

Figure 4: HSG65 Managing for health and safety – Plan, Do, Check, Act cycle[5]

The system is structured around the four key elements:

- Plan – determining policy and planning for its implementation;
- Do – risk profiling, organising to deliver the plan and implementing it;
- Check – measuring performance and investigating accidents, incidents and near misses; and
- Act – reviewing performance and acting on the lessons learned.

Plan

Determining policy

The policy sets the tone by clearly stating the organisation's aims and objectives for the management of health and safety. Of course, different organisations will feel differently about health and safety and will therefore want or expect different things from their policy.

The policy is a public demonstration of senior management's commitment to health and safety. It is best written by someone within the organisation, as it needs to reflect the organisation's values and beliefs. Organisations that are successful in achieving high standards of health and safety 'live out' their policies through the way they operate. Their policies shape the culture of the organisation while ensuring they meet legal requirements in a true way. Their policies aim to protect human resources and minimise financial loss and liability.

Planning for implementation

Planning is essential to bring the policy to life and to ensure risks are controlled, and that the organisation can respond to changing demands and maintain positive health and safety attitudes and behaviours. Successful planning requires organisations to identify where they are now, where they want to be and how they are going to get there. Once that is clear, the focus is on coming up with the system or approach to be used. That means setting up suitable arrangements and controls, operating and maintaining the system and linking it to other activities in the organisation.

The planning element includes the setting of specific objectives. The exact choice of objectives depends on how well developed the organisation's health and safety management is and what the organisation ultimately wishes to achieve. For example, the minimum standard to be achieved is to comply with the law; this may be enough for some organisations. However, other organisations may wish to have truly world class health and safety systems so that they can better reap the benefits of effective health and safety management. Therefore, there is no exact template for what plans should be made and it is left to the employer to decide exactly how safety and health are to be managed on a daily basis.

When planning for health and safety, organisations should always bear in mind minimum legal requirements. In addition, organisations may look to ensure that all their major hazards have been addressed and that any plans that are made are realistic when financial and technological constraints are considered.

> **CASE STUDY**
> **Achieving world class health and safety performance**[6]
>
> This case study is extracted from DuPont's sustainability report and sets out how DuPont works towards achieving world class health and safety performance.
>
> > At DuPont, we re-commit ourselves to our core value of health and safety every single day. This commitment is embodied in our company culture, our stakeholder engagement, and in the innovations we produce for the world.
>
> Safety and health is one of the company's core values and enshrined in the company code of conduct. Commitment to this is every employee's responsibility. Managers in each business are responsible for educating, training and motivating employees to understand and comply with this code.
>
> **Identifying hazards and risks:** Periodic safety perception surveys are sent to all contractors, subcontractors, and employees to assess the safety culture and invite feedback. The survey results inform the next steps needed by the site, the business or the corporate health and safety team. They create monthly corporate safety campaigns to address the top three causes of injuries and illnesses.
>
> Risk assessments and hazard evaluations are performed in a collaborative manner by cross-functional teams. A collaborative approach is also applied to develop prevention and mitigation strategies to reduce risks within the operational and business context.
>
> **Reporting and investigating incidents:** Employees report any work-related hazards, hazardous situations, near misses and incidents as a condition of employment, encouraged by open and proactive communication between workers and their line management. There are procedures in place and trained staff to report, classify and investigate near misses and incidents.
>
> **Communication and training:** Internal communication about health and safety is primarily conveyed through SharePoint sites, websites, digital signage, posters and team meetings. Sites also have systems in place to encourage and collect suggestions from employees on how to improve the safety and effectiveness of facilities and procedures.
>
> Each business within the company has ongoing training programmes that are designed to maximise the performance of its employees in meeting business objectives, including better health and safety outcomes.

Senior leaders and health and safety staff in each business review the company-wide health and safety performance on a weekly and monthly basis. Employees are updated on performance at quarterly global town hall meetings. Many teams hold monthly safety meetings to help reinforce core values, which include in-person training and updates on safety topics. All employees are encouraged to participate at least quarterly and are key participants in delivering the sessions. Meeting attendance and the topics covered are documented. Training and documented procedures are provided to ensure that workers are knowledgeable about the relevant safety and health risks of the facilities in which they work. Processes are in place to ensure understanding of these risks and competence of workers to effectively manage them in the completion of their work. Employees who cannot attend these meetings complete supplemental training.

Data analysis: Data collection and analysis are key to protecting employees and communities. A health and safety performance dashboard gives health and safety teams and corporate leaders a way to look at trends and to examine the main causes of injuries, illnesses and other incidents. This enables managers to create safety and health programmes that are better informed and tailored to their unique operations and working environments.

> **APPLICATION**
>
> Consider your organisation's health and safety policy. How does your organisation show it 'lives out' the policy?

Do

Risk profiling

Risk profiling allows the organisation to get a view of the nature and level of threats faced by the organisation. It looks at the likelihood of threats materialising, the level of disruption and costs and the effectiveness of control measures in place. It looks beyond health and safety risks and considers how issues in one area might impact on another. For example, a spillage of chemicals may have an environmental, service or quality impact in addition to a health and safety impact.

Organising

The 'organising' section identifies four key components that are essential for promoting positive health and safety outcomes:

1. Controls within the organisation, in particular:
    - the role of those in supervisory positions (leaders, managers and supervisors); and
    - the engagement and management of contractors;
2. Co-operation between workers, their representatives, and managers;
3. Communication throughout the organisations – worker and management behaviours, written material and in-person discussions; and
4. Competence developed through recruitment, selection, training and coaching and access to specialist advice. Individual competence allows people to carry out their work safely and without risk to their health.

Implementing the plan

This is the core of any health and safety management system. Once the principles have been set and roles and responsibilities identified, the hard work can begin. This part of the system is where day-to-day health and safety management happens.

It covers the major tasks performed by all those involved in health and safety. These include conducting risk assessments, writing safe systems of work, managing contractors, identifying training needs and delivering training.

Check

Measuring performance

Checking whether plans have been implemented and how well risks are being controlled is vital. An essential part of any management system is the need to look at past performance to see how well, or how badly, the issues are being managed and how well the system is performing. To ensure that this takes place, procedures and responsibilities for monitoring performance are required, together with suitable performance measures. Monitoring performance takes time and effort, and appropriate resources need to be allocated just as for any other aspect of the organisation's business. Selecting the right measures is key, as using the wrong measures will waste time and effort.

Monitoring

> **DEFINITIONS**
>
> *Active monitoring* involves monitoring the activities done to prevent accidents and incidents; for example, monitoring compliance with the health and safety management system and legislation.
>
> *Reactive monitoring* is about monitoring the deficiencies or failures in health and safety, such as accidents or incidents and other events with the potential to cause loss or harm.

Performance is typically monitored through active and reactive monitoring. Performance should be evaluated against predetermined standards to identify when action is needed to improve performance. Active and reactive monitoring aim to identify the causes of substandard performance and focus attention on areas of the health and safety management system requiring improvement.

Auditing may also be useful. At its most basic, the audit will seek simply to establish that the relevant elements are present; more sophisticated audits will make judgements and draw conclusions on the adequacy of what is being audited by assessing against standards. Internal auditing can be used to confirm the organisation is conforming to internal and legal standards. Developing an audit programme, including the scope, method and frequency of audits, is associated with achieving high standards of health and safety. You can find out more in 4.3: Health and safety auditing.

Investigating accidents and incidents

Even with good plans, accidents and incidents do happen. It is important that everyone in an organisation is prepared for these situations and knows what to do. So, plans to deal with an accident or emergency need to be regularly tested.

Findings from accident and incident investigations can inform action plans to prevent similar accidents or incidents from happening again and help identify areas of risk assessment that need to be reviewed.

Investigating near misses, where there has been no loss, is useful in preventing more serious incidents from occurring.

> **APPLICATION**
>
> What monitoring methods does your organisation use?

Act

Reviewing performance

Reviewing health and safety performance enables an organisation to confirm that its overall approach or system is effective and that the health and safety policy is valid. As well as closing the loop in the PDCA cycle, the outcome of the review forms the basis of the health and safety plan for the next cycle. It also provides an opportunity to celebrate health and safety successes with workers and their representatives. Some organisations also report their health and safety performance through their annual report or a separate corporate social responsibility report.

Additionally, comparing health and safety performance against other similar organisations with high standards for health and safety through benchmarking activities can provide ideas for improvement.

Take action on lessons learned

This section provides an opportunity within the health and safety management system to act on the findings of the 'review' stage; this will help to enhance performance and keep the safety management system up to date and fit for purpose.

Acting to remedy any shortcomings, such as an unidentified hazard that could lead to serious incidents and/or ill-health, and preventative action to deal with organisational vulnerabilities identified at the 'check' stage, allows for continual improvement of the system. An accident or incident may often reveal that, although systems, procedures and rules were in place, they were not complied with. Underlying causes can often be linked to arrangements that have not taken adequate account of human factors.

Organisational learning is important for health and safety management. If reporting and follow-up arrangements are not adequate, there can be a disincentive to report, resulting in valuable learning opportunities being lost. Without information about root causes, recurrence is more likely. Where departments in organisations operate in isolation, organisational learning will be inhibited.

2.1.2 The benefits of having a formal/certified health and safety management system

Having a formal health and safety management system in place, such as the two we have discussed, helps organisations prioritise the planning, organising, control, monitoring and review of measures. In so doing it supports a consistent approach with effective resource allocation. An informal approach may have some good practices, such as a policy, procedures and risk assessments, but is unlikely to offer the structured and consistent approach that can be achieved with a formal approach.

Many other aspects of an organisation's business are managed using well developed systems; adopting a similar approach for health and safety management helps to put health and safety on a level with other business issues. As the structure of other management systems, such as quality and environment, are similar, adopting a similar formal system for health and safety may allow the organisation to consider an integrated management system approach at some point in the future.

Implementing a health and safety management system also supports the achievement of legal compliance. This is through the identification of relevant legal requirements as well as other requirements that the organisation chooses to comply with. These other requirements could be things like industry best practice or formal management systems certification. Most organisations will achieve this by putting procedures in place that will ensure compliance with the requirements. This can also provide enforcement agencies with some reassurance about the organisation's approach.

The visible commitment of top management is a requirement of any management system, so adopting a health and safety management system ties top management into being actively involved and committed to it. As we have seen with ISO 45001:2018, this requirement is clearly spelt out. Most systems will also require the active engagement of the workforce too. So having a system provides a focal point for everyone in the organisation. It means that everyone will be clear on their role and responsibilities. Additionally, the health and safety of others, such as contractors, will also have to be considered.

Continuous improvement is a requirement of most management systems, including the ones discussed. Using audit and review, organisations can achieve incremental improvements in specific areas or in the management system itself identified through these processes.

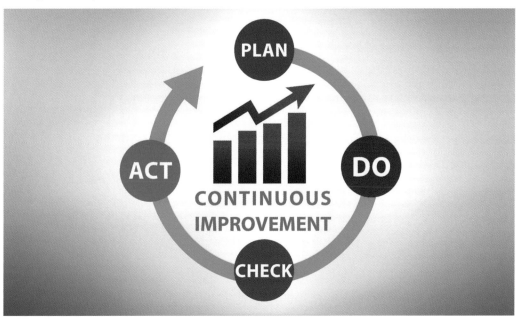

Figure 5: Cycle of continuous improvement

Demonstrating that health and safety controls are in place also supports the demands of corporate governance to demonstrate robust internal controls.

In summary, the potential benefits of formal health and safety management systems can be summarised as follows:

- evidence of corporate governance and meeting supply chain needs;
- alignment of health and safety with strategic direction;
- improved performance through proactive approach;
- increased efficiency and effectiveness;
- integration with other systems;
- structured and professional approach;
- timely identification of issues and opportunities;
- eliminated or minimised hazards;

- risks identified and managed;
- reduced errors, incidents and injury;
- developed learning organisation;
- improved culture and motivated workforce through engagement;
- increased leadership and worker involvement;
- reduced costs associated with health and safety failure and regulatory action;
- lower insurance premiums; and
- competitive advantage.

Some organisations desire written assurance by an independent third party that they are conforming to the requirements of a management system standard; this is referred to as 'certification'. ISO 45001:2018 provides for this third-party certification. Whether organisations seek certification may be influenced by the costs and the perceived benefits. Motivation to gain certification may come from internal and/or external stakeholders. Internal stakeholders (such as the leadership) may require assurance that the system meets a verifiable standard or may perceive that having certification adds value with clients or customers.

External stakeholders, such as investors or clients (existing or prospective), may require certification. In this case, having certification becomes an 'entry requirement' to operate or to bid for work. This is often the case when dealing with clients that are large multi-national organisations or those in the public sector.

Aside from the direct costs of certification, a significant amount of management time is needed to provide evidence of conformity and for the verification audits.

KEY POINTS

- Health and safety management systems provide a structured framework to help organisations manage the health and safety risks generated by their activities.
- Formal systems, such as ISO 45001:2018 and HSG65, are built on the elements of plan, do, check, act (PDCA), capturing the principle of continuous improvement.
- ISO 45001:2018 has built on experience of OSHAS 18001 and reflects the approaches of organisations that strive for and succeed at health and safety management.
- ISO 45001:2018 places increased emphasis on:
  - organisational context of the organisation's activities;
  - leadership and worker involvement;
  - documented information;
  - continual improvement;
  - hierarchy of control;
  - risk and opportunity management;
  - compliance status;
  - contractors, procurement and outsourcing; and
  - performance evaluation.

- HSG65 provides guidance on a framework for a systematic approach to effective health and safety management. It stresses that, for an effective health and safety management system, you need these core elements: leadership and management (together with good business processes), a competent workforce, an environment of trust and involvement of people. A prerequisite is that the organisation understands its risk profile. Like ISO 45001:2018, HSG65 also promotes the PDCA cycle to achieve sustainable continuous improvement.
- Formal health and safety management systems provide for the systematic, consistent, efficient and effective management of health and safety and its continual improvement. They can aid achievement of compliance, support integration with other systems, demonstrate adherence to corporate governance requirements and help sustain a positive health and safety culture.
- Certification of health and safety management systems may be required by internal or external stakeholders and provide assurance of conformity to recognised standards by an independent third party.

References

[1] International Organization for Standardization (ISO), 'ISO 45001:2018 *Occupational Health and Safety Management Systems: requirements with guidance for use*' (2018) (www.iso.org)

[2] Adapted from ISO 45001. See note 1

[3] HSE, *Managing for health and safety* (HSG65, 3rd edition, 2013) (www.hse.gov.uk)

[4] See note 3

[5] Adapted from HSG65. See note 3

[6] DuPont, 'Solution makers: 2020 GRI Index' (2020) (www.dupont.com)

2.2: What good health and safety management systems look like

> **Syllabus outline**
>
> In this section, you will develop an awareness of the following:
>
> - Statement of general policy – overall aims of the organisation in terms of health and safety performance:
>   - sets overall objectives and quantifiable targets (specific, measurable, achievable, reasonable, timebound – SMART principles)
>   - considers technological options
>   - proportionate to the needs of the organisation
>   - considers financial, operational and business requirements
>   - signed by top management
>   - important role in decision-making
> - Defined health and safety roles and responsibilities of people within the organisation:
>   - allocation of responsibilities; lines of communication and feedback loops; the role of line managers in implementing and influencing the health and safety management system and monitoring its effectiveness
> - Practical arrangements for making it work:
>   - the importance of stating the organisation's arrangements for planning and organising, controlling hazards, consultation, communication, monitoring compliance, assessing effectiveness
> - Keeping it current: when you might need to review the health and safety management system, including passage of time, technological, organisational or legal changes, and results of monitoring

As we have seen in 2.1: What they are and the benefits they bring are and the benefits they bring, most health and safety management systems are based on the principle of Plan, Do, Check, Act (PDCA). At the heart of an effective health and safety management system is a policy. This sets out a mission statement for health and safety and a mechanism for management control and accountability; it also sets a framework for implementing, monitoring, auditing and continuously improving.

The policy and the health and safety management system should evolve over time and allow for the continuous improvement of health and safety performance. Both should be subject to periodic management review. They should also be updated when other significant changes in the organisation, or the external environment, occur. We can best judge the quality of a health and safety management system by how effective it is in driving improvements in health and safety performance.

2.2.1 Statement of general policy – overall aims of the organisation in terms of health and safety performance

The organisation's policy should be a public declaration of commitment to high standards of health and safety management. It will show the organisation's commitment to ensuring that appropriate standards of health and safety are maintained. This is an important notion since organisations will need to live up to what they say in the policy. The wording is very important and must be carefully chosen to reflect the culture of the organisation.

The health and safety policy should state the overall aims in terms of health and safety and give a general statement of intent. Having clear aims and objectives makes it possible to build a system and aids in decision-making since the 'destination' (as identified by the aims) is clearly articulated and well known. If you know exactly what you are being asked to achieve then you can make suitable plans; but if your objective is unclear, or if it keeps shifting, then you will find it very difficult to plan. This will lead to a lot of waste in terms of time, effort and resources. A properly constructed and clearly expressed statement of policy should describe what the organisation ultimately wishes to achieve so that resources can be apportioned appropriately and with minimal waste. A well-worded policy statement allows for the effective management of health and safety risks.

If clear aims and objectives are stated, it will also be possible to more accurately allocate roles and responsibilities that will make the best use of people's individual skills and abilities. This will involve as many people as possible in a combined effort to manage health and safety risks.

Having a health and safety policy is a legal requirement under the Health and Safety at Work etc Act 1974 (HASAWA). The Management of Health and Safety at Work Regulations 1999 (MHSWR) also amplify this requirement. Additionally, it is an integral component of safety management systems like ISO 45001:2018. Therefore, if the law is to be complied with, or an effective system for the management of health and safety is to be implemented, it is essential to devise a policy.

The health and safety policy underpins everything else, so its importance cannot be overestimated. An effective health and safety policy should be specific to the organisation and its activities. It should be developed in consultation with the management and workforce. It will need to be effectively communicated, monitored and revised as appropriate.

The health and safety policy usually consists of a policy statement and a policy document. The policy statement is usually no more than a page long, containing a brief statement of the organisation's attitude towards, and commitment to, health and safety. The best policy statements are personal statements of commitment from the most senior manager in the organisation. The policy statement does not go into detail but gives sufficient information to establish the standard to be achieved.

The policy document is a much more detailed document that includes the policy statement but adds the 'Organising' and 'Arrangements' sections to create a complete reference for use by employees and others. These are outlined in Figure 1.

Figure 1: The components of a safety policy document

ISO 45001:2018 (clause 5.2) also provides a steer on the contents of the policy.[1] It states that the top management in the organisation must take the lead on setting up the policy, implementing it and maintaining it. The policy must be fit for purpose, taking account of the organisation's size, context, role and the risks and opportunities the organisation presents. It should set aims and objectives for health and safety and a commitment to:

- a safe and healthy working environment for preventing injury and ill-health;
- meeting legal and other requirements;
- eliminating hazards and reducing risks;
- continually improving the system; and
- consulting and engaging workers and their representatives in dialogue about health and safety matters that affect them.

CASE STUDY
Example health and safety policy statement

The ABC Group policy aims to prevent incidents and ill-health by documenting, implementing, maintaining and continually improving an occupational health and safety management system. The system is to control health and safety risks and performance arising from all of our work activities.

Our health and safety management system meets the requirements of ISO 45001:2018 in respect of activities carried out by ABC Group companies.

The purpose of this policy is to set out the framework for setting objectives and the processes and procedures that, through correct delegation of duties, will as far as reasonably practicable:

- Provide and maintain safe and healthy working conditions for the prevention of work-related injury and/or ill-health.
- Provide adequate welfare facilities.
- Eliminate hazards and reduce occupational health and safety risks using the hierarchy of controls.
- Provide the essential, relevant and appropriate level of health and safety information.

- Provide instruction and training for all workers and other persons working under the control of ABC Group.
- Fulfil legal and other requirements.
- Ensure consultation with, and participation of, workers and our health and safety representatives on matters affecting their health and safety in the workplace.
- Ensure health and safety is not compromised by other business objectives.
- Provide an open-door policy to talk to a safety or HR representative.
- Ensure continual improvement of the occupational health and safety management system.

The ABC Group's senior management give their full support and commitment on all occupational health and safety issues and will provide sufficient funds and resources to enable correct implementation of this policy, while meeting all expectations of our stakeholders.

The successful implementation of this policy relies on all duties and responsibilities being fulfilled, including compliance with applicable legal requirements, and with other requirements that relate to ABC Group's occupational health and safety hazards, with the aim of continual improvement through programming, monitoring and periodic reviews. Objectives will be set at the management review meetings.

This document is reviewed at least annually or when changes occur, and a revision may be required.

A N Other
Chief Executive
ABC Group plc

SIGNED:
DATE:

This policy is displayed in all offices and is available to all workers via the ABC Group intranet. All workers are encouraged to read it and communicate any queries to a director.

A copy of this policy can also be obtained on request from the Group Health and Safety Director (to any interested parties) and is available on the ABC Group website www.abcgroup.com

We will look at the organisation and arrangement elements later (see 2.2.2: Defined health and safety roles and responsibilities of people within the organisation).

APPLICATION

Compare your organisation's health and safety policy against the content that we have discussed. Are there any differences? Can you identify any improvements?

Objectives and quantifiable targets

Objectives are often used to set the health and safety policy.

The organisation's aims in the health and safety policy may endure over time; for example, the aim to prevent work-related injury, ill-health and incidents. These set the overall framework for the objectives, which may change annually.

The health and safety objectives should be directly linked to the work of the organisation and its overall objectives. They need to reflect the nature of the hazards and risks presented by the work that the organisation does. Having objectives linked to specific risk issues helps to ensure adequate focus on managing these issues. For example, in a transport and distribution company, there may need to be a focus on introducing improved layout and lighting of the parking bays to help reduce the risks of accidents due to reversing.

Objectives should also consider available technologies and any advances in these technologies since the previous review of objectives. For example, the introduction of vehicle-reversing alarms and in-vehicle cameras. As well as linking to the overall organisational objectives, health and safety objectives should also avoid conflicting with other financial, business and operational objectives to avoid goal conflict. For example, an operational objective to increase production in isolation of health and safety considerations could introduce conflict. Incorporating health and safety objectives linked with wider objectives into individual managers' objectives helps ensure focus and overall alignment.

Setting quantifiable performance targets for health and safety objectives allows for the monitoring and measurement of progress towards the target. It also makes clear what is to be achieved and by when. If an objective is insufficiently defined it is unlikely to result in the necessary action to achieve it. So 'preventing accidents' or 'increasing worker involvement' are too vague or undefined. Well-designed and expressed targets need to have certain attributes. The SMART approach to target setting can be used to ensure that this happens. Objectives need to be Specific, Measurable, Achievable, Realistic and Time-bound:

- Specific – to make an objective specific, consider who is involved, what needs to be done, where and by when, taking account of any requirements, and why.
- Measurable – measurable objectives should have quantifiable indicators such as a number or a percentage to achieve.
- Achievable – an achievable objective considers the reality of achievement in the time available, which means there needs to be consideration of any barriers to success.
- Reasonable – reasonable objectives should motivate those who have to achieve them to do so. They need to link to the overall organisational purpose and relate to health and safety management.
- Time-bound – a time-bound objective includes a timescale for achievement of the objective. This is more likely to lead to planning to achieve the objective in the time frame than an open-ended objective.

The objective 'all workers will receive a one-day course on working safely within three months of joining the organisation' is specific, measurable and time-bound, and achievable and realistic for the organisation.

ADDITIONAL INFORMATION

Objective

All workers will receive health and safety training relative to their role.

SMART objective

All workers will receive a one-day course on working safely within three months of joining the organisation.

If a target is set over a long period of time, such as a year or longer, it might be useful to set intermediary targets that lead to the overall target. So quarterly targets may be appropriate. This also helps improvement across the period and avoids the need to rush to achieve the target close to the end of the year.

Targets need to be linked to the objectives but can cover a range of topics, such as:

- reductions in accidents/incidents and work-related ill-health;
- improvements in the number of workplace inspections, workers trained or completed risk assessments; and
- improvement in the extent of compliance with statutory requirements or audit recommendations closed out on time.

Objectives and targets need to be documented and communicated. Responsibility for achieving objectives should be assigned to managers. Managers with responsibility for meeting objectives should monitor progress, which should be reported periodically. This will help identify any problems in fulfilling the objective and give the chance for plans to be amended. Objectives and targets are usually reviewed annually as part of an annual planning process. This helps to ensure objectives remain relevant and support continual improvement.

> **APPLICATION**
>
> Compare the health and safety objectives of your own organisation with the SMART criteria. Can you identify any improvements?

Proportionate to the needs of the organisation

While having a policy should be a common feature of most organisations, the exact contents will differ according to the organisation's aims, risk, structure and resources. Policies should evolve as an organisation grows or changes and its approach to health and safety management changes. So, it is important that the appropriateness of the policy is monitored over time.

A small organisation with low risks may have a relatively simple policy compared with a larger, more complex organisation with high levels of risk. A larger organisation will need a policy with wider scope and much greater depth. Essentially, the policy needs to be proportionate to the needs of the organisation and written using language and tone that will be suitable for the reader. This means avoiding complex technical or legal terms or the inclusion of unnecessary information.

Signed by top management

Whatever the size of the organisation or its complexity, the policy should always be signed by the most senior accountable person in the organisation. Not only does signing the document show that the most senior person has accepted their responsibilities, it also gives authority to the policy and demonstrates senior management commitment to it. It emphasises that health and safety is as important as other business objectives. Adding a date to the statement indicates its currency and the last time it was reviewed; this can give impetus to ensuring it is kept up to date.

Supports decision-making

'Policy' is a generic term used to describe a set of rules that guide decision-making. In the context of health and safety, the term 'policy' can be used to describe the organisation's aims or the parameters within which decisions are made.

For example, Company A's policy might be to 'comply with health and safety law', whereas that of Company B might be to 'put the safety, health and welfare of workers before commercial objectives'. The standard set in the policy of Company A is a low one: simple legal compliance. However, the policy expressed by Company B sets a much higher standard and may even run the risk of creating a risk-averse environment. So, the policy sets the overall tone of how the organisation operates.

Although policies have common components, they must reflect the context, needs and culture of the organisation. To be effective, the policy requires communication to all members of the organisation and external interested parties, such as contractors, neighbours and suppliers. An effective communication plan needs to accompany a health and safety policy – simply putting it on a noticeboard and hoping it will be seen, understood and implemented is not enough. Running briefing sessions, making the policy widely available across the organisation (such as through the intranet) and using formats and languages suitable for the workforce are necessary for effective communication. For new workers, the policy can be included in the induction process.

2.2.2 Defined health and safety roles and responsibilities of people within the organisation

Everyone in the organisation has a role and responsibilities to achieve the aims set out in the policy statement. So, it is essential that these roles and responsibilities are clearly defined and people are made aware of them. The policy should clarify that the fulfilment of the policy requires workers to accept responsibility for health and safety within their control (ISO 45001:2018 clause 5.3). Note, however, that although responsibility can be assigned, top management are ultimately accountable for the health and safety management system.

Employers

Employers must allocate responsibility, accountability and authority for the development, implementation and performance of the health and safety policy. Roles with responsibility, authority and accountability for identifying, evaluating and controlling risks must be established. This includes the managing director, the directors, line managers and employees, plus those with specific safety-related jobs such as safety representatives, first-aiders and fire wardens.

It is also sensible to nominate a director who has overall responsibility for health and safety in the organisation. This sends a clear message that health and safety issues are taken seriously at board level and facilitates top-level decision-making.

Line managers

Line managers are an important group, since they play a pivotal role in influencing attitudes to health and safety among the workforce. Line managers occupy the space between senior management and 'shop floor' workers. They are ideally positioned to see health and safety from both sides. They are the people who many 'shop floor' workers will look to for guidance and leadership since they are in a position of responsibility (being the first tier of management). However, they are still close enough to the work on the 'shop floor' to be able to understand and empathise with the workers' point of view. The example they set will often be a crucial factor in setting the tone for compliance with health and safety best practice. This will ultimately establish the nature of the safety culture in the organisation.

2.2 What good health and safety management systems look like

Line managers are also ideally positioned to monitor the effectiveness of safety initiatives and controls. Without their input, it may be that problems with the implementation of some safety initiatives go unnoticed. This may mean that the more senior managers are unaware and assume that things are proceeding as planned when the opposite is true.

All workers

Clearly, the reference to 'all workers' shows that everyone has some degree of responsibility for managing health and safety at work. There are some responsibilities that are common to all workers.

Examples of roles and responsibilities for a health and safety adviser and workers are shown here (this is not an exhaustive list).

Health and safety adviser
(a) Shall supervise the company health and safety programme
(b) Shall regularly inspect the plant to ensure that the programme is being complied with and make recommendations directly to the executive concerned on matters relating to health and safety
(c) Must monitor that action on all work necessary to ensure health and safety, as agreed by management, is being taken and findings reported back as appropriate
(d) Shall, in conjunction with the human resources department, arrange adequate material and publicity for the health and safety programme
(e) Shall investigate all accidents and recommend corrective action
(f) In conjunction with the works engineer, shall inspect new and temporary equipment and processes for potential hazards
(g) Shall recommend any necessary safety rules
(h) In conjunction with the works engineer, shall inspect and investigate all new plant and plant layouts before they are commissioned
(i) Shall be responsible for arranging instruction courses

Workers
(a) Shall make themselves familiar with the departmental safety manual
(b) Shall observe and conform to all safety rules at all times
(c) Shall wear appropriate safety clothing and safety equipment and use appropriate safety devices at all times
(d) Shall conform to all advice given by the safety officer and instructions of others with a responsibility for health and safety
(e) Shall report all accidents and damage to the departmental or section supervisor whether workers are injured or not
(f) May make suggestions to improve health and safety in the workplace
(g) Must report all hazards to supervisors

> **APPLICATION**
>
> Consider the policy document for your own organisation. Are all the relevant roles and responsibilities adequately defined? Can you identify any improvements?

2.2.3 Practical arrangements for making it work

This is all about how the aims are put into practice. The 'arrangements' section of the policy can be seen as a guidance document on how health and safety is managed in the organisation – a 'safety handbook' perhaps. In theory, if a new worker wishes to find the answer to any health- or safety-related question, they should be able to find it in the 'arrangements'. If they cannot, then the answer should probably be added at the next policy review.

There are no hard and fast rules on how the 'arrangements' section of a policy document should be structured. It does need to reflect the context, nature and size of the organisation. Some arrangements may be common across organisations, although there would be a distinct difference between the specific arrangements for a utility company and those of a hospitality organisation. Arrangements can be divided into general arrangements and specific arrangements.

General arrangements need to be clarified for such things as planning and organising, controlling hazards, consultation, communication, monitoring and assessment of the effectiveness of the system. This is because the policy document will be used as the main source of reference for 'how we do things round here' and so it becomes a key element in establishing consistency of approach, helping to ensure that nothing is missed and that safety is managed consistently across the organisation.

Specific arrangements take each health and safety issue separately and explain how that issue is to be managed by the organisation. An 'issue' might be:

- a hazard or hazardous activity (eg noise, work equipment, hazardous substances, electricity, fire, confined spaces, work at height, manual handling);
- a control measure (eg PPE, training); or
- an administrative procedure (eg risk assessment, monitoring and review, consultation, contractor safety, incident reporting, emergency situations).

For each issue, the arrangements section should normally specify who is responsible for doing what and when. For instance, under 'consultation', the arrangements might say who chairs the safety committee, who is represented on it and how often it meets. Similarly, under 'electricity' the arrangements might state how often portable appliance testing (PAT) is carried out, how the results are recorded and the procedures for identifying and isolating any electrical equipment found to be dangerous prior to repair or disposal. Alternatively, the same information might be included under the heading of 'maintenance'. If detailed work procedures, manuals or organisational codes of practice exist, then there is no need to repeat the details here, but a clear cross-reference to the other document should be provided.

The extent to which arrangements are documented will depend on the size, nature and complexity of the organisation, the need to demonstrate compliance with legal and other requirements and the competence of workers. Nevertheless, documentation does need to be a consideration, as set out in ISO 45001:2018 clause 7.5. The extent and means of documentation will need to take account of what is necessary for health and safety.

With large organisations, most of the detailed arrangements are likely to be contained in separate documents and/or are developed by individual operational units rather than at corporate level. However, the organisational policy should adequately describe the corporate arrangements for issues such as consulting with employee representatives and reviewing performance.

2.2.4 Keeping it current: when you might need to review the health and safety management system

The health and safety management system will need to be maintained to ensure its ongoing effectiveness. Both the system and the policy document are living things, in the sense that they are constantly evolving to meet changing organisational requirements and the demands of the outside world. To ensure that the right changes happen at the right time, the system will need to be reviewed. We will now look at the circumstances when a review might need to take place.

Periodic review

Reviewing the system periodically, perhaps on an annual basis, will provide a 'sense check' that it remains fit for purpose and will help maintain its profile with the management team. ISO 45001:2018 clause 9.3 indicates that the management review should consider:

- progress with actions from previous management reviews;
- any changes in the external and internal environment that are relevant to the management system, such as changes in legislation or industry standards or an organisational restructure;
- the extent to which the policy and objectives have been met;
- health and safety performance information;
- adequacy of resources for maintaining the effectiveness of the system;
- communications with interested parties; and
- opportunities for continuous improvement in health and safety performance or the performance of the system itself.

The health and safety management system review is the ideal time to conduct the annual review of the policy, since the two are inextricably linked. Therefore, even if there are no other obvious reasons for reviewing the policy (such as those outlined), it should still be reviewed periodically. The exact frequency of the review will be determined by the organisation.

Changes in the organisation

One of the most important aspects of the policy is that it must be signed by the most senior person in the organisation. But what if they leave? The new senior person may have differing thoughts about safety and health management so these will need to be clearly expressed in a change to the policy. Even if the new senior person agrees completely with what their predecessor has said, they will still need to sign a fresh copy of the policy to emphasise to stakeholders such as workers and customers that health and safety is still of paramount importance.

If the structure of the organisation changes, then this may affect the way in which communication flows around the organisation and, with it, how safety is managed. Again, the policy document (in particular the 'organisation' element) will need to be updated to reflect such changes.

Technological changes

As new technology is introduced into a workplace, new hazards may well be created or the method of work altered in a way that might create risks. This may well lead to a need for changes to be made in the arrangements, perhaps with reference to such things as machinery maintenance, use of guarding, competence requirements, lock-off systems and so on.

Changes in legislation

These will occur periodically, and it is essential to keep up with these changes. A failure to do so may mean that the organisation is falling short of its legal obligations and may expose itself to legal action. Fortunately, however, it is likely that changes in the law will not just suddenly occur without warning. There is usually a period of public consultation before a new law is introduced or an old one altered. It is also quite normal for there to be a 'transition period', allowing time for organisations to make the necessary arrangements to comply with new or amended legal requirements.

Therefore, it is important for an organisation to have some means by which changes in the law can be brought to its attention at an early stage so that it has plenty of time to adapt.

Following the result of monitoring

Monitoring is an essential element of any health and safety management system. Monitoring may from time to time uncover areas that are not as well managed as they might be, in which case improvements may be considered necessary. For example, regular inspections may identify shortcomings in fire safety caused by poor housekeeping; the arrangements for housekeeping can therefore be strengthened to help address this problem.

It is not just internal monitoring that can throw up a need to review the system; visits from enforcement officers and insurers can also flag up issues that may have gone unnoticed. This could be because workers are too close to the system and an impartial third-party view can see something that workers cannot.

KEY POINTS

- Effective health and safety management is based on the PDCA cycle and built on a policy that sets out a clear statement of commitment and vision, which creates a framework of accountability.
- A good system is effective in driving continuous improvement in health and safety performance.
- The policy should be a public declaration of commitment to high standards of health and safety management in an organisation. It sets out the aims and objectives to be achieved.
- The policy needs to take account of the needs of the organisation – its context, nature, size and complexity. It needs to be effectively communicated to all members of the organisation and interested parties.
- Objectives need to be specific, measurable, achievable, realistic and time-bound (SMART).

2.2 What good health and safety management systems look like

- All workers at all levels of an organisation have responsibility for health and safety; however, some, such as top management, have more responsibilities and are accountable. Other roles, such as health and safety practitioners or first-aiders, have specific responsibilities.
- The arrangements section of the policy document sets out 'how' health and safety is going to be managed. Usually it includes general arrangements and specific arrangements.
- General arrangements cover things like planning and organising, controlling hazards, consultation, communication, monitoring and assessment of the effectiveness of the system.
- Specific arrangements take each health and safety issue separately and explain how that issue is to be managed by the organisation.
- The health and safety management system requires maintaining to ensure its effectiveness. Circumstances such as the passage of time, technological, organisational or legal changes and results of monitoring would warrant a review of the system.

Reference

[1] International Organization for Standardization (ISO), 'ISO 45001:2018 *Occupational Health and Safety Management Systems: Requirements with guidance for use*' (2018) (www.iso.org)

ELEMENT 3

MANAGING RISK – UNDERSTANDING PEOPLE AND PROCESSES

3.1: Health and safety culture

Syllabus outline

In this section, you will develop an awareness of the following:

- Meaning of the term 'health and safety culture'
- Relationship between health and safety culture and health and safety performance
- Indicators of an organisation's health and safety culture:
  - incidents, absenteeism, sickness rates, staff turnover, level of compliance with health and safety rules and procedures, complaints about working conditions
- Influence of peers on health and safety culture

In Element 2: How health and safety management systems work and what they look like, we discussed the importance of effective occupational health and safety management systems built on a policy that sets a clear vision. However, occupational health and safety management systems are not effective in workplaces with an unsatisfactory health and safety culture. To be fully effective, a health and safety management system must be supported by a positive health and safety culture.

3.1.1 Meaning of 'health and safety culture'

The term 'safety culture' first came to prominence following major accidents in the late 1980s, including the Chernobyl nuclear power plant disaster in 1986 and the Piper Alpha oil rig explosion and fire in the North Sea in 1988. The term appeared in the International Atomic Energy Agency summary report into the Chernobyl disaster to describe how the thinking and behaviours of people responsible for safety had contributed to the accident. From then onwards, the importance of creating a corporate atmosphere or culture in which health and safety is understood to be, and is accepted as, a priority, has been recognised.

3.1 Health and safety culture

> **DEFINITION**
>
> *The Health and Safety Commission's Advisory Committee on the Safety of Nuclear Installations (ACSNI) has given this definition of **safety culture**:[1]*
>
> *The safety culture of an organisation is the product of individual and group values, attitudes, perceptions, competencies, and patterns of behaviour that determine the commitment to, and the style and proficiency of, an organisation's health and safety management.*

Early definitions of culture focused on people's beliefs, attitudes and behaviours rather than the wider context of the organisation, such as the physical infrastructure and the management systems in which people work. More recently, safety culture is considered to take a more holistic approach that considers an organisation's total approach to health and safety management, including physical and management controls:[2]

> Safety culture can be defined as those sets of norms, roles, beliefs, attitudes and social and technical practices within an organisation which are concerned with minimising the exposure of individuals to conditions considered to be dangerous.

A simpler and less formal definition that is often used is that safety culture is 'the way we do things round here'.[3]

The term 'safety culture' is now more commonly known as 'health and safety culture' to indicate interest in work-related health.

Health and safety culture can be made up of three layers:[4]

- Outer layer (actions) – the way people and the organisation itself behave are the most observable level of health and safety culture; for example, whether people keep the workplace tidy, managers lead by example and workers wear personal protective equipment (PPE).
- Middle layer (feelings) – the values, beliefs and attitudes that people and the organisation hold underpin, and are a strong determinant of, the observable level; for example, whether health and safety is seen as a matter of legal compliance or as good business sense.
- Central core (principles) – the fundamental principles that drive how the organisation operates, ie its ethos.

3.1.2 The relationship between health and safety culture and health and safety performance

Although it is probably true to say that good health and safety cultures often appear in well-run organisations, it would not be right to say that if the health and safety culture is good then the organisation will be successful. However, there is certainly a link between health and safety culture and health and safety performance.[5] Having a more positive health and safety culture is linked to a better safety performance. Given the significance of organisational health and safety culture to health and safety management and the prevention of injury and work-related ill-health, assessing culture as part of continuous improvement is advised. Culture manifests itself in several ways and so there is no single measure. It requires a range of indicators linked to the outcomes of, and the inputs to, health and safety management.

Decisions in organisations are influenced by their culture and this influence may be positive or negative. Some organisations may have a health and safety culture that is focused on minimum compliance with national laws while others may be focused on taking an approach of continuous improvement, and there will be organisations with approaches in between these extremes. So, we can say a health and safety culture has different levels of maturity.

> **TIP**
>
> **Creating a positive health and safety culture**
>
> Organisations with strongly positive health and safety cultures will have effective health and safety policies in place:
>
> - Clear leadership and direction from the senior management team will back up the health and safety policies.
> - All workers will share similar positive ideals and values about safety and health and will therefore work together as a team to create and maintain a positive health and safety culture.
> - Workers will look out for the safety of their co-workers and others.
> - There will be adequate investment in health and safety training, advice and equipment.

With all workers thinking and acting safely, the accident rate is likely to be low. This means that less money will be wasted through the costs of accidents, which translates to lower losses from the organisation's 'bottom line'. Senior management will see the benefits in a happier and better motivated workforce, who not only have fewer accidents but also work more efficiently and are more productive.

CASE STUDY 1

A gas utilities group wanted to further reduce injuries, illnesses and unsafe behaviours. To increase focus on health and safety it incorporated its health and safety strategy into the overall business management plan and set about increasing the involvement and participation of workers and their health and safety representatives. It did this by, for example:

- providing everyone with informative and visual monthly updates on issues and improvements;
- ensuring that a manager and safety representative investigated lost-time injuries on the day of the incident, and that any lessons learnt were quickly communicated to staff;
- defining the roles and responsibilities of line managers and supporting them through a staff performance review process and safety management training;
- making it easier for staff to report incidents and hazards through a 'hotline';
- making sure health and safety was the first item on the agenda of all management meetings;
- having frequent meetings between safety advisers, safety representatives and managers; and
- ensuring that directors supported the collaborative approach through good communication, attending management meetings and meeting staff members.

> To support the new ways of working there was significant investment in safety management and behaviour training. As a result, the company saw the following improvements:
>
> - an improved safety culture – including ownership at all levels, with commitment and competence improvements;
> - improved incident investigation and procedures to help prevent incidents happening again;
> - increased reporting and resolution of hazards and near misses;
> - staff development of health and safety leadership skills transferable to other business performance areas;
> - improved staff morale and pride because of management acknowledging their performance;
> - improved reputation with stakeholders; and
> - a reduction in accidents, incidents and injuries of over 80% and in lost-time injuries from 35.5 per 1000 staff to 6.6 over five years.

An inadequate or negative health and safety culture will be characterised by an organisation that fails to appreciate and value health and safety practice. Workers will not be interested in health and safety and general management and leadership will be unsatisfactory. There is likely to be a culture of blame, whereby workers are blamed if they have accidents or inadvertently damage equipment. Safety and health will be a much lower priority than other areas of business risk. The net effect will be that there will be less investment in health and safety.

Less investment in health and safety means that workers will be less health and safety conscious as they will not have had the necessary training. Tools and equipment may be badly maintained and so may be more dangerous as a result. Guards on machines may be broken or missing. The culture of blame will lead workers to keep quiet about shortcomings that may affect their health and safety, for fear of being disciplined. A situation such as this is likely to see a high accident rate and a reduction in morale and will probably result in workers leaving the organisation.

The first workers to leave will be those who feel they can get a better, not to say safer, job elsewhere; these are likely to be the more highly skilled, competent workers who are confident in their ability to secure alternative employment. Once they leave, average skill levels will reduce, as will efficiency levels. Less experienced workers will now do more hours. It follows that they will almost certainly have more accidents, leading to a further drop in morale. The cycle continues.

3.1.3 Indicators of an organisation's health and safety culture

Developing and encouraging a positive health and safety culture is key to effective health and safety management and performance. As discussed, health and safety culture is made up of a number of facets (aspects), so to assess or measure this requires the use of a broad range of indicators. This means identifying tangible indicators of performance that reflect some of the underlying aspects of health and safety culture.

Typical indicators of health and safety culture include:
- incident and work-related sickness rates;
- absenteeism rates;
- staff turnover;
- level of compliance with health and safety rules and procedures; and
- complaints about working conditions.

Incident and work-related sickness rates

This can be looked at in two ways. The obvious way is to say that an organisation having more incidents and accidents is likely to be less safe and so may have an ineffective health and safety culture. However, this may not necessarily be the case. An organisation with lots of reported accidents may have a good and well-used reporting system, which indicates an improving health and safety culture, whereas one with seemingly low incident, accident and work-related sickness rates may have an inadequate reporting health and safety system and/or a blame culture.

Absenteeism rates

One facet of a good health and safety culture is that workers tend to be happier and better motivated, with higher morale. This translates to lower levels of absenteeism. Contrast this with an organisation that has an unsatisfactory health and safety culture – in an extreme situation, workers may take every opportunity they can not to go into work for fear of being injured or made ill. It may also be the case that workers are unable to come into work because they have been injured or made ill by their work. Therefore, relatively high absenteeism rates should be investigated as they may reveal an ineffective health and safety culture.

Staff turnover

Relatively high levels of staff turnover tends to be linked to some sort of disquiet among the workforce, one possible reason being low morale caused by a perception that 'management' do not care about the safety and health of the workforce. Alternatively, those organisations with a positive health and safety culture tend to be better places to work, with correspondingly higher levels of staff morale and lower staff turnover.

Level of compliance with rules and procedures

Workers who are well motivated and have good morale will tend to follow rules and procedures more readily than those who are not well motivated and have low morale. Even though an organisation has plenty of rules and procedures, we cannot say that it has a good health and safety culture until we take the time to assess the degree of compliance with those rules. If an organisation is not compliant with its own health and safety rules and procedures, it is not taking the management of health and safety issues seriously, indicating a health and safety culture that is liable to be unsatisfactory.

Complaints about working conditions

Organisations with positive health and safety cultures genuinely want to know when there are shortcomings with health and safety management arrangements so that action can be taken to continuously improve standards. The fact that the health and safety culture is positive will probably mean that comparatively few serious issues are raised, but when they are brought to the attention of the management team they get dealt with efficiently and effectively.

Once again, organisations with negative health and safety cultures will do the exact opposite. They may not have any effective system for reporting deficiencies in working conditions, or they may even actively discourage reporting, perhaps branding workers who do report problems as 'troublemakers', even though the complaints that are being made are perfectly legitimate.

Other indicators used to assess health and safety culture relate to what the organisation is doing to manage health and safety. When an organisation has a range of indicators assessing its inputs to health and safety management there is likely to be a positive health and safety culture. Indicators of this type could include: the extent of visible leadership and commitment; inclusion of health and safety in the annual report; the amount and effectiveness of health and safety training; and completion of actions following inspections, audits and investigations.

When carrying out an assessment of health and safety culture it is good practice to consider indicators that reflect both the outcomes of and inputs to health and safety management.

Having noted some typical indicators of a health and safety culture, we should also take a moment to consider those features that may cause a health and safety culture to decline. A health and safety culture may decline due to the following:

- Workplace reorganisation and the uncertainty that such changes create. This may have an adverse effect on morale, which may result in a reduced level of commitment.
- Inadequate control exercised by senior management, which may manifest itself through inconsistent behaviour (eg not complying with their own rules, such as wearing PPE), or unsatisfactory standards of supervision (eg ignoring breaches of the rules).
- Insufficient resources, such as money, time and competent personnel.
- Bad relationships between managers and workers, leading to low levels of co-operation.
- Insufficient risk assessment, leading to unsatisfactory standards of decision-making and inadequate controls.
- Not enough awareness and/or training, resulting in low levels of competence.
- Inadequate communication and understanding of priorities eg conflicts between productivity and safety.

APPLICATION

- What indicators of a positive health and safety culture does your organisation show?
- What indicators of a negative health and safety culture does your organisation show?

3.1.4 The influence of peers on health and safety culture

When workers come together in groups, acceptable rules of behaviour, dress etc will be established. These rules are frequently referred to as 'norms'. Most people will feel a strong desire to conform to these 'norms'.

In any group, some people will have more influence than others. Interestingly, this may be for a variety of reasons, and not simply because a particular person has been given the title of 'manager' or 'supervisor'; sometimes relatively junior workers may be very influential. The influential people will establish the acceptable patterns of behaviour, which others will then follow. This will develop to the point where a worker might consider doing one thing but will end up doing what they think will be acceptable to the group. This is known as 'peer pressure' and it can be positive or negative.

If the influential person in the group accepts the need for high standards of health and safety, then it is probable that the rest of the group will do the same, the outcome being a positive health and safety culture. This is one of the main reasons why we have stressed the importance of health and safety management being led by the most senior people in the organisation.

Conversely, if the most influential person in the group views health and safety as a waste of time, a negative health and safety culture is likely to result.

Case study 2 illustrates the effect that peer pressure can have.

CASE STUDY 2
A young worker has just left college, where he has been taught how to dress correctly in his overalls and how to work safely when using machines and hand tools. He arrives for his first day with his new employer, keen to make a good impression and 'fit in'. When he starts to work, he sees that the older workers are wearing what he knows to be inappropriate clothing and are not observing the same safety precautions that he has been taught. As he wants to fit in with the group, he will feel an urge to model his behaviour on theirs, so succumbing to peer pressure.

When considering how best to develop a positive health and safety culture, a useful technique is to identify the most influential person in a group and get them to agree with your point of view. If this is done successfully, then the rest of the work group will follow their lead. Other work groups will then notice the success and will start to do the same, thus spreading the positive health and safety culture throughout the organisation.

KEY POINTS

- 'Safety culture' first came to prominence in the later 1980s following a series of major disasters.
- Health and safety culture is the product of norms, roles, beliefs, attitudes and social and technical practices within an organisation, aimed at minimising health and safety failures.
- A less formal definition of safety culture is the 'way we do things round here'.
- Health and safety culture can be made up of three layers. The way people and the organisation behave (outer layer – actions) is much more visible than their values, attitudes and beliefs (middle layer – feelings) and the ethos of the organisation (central core – principles).
- Health and safety cultures have different levels of maturity (for example, minimum compliance with the law versus continual improvement).

- There is a link between health and safety culture and health and safety performance. Organisations with a health and safety culture that promotes a preventative approach are more likely to have a better health and safety performance than those that underinvest in health and safety.
- A broad range of indicators is required to assess health and safety culture. Typically, these will relate to health and safety outcomes such as incidents, absenteeism, sickness rates, staff turnover, level of compliance with health and safety rules and procedures and complaints about working conditions. Others relate to the inputs to managing health and safety, such as visible leadership and effectiveness of training.
- The influence of peers can positively or negatively affect health and safety culture. Influential people in a work group can be used to support and promote the development of a positive health and safety culture.

References

[1] Health and Safety Commission, *ACSNI Study Group on Human Factors*, Third Report (HSE Books, 1993)

[2] B Toft and S Reynolds, *Learning from Disasters: A Management Approach* (Perpetuity Press, 1999)

[3] Confederation of British Industry, *Developing a Safety Culture: Business for Safety* (Confederation of British Industry, 1990)

[4] FW Guldenmund, 'The nature of safety culture: a review of theory and research' (2000) 34 *Safety Science* 213–257

[5] A Smith and E Wadsworth, 'Safety Culture, Advice and Performance, Report submitted to IOSH Research Committee' (IOSH, 2009) (iosh.com/safetyculture)

3.2: Improving health and safety culture

Syllabus outline

In this section, you will develop an awareness of the following:
- Gaining commitment of management
- Promoting health and safety standards by leadership and example and appropriate use of disciplinary procedures
- Competent workers
- Good communication within the organisation:
  - benefits and limitations of different methods of communication (verbal, written and graphic)
  - use and effectiveness of noticeboards and health and safety media
  - co-operation and consultation with the workforce and contractors, including:
    - appointment, functions and entitlements of worker representatives (trade union appointed and elected)
    - benefits of worker participation (including worker feedback)
    - the role of health and safety committees
- When training is needed:
  - induction (key health and safety topics to be covered)
  - job change
  - process change
  - introduction of new legislation
  - introduction of new technology

Having set out what we mean by 'health and safety culture' and its importance to an organisation's health and safety performance, we turn our attention to how the health and safety culture of an organisation can be improved.

In general, health and safety culture can be improved through a range of measures, including:

- securing the commitment of management, who must lead by example;
- appropriate use of disciplinary procedures;
- use of competent workers;
- effective communication;
- effective worker co-operation and consultation; and
- training.

We have seen already that all these measures are important for an effective health and safety management system. Indeed, organisations that have an effective health and safety management system that genuinely drives improvement in performance are likely to have a positive culture. However, even where a positive culture exists there is likely to be scope for improving it. A health and safety culture will grow and mature over time with attention to the measures we have outlined.

3.2 Improving health and safety culture

> **TIP**
>
> The legislation relating to worker consultation and provision of information in the UK is contained in the following regulations:
>
> - the Safety Representatives and Safety Committees Regulations 1977, which apply to workplaces where the employer recognises trade unions and where trade unions are recognised for collective bargaining purposes;
> - the Health and Safety (Consultation with Employees) Regulations 1996, which apply to workplaces where employees are not in a trade union and/or the employer does not recognise the trade union, or the trade union does not represent those employees not in the trade union; and
> - the Health and Safety Information for Employees Regulations 1989.
>
> Guidance on these regulations can be found in the following Health and Safety Executive (HSE) publications:
>
> - *Consulting workers on health and safety. Safety Representatives and Safety Committees Regulations 1977 and Health and Safety (Consultation with Employees) Regulations 1996 (as amended). Approved Codes of Practice and guidance* (L146);[1] and
> - *Involving your workforce in health and safety: Guidance for all workplaces* (HSG263).[2]

3.2.1 Gaining commitment of management

This starts with the most senior managers, at board level, and cascades down through the various layers of management, so that all managers show that they understand the need for, and are committed to, effective management of health and safety in their respective departments. The absence of management commitment can lead to a lack of support for health and safety from the workforce.

A policy that is written as a personal statement from a managing director will be far more effective than one that is simply copied from another source and has little apparent relevance to the organisation.

Securing overt and sincere commitment from all management helps to cement health and safety as a core value of the organisation.

3.2.2 Promoting health and safety standards by leadership and example and appropriate use of disciplinary procedures

Although it is important, the policy should not be the only way in which senior managers show their commitment. Other techniques that have been shown to work include:

- direct involvement in safety tours eg senior managers making a point of doing a brief safety walk round every time they arrive on a new site;
- discussing safety at first hand with the workforce;
- attendance at safety committee meetings;
- having health and safety as a standing agenda item at board meetings, and dealing seriously with it;

- complying fully and consistently with all site safety rules; managers are also workers after all; and
- making sure that others adhere to company safety rules and taking appropriate action when they do not.

In a fully engaged and committed workforce, the need to take disciplinary action against workers who break the safety rules should hopefully be rare. However, this does remain an option and employers should not be afraid to use it when appropriate. The key here, as with any disciplinary matter, is to act consistently. In that way, the workforce will quickly recognise that this is an important issue and that there will be consequences of not following the rules or otherwise behaving in an unsafe manner.

Of course, the nature of the disciplinary action will vary according to the degree of non-compliance with the rules. Some acts, such as those that seriously endanger the safety of co-workers or others, could result in dismissal, whereas less serious breaches, for example, 'horseplay', might only result in a verbal warning.

3.2.3 Competent workers

Everyone should have at least some level of health and safety competence to help keep themselves and others safe and healthy at work. Their appropriate level of competence will ultimately depend on their role and responsibilities.

All workers need at least a basic understanding of health and safety in relation to their role, so that they can work to the requirements of their organisation's health and safety policy. For example, they need to at least understand the specific safe working procedures for their role, including general health and safety requirements, such as evacuation procedures.

Those in line management positions ought to have a greater understanding of their health and safety responsibilities. For example, they should know about and manage the risks associated with the work they oversee and how health and safety legislation applies to them, as well as the expectations of their organisation's health and safety policy. They also need to be sufficiently knowledgeable about the work that the people they manage are doing to be able to identify poor or unsafe practices.

Senior managers and directors need to know their legal accountabilities in terms of health and safety law and any specific requirements of their sector.

If an organisation appoints highly competent, skilled workers, the potential for accidents will be far less than if the workforce is made up of less well trained and less competent workers.

3.2.4 Good communication within the organisation

Effective communication in an organisation helps ensure the correct health and safety behaviour takes place. Studies of major accidents show that, without the correct communication, errors or violations can occur, with disastrous consequences.

If you think of an organisation as a living entity, communication can be likened to the blood flow that keeps the organisation alive and allows it to grow and develop.

To make communication effective, its purpose, content and the method used to convey the message all need to be considered, along with the needs, capabilities and expectations of the receiver of the message. Communication needs to flow up, down and laterally across an organisation. When the purpose of the communication is intended to bring about a change in behaviour, as is often the case with health and safety, it needs to provide information on why a certain behaviour is required. This is important for managers when communicating changes in procedures or reinforcing a requirement.

Managers also need to use praise when they see examples of good health and safety behaviours. When done sincerely, this technique is powerful in reinforcing safe behaviour.

Workers need opportunities and mechanisms to engage with managers to raise concerns, give feedback and be consulted on health and safety matters.

Benefits and limitations of different methods of communication (verbal, written and graphic)

Communication can be considered as a process that occurs between a sender ('transmitter') and a 'receiver' involving a message and a channel through which the message is transmitted, for example, a face-to-face conversation, email or presentation. The receiver should be able to send feedback to show that they have received and understood the message and to respond to it.

This suggests that the communication process is more than transmitting information. It is concerned with an exchange of understanding and shared meaning. Successful communication helps to build co-operation, shared values and improved practices. As the conversation or the presentation continues, for example, so the transmitter–receiver process is repeated.

Figure 1 shows a simple communication process (simplified from Schramm's model of communication).[3]

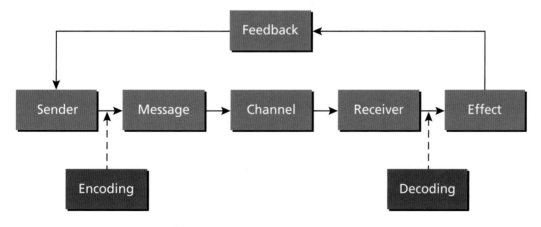

Figure 1: Simple communication process

The effectiveness of the communication process can be impacted by various sorts of noise or interference, including:

- distractions caused by medication or feeling unwell, tired or hungry;
- environmental issues, such as background noise, unsuitable lighting or temperature;
- lack of attention by the receiver due to preoccupation, impairment or lack of trust in, or respect for, the sender;
- use of technical jargon and abbreviations that make the message more complex;

- the sender having a dialect that makes the message hard to understand;
- complex, illogical or ambiguous messages;
- timeliness of communication – if the receiver gets too many communications at the same time, the message may get missed; and
- use of inappropriate methods to send the message.

For example, consider a situation where a group of specialists, perhaps engineers, get together; they will use technical terms and jargon they all know and understand.

Problems may arise when one of the engineers uses the same jargon and technical terms ('symbols') to communicate with someone who does not fully understand these but is perhaps hesitant to express their confusion.

The responsibility for successful message transmission rests with the sender; this can only be achieved through some form of feedback to check that the message has been correctly received.

One of the main advantages of face-to-face spoken communication is that immediate feedback is possible. However, if the spoken communication is over the phone, for instance, feedback via body language symbols (eg shrugs, signs of boredom and lack of attention) is lost.

Table 1 gives some examples of communication activities and methods and their associated problems.

Communication activity	Method of communication	Potential problems
Maintenance of stock levels and stock reordering	Paper/computer	Poor handwriting, mistyping, mistakes in quantities being reordered
Instructions to dumper driver	Verbal	Driver mishears because of ambient noise
Assembly instructions for a piece of equipment	Information sheet	Confusion because the instructions have been badly translated and the sheet has the appearance of looking like it has been copied over and over
Mixing cement for floor tiles	Information is printed on the cement bag	Mixing ratios are printed on a panel at the very top of the bag, which may be cut off and thrown away when the bag is opened
Training course for new workers	Lectures and demonstrations	Some workers do not have English as their first language, particularly technical English

Table 1: Examples of communication

In each case, there is appropriate feedback that might prove beneficial. For example, in the case of stock control, orders above a certain level may need to be countersigned by a second person; alternatively, a computer-based system could generate an alert if, for example, the level of stock reordering falls well outside of the usual monthly pattern.

3.2 Improving health and safety culture

> **APPLICATION**
>
> Consider a range of situations in your organisation requiring information to be communicated and the methods used to communicate.
>
> What are the potential problems and how could they be overcome?

So, if you want your communication to be effective, what do you have to consider?

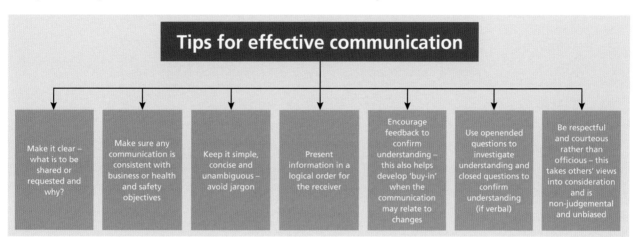

Figure 2: Tips for effective communication

Choosing the correct method or channel for the communication is key to making it effective. You should consider the principles of good communication and the benefits and limitations of various methods, whether verbal, written or graphic.

1. Verbal – involves direct conversation between individuals in person or through another medium, eg telephone or online, audio or web conferencing platforms. It allows for two-way communication, which can be useful in checking that a message has been understood; this is sometimes very important. The barriers to effective communication mentioned earlier are all pertinent to verbal communication. When information is more complex, the written form may be better.

2. Written – involves communication to individuals and groups using various media, including letters, emails, reports, written procedures and handouts. Increasingly these types of communication are transmitted electronically rather than in hard copy. The receiver can retain the information for reference and can read and process it when it suits them. However, written communication can often be one-way. It allows information to reach large audiences or a range of people at the same time, but this presents challenges, firstly in ensuring that the message has been understood and secondly in obtaining feedback. An example of this would be communicating a new health and safety policy. In this case you could consider multiple forms of communication, such as first-line managers following up written information with team briefings. Language, structure, style and tone can be barriers, so it is worth getting a few people to read your communication before circulating it.

3. Graphic – involves the use of charts, diagrams, infographics, signs and photos and can be used in conjunction with other forms of communication or instead of them. Graphic communication can often clarify more complex concepts or relationships. A safety sign is a good example of graphical communication. The sign's colour and pictogram convey the hazard and/or the control measures, and are internationally recognised. Incorporating a visual element into other health and safety communications, such as procedures, can help to reinforce messages.

Some of the advantages and disadvantages of the three communication methods are outlined in Table 2.

Method	Good points	Potential problems
Verbal, eg meetings, presentations, workshops, informal/formal conversations (face-to-face, phone/video conferencing)	• Builds rapport and trust • Generates high levels of understanding • Rapid feedback facilitates quick decisions • Best for conveying private, confidential and sensitive information	• Cannot be relied on for most critical health and safety communications except in highly trained teams • When agreement is important, it should be followed up by written confirmation • Requires a greater level of attention and receptiveness from the listener(s) • Misunderstandings can occur
Written, eg letters, newsletters, annual reports, procedures, posters, handouts, email, social media	• Can be read at a convenient time • Edits and revisions are possible to keep content relevant • Provides a permanent record for later reference • Recipients can review • Useful for complex messages and facts and figures	• Limited immediate or direct feedback can lead to assumptions of acceptance and delays in finding out that the message has not been understood • Can take many drafts to compose • Badly composed messages can distort meaning
Graphic, eg charts, diagrams, drawings, tables, graphs, displays, videos, photographs, pictograms, infographics	• Strong, relevant imagery is a powerful communication tool • Combining words and images can have greater impact • Meaningful visual aids in presentations can result in greater attention to, and retention of, information • Charts and graphs can increase understanding and encourage debate	• Poor design/presentation can detract from the message

Table 2: Summary of three communication methods

It is important to remember that each person will have communication preferences. Some people will prefer a written communication while others will prefer a verbal communication. Sometimes it may be appropriate to have a combination of communication methods for any given message. Take this resource, for example;

there is lots of written text (as you would expect) but this is supplemented by diagrams, pictures etc. Preferences really depend on an individual's learning style. So, you should consider the strengths and weaknesses of each communication method when selecting which to use.

Use and effectiveness of noticeboards and health and safety media

Noticeboards

Noticeboards are a long-standing health and safety communication method. When located in work areas where there is a high level of footfall, they can provide a cheap and effective means of communicating policy statements, current health and safety information and any proposed changes, and also provide key organisational contacts, including emergency information. Noticeboards rely on workers reading them, so when it is essential that everyone understands a procedure, for example, you should also use another way of communicating it, such as a team briefing. Information on a noticeboard needs to be kept up to date, legible and maintained to be effective. Out-of-date information should be removed. Assigning responsibility for maintaining the noticeboard will help to ensure this happens.

The use of large display boards at site entrances conveys a powerful message. They are a clear statement that the company is prepared to subject its safety record to public scrutiny, and show a clear commitment to workplace health and safety.

Posters

Posters are generally cheap, cause minimal disruption in the workplace and can be changed quite easily, according to the topic or aspect of health and safety that needs to be promoted. Refreshing posters periodically ensures that they remain prominent and do not disappear as 'part of the wallpaper'.

Posters are most effective when they convey positive messages, such as the benefits of working safely. Poster location and content also need to take account of the target audience; for example, a poster aimed at encouraging the wearing of eye protection in a workshop needs to be located close to the machinery. The messages posters contain need to resonate with the audience – use of inappropriate language or images could in fact dissuade workers from taking the message on board.

Customised in-house posters may convey a more powerful message than generic ones and can target specific workplace needs and performance. A consistent house style for posters, noticeboards and the materials on display will provide a strong unifying influence.

The aim is to make sure that the different media cross-refer and play to their own strengths. For example, posters announcing that new procedures have been introduced could point workers to leaflets contained in a pocket below the poster where further details can be found. Alternatively, the poster and leaflet could be displayed on the company's intranet.

There is a legal requirement under the Health and Safety Information for Employees Regulations 1989 to display the HSE-approved poster containing information about employer and worker obligations in relation to health and safety (see Figure 3). Alternatively, employers may provide individual leaflets to workers containing the same information.

Figure 3: HSE poster on employer and worker obligations

Videos/moving images
Films can be used to reinforce messages and raise awareness of health and safety topics to motivate workers to behave safely. They can be shared with large audiences, who are not necessarily on site if working remotely. Visual impact can maintain the viewer's attention during the communication so that key points can be conveyed, for example, during a site induction for new starters and contractors.

Team briefings/toolbox talks
Short talks on hazards, risks and control measures, given by a supervisor to their team, are a useful way to raise awareness of, explain and remind workers about issues relevant to their work. A session would usually cover just one topic at a time to increase focus and avoid overload; this will also encourage workers to ask questions and share their views. Used frequently, such talks help promote consistent safe and healthy behaviours.

Handbooks
Some organisations issue worker handbooks (electronically or in hard copy) containing policies, rules and procedures, including emergency information. To be effective, handbooks need to be kept up to date, which can be a considerable task,

more so if the handbook is published in hard copy. Many organisations will have an electronic version accessible via their intranet. So that workers are kept up to date, changes in procedures may need to be communicated through a team talk before the handbook can be updated.

Co-operation and consultation with the workforce and contractors

Co-operation between the management team and the workforce is essential if health and safety issues are to be managed effectively. Organisations with positive health and safety cultures will have mechanisms in place for engaging the workforce in open and honest dialogue, which in turn develops trust.

Good co-operation can be achieved in several ways, including:

- consulting with workers directly or indirectly;
- consulting with worker health and safety representatives; and
- through health and safety committees.

Consultation is a two-way process. It involves a flow of information and allows both sides to contribute to the discussion and share ideas before any decision is made or change introduced. Consultation is distinctly different from 'informing', which is a one-way process involving the communication of a decision that has already been taken. This can result in resistance to change from the workforce.

Requirements for consultation and co-operation are included in s2(7) of the Health and Safety at Work etc Act 1974 and expanded on in regulations. There is also a requirement included in ISO 45001:2018. Employers are required to consult workers on health and safety issues that affect them and before any changes are made, for example:

- changes in work procedures, types of work, equipment and working hours;
- information provided on workplace hazards, risks and precautions; and
- incident investigation reports, risk assessments, emergency arrangements and plans.

Consultation can be carried out by consulting individuals directly or indirectly. Direct methods include meetings, tours and performance appraisals. Indirect methods include the use of surveys and suggestions schemes. Another way to consult is through worker representatives, who participate in consultation with management on behalf of the work group they represent, either as a single representative or with other representatives at health and safety committee meetings.

Appointment, functions and entitlements of worker representatives

The Safety Representatives and Safety Committees Regulations 1977 (SRSCR) apply to workplaces where the employer recognises trade unions and trade unions are recognised for collective bargaining purposes. The Health and Safety (Consultation with Employees) Regulations 1996 apply to workplaces where employees are not in a trade union and/or the employer does not recognise the trade union, or the trade union does not represent those employees not in the trade union. The table below identifies the main functions of representatives under the two sets of Regulations:

Safety Representatives and Safety Committees Regulations 1977	Health and Safety (Consultation with Employees) Regulations 1996
Representatives:	
Appointed in writing by a trade union recognised for collective bargaining purposes.	Elected by the workforce, where the employer has decided not to consult directly.
Title/position:	
Safety representatives.	Representatives of employee safety.
Functions:	
Investigate potential hazards and dangerous occurrences at the workplace, complaints by an employee relating to health, safety and welfare at work, and examine causes of workplace accidents.	
Representation to the employer on the above investigations, and on general matters affecting the health and safety of the employees they represent.	Representation to the employer on: • potential hazards and dangerous occurrences; • general matters affecting the health and safety of the employees they represent; and • specific matters on which the employer must consult.
Inspect the workplace.	
Represent employees in dealings with health and safety inspectors.	Represent employees in dealings with health and safety inspectors.
Receive certain information from inspectors.	
Attend health and safety committee meetings.	

Table 3: Functions of health and safety representatives[4]

Both appointed safety representatives and elected representatives of employee safety have rights to the paid time off necessary to carry out the functions outlined in Table 3 and to undergo any training needed to carry them out. In the case of appointed safety representatives, the Trades Union Congress (TUC), or the trade union concerned, will offer and usually meet the costs.

Benefits of worker participation

Worker participation in health and safety provides the employer with a wider perspective of how hazards and risks affect workers, how control measures are working in practice and the implications of proposed changes in work. Drawing on worker experience helps to improve the effectiveness of health and safety measures. If workers are actively involved in the decision-making process, then they will be far more likely to accept change as and when it occurs.

3.2 Improving health and safety culture

It is important to note, however, that consultation does not mean that management must always do what the workers want. There must be an understanding that workers will have the opportunity to raise issues they see as being important, but that not all requests will be actioned. Equally, not all issues are worthy of consultation and doing so would be impractical.

For most organisations, this is an acceptable approach. Workers feel they have had the opportunity to 'have their say' and management pick up on the general mood of the workforce while also gathering valuable information on improvements that can be made. Such an approach is good for morale, which in turn is good for productivity and allows management to demonstrate commitment to health and safety.

CASE STUDY
A participative approach to mental health

A large international telecommunications company with over 100,000 workers in 170 countries wanted to address health and wellbeing issues affecting its workforce by getting them involved.

Mental health was an issue of growing concern and stress was a significant cause of time away from work. The company considered health and wellbeing issues important for the workforce, the business and society. The aim was to increase worker involvement in managing health issues to improve the health of staff and facilitate joint problem solving.

The company's health and safety representatives worked with managers to form and promote a strategy on tackling health issues. To get the workforce involved, wellbeing and communications leads and health and safety specialists worked together to tailor messages about the importance of mental health issues and cascade these to the workforce to raise awareness. One part of the organisation used its radio channel and the plasma screens in its contact centres to do this, and another part used its in-house magazine.

Three months into the initiative, a follow-up online survey found that, of those who had accessed the materials:

- 68% learned something new about how to look after their mental health;
- 56% put in place some of the recommended practices; and
- 51% of those who had made changes noticed improvements in their mental health and wellbeing.

Over the longer term (4–5 years), the sickness absence rate due to mental ill-health fell by 30% despite challenging market conditions. The company got nearly 80% of people who had been off for more than six months with mental ill-health back into their own jobs.

A participative approach that was 'done with' workers rather than one 'done to' them was more effective. Having senior business champions providing high-level visibility and promotion of the initiative was also beneficial.[5]

APPLICATION
How effective is worker participation in health and safety in your organisation?

The role of health and safety committees

Formation of a health and safety committee may be requested by representatives (based on reg 9 of the SRSCR). In any case, it is a good practice that helps with consultation and fosters worker participation and a collaborative approach.

For such a committee to be effective, it is important to set out its objectives, functions and composition. Objectives might include:

- monitoring trends in accidents and ill-health so that reports can be made to management on unsafe/unhealthy conditions and practices, together with recommendations for corrective action;
- examination of safety audit reports;
- consideration of reports and information provided by inspectors from the enforcing authority;
- consideration of reports that safety representatives may wish to submit;
- assistance in the development of works safety rules and safe systems of work;
- monitoring the effectiveness of the safety content of worker training; and
- monitoring the adequacy of safety and health communication and publicity in the workplace.

Composition

The membership and structure of the committee should be agreed in consultation between management and workers/their representatives. Good practice is to keep the total size as reasonably compact as possible, with management representatives not exceeding the number of worker representatives. Management representatives should include a good cross-section of line management and functional responsibilities, with sufficient expertise to provide information on company policy and technical matters, and adequate authority to act on recommendations. Specialists such as health and safety practitioners, occupational health physicians or nurses and occupational hygienists should be members of the safety committee by virtue of their position of employment. Other specialists should be invited to attend committee meetings when the need arises.

Operating the committee

Meetings should take place as often as necessary, depending on the nature of the business, size of the workplace, numbers employed and the degree of inherent risk. There should be sufficient time to ensure full discussion of all agenda items. It is advisable not to cancel or postpone a meeting except in very exceptional circumstances, when a new date should be agreed and publicised without delay. Meeting dates should be published well in advance of the date, making use of internal communication channels such as health and safety noticeboards, the intranet and team briefings, so that as many workers as possible are aware of them.

Sending copies of the agenda and any accompanying papers to all committee members at least one week before each meeting allows them sufficient preparation time. Minutes should be taken and agreed, and copies sent as soon as possible after the meeting to each member of the committee, each worker representative appointed and the most senior executive responsible for health and safety. Displaying the minutes where they can be seen by all workers is also important, as is keeping top management informed.

Meetings should be conducted with a degree of formality but informal enough so that people do not feel intimidated or put off attending. An effective chairperson can ensure the smooth running of the meeting and encourage all members to participate.

Effectiveness

For the health and safety committee to work effectively there must be good communication flows between management and the committee and between the committee and the workers. Commitment by all parties is essential. Management must genuinely want to use the workers' knowledge and experience and the workers must be similarly committed to improving health and safety standards.

> **TIP**
>
> To maintain the momentum of work and the effectiveness of the committee:
>
> - meetings should be held regularly;
> - minutes should be published as soon as possible after the meeting to provide evidence of discussions and recommendations; and
> - management decisions based on the recommendations should be prompt and translate into action.

3.2.5 Training

A good training programme helps to develop the required competencies so that workers can perform their work to a safe and adequate standard. Training should start from the moment a new worker commences work and continue throughout their career.

Training programmes help to improve human reliability by developing competencies at a faster rate, and more safely, than might otherwise be the case if workers were left to learn by themselves through trial and error. The training programme can incorporate 'lessons learned' from previous incidents, so that the same errors are less likely to recur. If done well, this will lead to a reduction in injuries and other forms of loss, so the investment is almost always worthwhile.

Many organisations will have what is known as a 'training matrix'. This identifies what skills a person in a particular job role should have, so that new workers can be brought up to the required standard. This training matrix will be based on a training needs analysis. This analyses the skills needed to perform a particular task and then maps them against skills that workers are likely to have. The gaps between available and required skills are the training requirements that will then be incorporated into a training programme.

A variety of methods can be used to deliver effective training; much depends on the facilities available and the nature of the topic to be covered. For example, some training, such as resuscitation training, requires hands-on practice to do confidently, quickly and with skill. Other types of training, such as induction training, tend to be more informational in nature (ie fact-based) and do not really require practice.

In fact, good training usually makes use of a range of methods. This helps to break the sessions up and will also help to ensure that there is something for everyone, since different people learn in different ways.

> **TIP**
>
> Typical methods used in training sessions include:
>
> - presentations – assisted by visual aids and with the opportunity to ask questions;
> - group work – a trainee-centred approach that allows people to learn under supervision by carrying out tabletop exercises and sharing ideas and experiences;
> - role play – putting trainees in a realistic situation, such as in first-aid training; and
> - discussion – a useful way for trainees to gain clarification and to promote understanding of key points.

Training is needed in many different circumstances.

Induction

When people join a new organisation, they usually undergo induction training to orientate them to their new environment and to help them start to understand the culture, site rules and so on.

An induction session might cover:

- the organisation's health and safety policy;
- key people with responsibilities for safety management who the new worker can go to for advice or to report a shortcoming in safety arrangements;
- specific risks and controls;
- emergency procedures;
- first-aid provision; and
- welfare facilities.

In some large organisations, the induction may be spread out over several days or even weeks, and each session will focus on a different aspect of the organisation's operations. In smaller organisations, a relatively brief session will normally be enough.

Job or process change

Training and development should continue throughout an individual's employment with an organisation and throughout their career if they move to other organisations. Training may be needed when people move to a new role within the organisation, for example, being promoted into a management or leadership position. Changes in workplace processes, equipment or substances may also demand update training so that workers are kept fully aware of the changes that might affect their health or safety. The absence of training could result in incorrect or unsafe behaviours and risk of harm.

New legislation or standards

The introduction of new legislation or standards could impose new requirements on the organisation for which training needs to be implemented. For example, if there is new hazard information issued with substances used in a motor repair workshop, the processes involving this substance would need reviewing and amending accordingly, and workers would need to be retrained.

New technology

New technology can also mean changes in work practices that lead to a training requirement. For example, the mechanisation of a manual handling process to reduce injury risks would require equipment users to learn new skills to operate and maintain the equipment safely.

Training should not be a one-off; it should be refreshed periodically. This is particularly important when a relatively high degree of skill is needed, such as driving a forklift truck. Refresher training offers an opportunity to address bad habits and to communicate new techniques. Refresher training is usually required when the impact of not working in a particular way could lead to high risks. One example of this is the annual requirement in many countries for first-aider refresher training. Other examples might be when accident/incident investigations or audits indicate there may be a specific competence issue or when someone returns to work after a period of absence.

KEY POINTS

- It is good practice to look for opportunities to improve health and safety culture.
- It is important to gain the commitment of all management, including top management, promote health and safety standards by leadership and example, and use disciplinary procedures when appropriate.
- Everyone in the organisation should have an appropriate level of competence for their role.
- Mechanisms for health and safety communication throughout the organisation should be developed, including to ensure the most appropriate method is used for the situation.
- Communication methods are verbal, written and graphic, and the appropriate method should be selected for the given situation. Choosing carefully can be the difference between a message being transmitted and understood, thereby mitigating the risk, or causing a misunderstanding that could lead to increased risk of harm.
- Noticeboards, posters, team briefings and video can all play a part in communicating health and safety information, but each must be reviewed to ensure they remain relevant and current.
- Training is an essential tool in developing the competence of managers and workers throughout an organisation. Training should start from the beginning of employment with induction training and be continued throughout employment, particularly when the individual's job changes, there is a change in process and when new legislation or technology is introduced, all of which have a potential impact on the risk of harm.

References

[1] HSE, *Consulting workers on health and safety. Safety Representatives and Safety Committees Regulations 1977 and Health and Safety (Consultation with Employees) Regulations 1996 (as amended). Approved Codes of Practice and guidance* (L146, 2nd edition, 2014) (www.hse.gov.uk)

[2] HSE, *Involving your workforce in health and safety: Guidance for all workplaces* (HSG263, 2015) (www.hse.gov.uk)

[3] Schramm, W, 'How communication works' in: Schramm W (ed), *The process and effects of mass communication* (University of Illinois Press, 1954) pages 3–26

[4] Source: HSE, *Consulting employees on health and safety: A brief guide to the law* (INDG232, rev 2, 2013) (www.hse.gov.uk)

[5] Adapted from HSE Case study: BT Group plc (www.hse.gov.uk)

How human factors influence behaviour positively or negatively

By the end of this chapter, you will develop an awareness of the following:

- Organisational factors, including: culture, leadership, resources, work patterns, communications
- Job factors, including: task, workload, environment, equipment, display and controls, procedures
- Individual factors, including: competence, skills, personality, attitude and risk perception
- Link between individual, job and organisational factors

When an accident happens, it may be tempting to look at the person who had the accident, or who caused it, and say it was their fault and only their fault. Similarly, when people behave in an unsafe way, it may be tempting to blame them and not consider the circumstances of their work or the way the job was designed.

This is rather short-sighted as it fails to consider important factors that may lead to a better understanding of how accidents can be prevented. Accident investigations usually reveal multiple causes of these events, and common failings are organisational design, management and decision-making.

The study of 'human factors' helps us develop a better understanding of the many influences on people's behaviours and how these contribute to health and safety failings. Conversely, armed with this understanding, we can also identify how health and safety behaviour at work can be improved.

DEFINITION

Human factors refer to environmental, organisational and job factors, and human and individual characteristics which influence behaviour at work in a way which can affect health and safety.[1]

There are three key human factors that influence behaviours at work:

1 Organisation, including the organisation's culture, its leadership, resources, work patters and communications.

2 Job, including the task, workload, environment, displays and controls, and procedures.

3 Individual, including competence, skills, personality, attitude and risk perception.

These three factors are not mutually exclusive; they interact with each other and collectively influence behaviour, as illustrated in Figure 1.

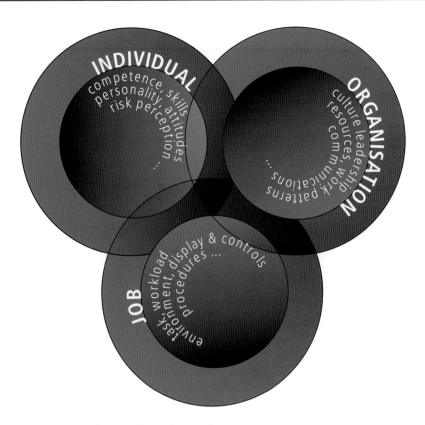

Figure 1: How human factors interact[2]

3.3.1 Organisational factors

These are the attributes of an organisation that influence the way in which people within it behave. These include:

- the culture of the organisation;
- commitment and leadership;
- resource availability;
- working patterns; and
- communications.

Culture

We discussed culture in 3.2: Improving health and safety culture. It can be positive or negative and people who work in the organisation will tend to behave in a way they feel is in line with that culture. If there is a positive culture, with a high degree of engagement from workers and management in the organisation's activities, this is likely to lead to engagement in health and safety matters and to safety-conscious behaviour from individual workers.

Conversely, a negative organisational culture will result in insufficient engagement generally and in health and safety and has the potential to result in unsafe or undesired behaviour from workers. Striving for a culture that promotes management and worker engagement in health and safety and a belief that deviating from health and safety practices is unacceptable will minimise the influence of this factor.

Commitment and leadership

Leadership demonstrated by top-level management may have a positive or negative impact on health and safety behaviours. If they do not demonstrate commitment, then this in turn will influence the layers of management beneath them and middle and first-line managers may not show commitment or be reluctant to challenge unsafe behaviours. Crucially, all those in management positions must lead by example.

Imagine a situation where company management have set out clear rules that say eye protection must be worn in an area. Imagine that all workers follow this directive, believing it to be made for all the right reasons. Now imagine that a couple of weeks later the manager who set the policy is seen walking through that area without any eye protection. The implications are clear, and the effect will be to instantly undermine everything that has been done up to that point with respect to the management of health and safety in that organisation. And that really is an important point – it is not just the wearing of eye protection that will become controversial, it is the workforce's view of the importance of all safety precautions. If the workforce starts to question the relevance of controls, they will stop using them and accident rates will start to rise.

Leading by example will help to ensure that such a situation never arises. Ensuring responsibilities are clarified, that health and safety management is co-ordinated throughout the organisation and that everyone knows 'what good looks like' will support managers in providing a positive example.

Resources

The organisation can influence safety behaviours by providing adequate and appropriate resources, such as equipment and training. If these are provided, then the workers will have the right tools for the job and will be less likely to improvise, which could lead to accidents. Insufficient or inappropriate resources could lead to workers developing alternative and potentially unsafe shortcuts.

Work patterns

Many people work shifts, which may include working at nights; others work very long hours or have peaks in demand. Organisational work routines of this type will increase the likelihood of workers becoming fatigued, which will lead to errors in judgement and to mistakes.

Figure 2: Night workers may make errors due to fatigue

Communications

To manage anything effectively, an organisation will need to have good methods of communication. Communication must be two-way and should allow for checks to be made that a message has been received and clearly understood. Those organisations with unsatisfactory communications are unlikely to manage safety and health issues well, since workers will be less likely to know about hazards, risks and what is expected of them.

Organisations that rely on one-way communication from managers to workers will miss the opportunity to gain feedback from workers on health and safety matters on the front line as workers are not encouraged to report issues or concerns. This could lead to bad practices becoming the accepted norm.

> **TIP**
>
> What organisations need to do to manage organisational human factors:
>
> - Get clear commitment from everyone to promote a culture for health and safety.
> - Provide positive leadership from top management.
> - Identify responsibilities and co-ordinate effort effectively.
> - Provide adequate resources – time, money and effort, including management time and supervision.
> - Effectively communicate standards of behaviour and their importance.
> - Plan work patterns to avoid fatigue and stress.

3.3.2 Job factors

Job factors are the aspects of the job or task that dictate the way in which workers will carry it out. Job factors are closely linked with organisational factors and may result from them.

Job factors that may influence behaviour are illustrated in Figure 3.

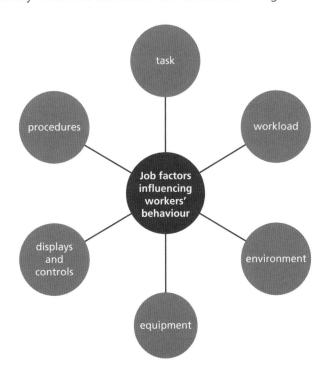

Figure 3: Job factors influencing workers' behaviour

Task

Tasks can be physically or mentally demanding, or both. For example, if the task is to be carried out in an awkward or uncomfortable position, workers will probably try to complete it as quickly as possible, which may compromise safety. Alternatively, there may be a perception that the job will only take a few minutes or is 'safe' – this will lead to fewer precautions being taken.

Workload

When workloads are easily manageable, workers can afford to do everything according to best practice. However, at times of high demand, when timescales are short, the temptation to cut corners to meet the deadline is great. Conversely, if the work is monotonous and boring, workers will be prone to 'switch off', leading to slips and lapses that can result in injury.

> **CASE STUDY**
>
> Train driving is a safety-critical task. High mental or physical workload ('overload') can cause people to make mistakes. Many improvements have been made to minimise distractions and the driver's cab is more comfortable and train traction systems are smoother. Some train-driving tasks, such as speed control and braking, can be automatically controlled.
>
> However, these conditions can lead to a state of cognitive 'underload'. This occurs when an individual's mental workload is too low, their arousal levels drop and performance suffers. To prevent incidents involving underload from occurring, train drivers need to be able to actively monitor and respond to their surroundings when required.
>
> A project to find ways to address underload for train drivers revealed that most drivers experienced underload on a regular basis. Drivers used a range of techniques to overcome this, such as self-discipline before the journey started, making sure to get enough rest before work, turning the heater on or off, breaking up the journey in their mind into shorter journeys and speaking aloud. Based on the drivers' input, a toolbox of mitigation techniques was created by drivers for drivers and accessible to all drivers to self-manage their workload levels. After a trial period, drivers reported feeling better equipped to manage underload using the toolbox techniques.[3]

Environment

If the workplace environment is uncomfortable, workers are likely to try to complete the job as quickly as possible or may adopt a potentially unsafe system of work to overcome the difficulties they encounter. Other environmental elements that may influence safety behaviour include lighting levels, noise, heat and humidity.

For instance, if the workplace is very noisy, workers may misunderstand instructions, which will lead them to do a job in a potentially unsafe way. Insufficient lighting may mean that workers are unaware of hazards and therefore fail to account for them when doing the job. High levels of heat and humidity can be distracting and can lead to dehydration, which may compromise decision-making, leading to mistakes and subsequent accidents.

Equipment

The employer should ensure that the right tools are provided for the job. An ergonomic approach to tool design and selection will include consideration of any limitations of those people doing the work so that chosen tools minimise effort, allow efficient working and reduce the risk of injury. If the right tools are not provided, then the worker may decide to use an inappropriate tool to get the job done. Badly designed and maintained tools result in frustration and fatigue, which can lead to unsafe behaviours.

Displays and controls

Ergonomic principles should also be considered when designing information displays and controls. Well-designed displays that correspond to control inputs help to reduce human error by giving a clear indication of the effect of a particular action. On the other hand, displays that are not well designed or positioned, or controls that are difficult to reach or operate, may increase the potential for unsafe behaviours.

Procedures

If procedures are clear, relevant and well written, then workers will be more likely to follow them and will be less prone to behaving unsafely. Inadequate procedures that are not relevant enough and are badly written will probably be ignored by most workers, who will do things their own way, which may not be the safest way. Methods used to communicate procedures are equally important and should take account of individual differences. Two-way communication allows for understanding to be checked, which is essential for safety-critical procedures.

> **TIP**
>
> What organisations need to do to manage organisational human factors:
>
> - Assess tasks and the likely errors involving manager and worker.
> - Apply ergonomic principles to the design of interfaces between people and equipment, tools, displays and controls.
> - Take account of maintenance requirements of equipment.
> - Be consistent in how procedures are presented and support this with effective communication.
> - Plan workload to minimise fatigue and stress. Consider absence cover and emergencies. If work needs concentration to avoid errors, make sure it can happen without interruption.
> - Control the working environment eg lighting, temperature, noise.

3.3.3 Individual factors

In addition to organisational and job factors, individual factors play an important part in health and safety behaviours. Everyone is different. Individuals have differing physical and mental characteristics. Mental characteristics – attitude, motivation, perception and mental capability – are shaped by inherited characteristics and life experience. Individuals have physical characteristics such as size, age and potentially physical disability or limitation, all of which can have a health and safety impact. For example, a person's size might restrict working in a confined space and colour vision deficiency (so-called 'colour blindness') may interfere with identification of colour-coded controls on a control panel. It is important to consider the job requirements so that effects of mental and physical characteristics that cannot be changed are minimised. This is particularly important for safety-critical roles in high hazard environments. Job or task analysis (see 3.4: Assessing risk) will be useful.

You should also be aware that the same person may behave in different ways at different times due to fatigue or variances in attitude.

Typical individual factors include:

- competence;
- skills;
- personality;
- attitude; and
- perception of risk.

Competence

We recognised the important contribution competent workers make to the health and safety culture of an organisation in 3.2: Improving health and safety culture. Competence can be viewed as a combination of different things, such as qualifications, training, knowledge, experience and other qualities that are needed for the task. When appointing a person to a role, the employer should consider these factors to ensure that they are choosing the best person for the job. Just because someone has a qualification does not automatically make them competent – that comes with the benefit of experience and may take years to develop after the initial qualification has been obtained.

Think about a teacher – they may have all the qualifications in a subject that it is possible to have, but if they lack the basic ability to explain things clearly in a language appropriate to their audience, then they cannot be described as a competent teacher.

If workers are given tasks that are beyond the limits of their competence, they will be at greater risk due to their absence of awareness of existing or potential hazards. For this reason, it is essential that the employer has some mechanism in place for assessing workers at the recruitment stage and periodically throughout their employment.

Ensuring competence of managers and workers increases the likelihood of safe behaviours being followed, as there will be greater understanding of why things need to be done in a certain way. The decision-making of competent managers is more likely to show regard for health and safety and encourage workers to do the same. Knowing the limits of your competence is equally important. While it may never be possible to rule out errors, regardless of how knowledgeable, skilled and experienced you are, being aware of any limitations in your competence can potentially reduce the chances of making mistakes.

Good practice suggests that organisations should define the health and safety competences required for each specific role. This provides clarity for the job owner and can help with competence assessment and assurance processes.

Skills

This is closely related to competence. Someone who is 'skilled' can work more accurately and quickly and is less prone to making mistakes.

Knowledge of how to do something only takes a person so far; being able to do it properly requires practice – sometimes a lot of practice. Skill levels improve as you gain experience and have the opportunity to practise. Driving is a good example of this.

If a person has some driving lessons, then passes their driving test, they will have shown that they have attained a sufficient level of competence to be allowed to drive alone and not cause harm to others through bad driving. Notwithstanding this, you probably would not refer to them as a 'skilled driver' until they had a lot more experience and, perhaps, had received advanced training.

Personality

Personality usually refers to traits, sentiments, attitudes, unconscious mechanisms, interests and values that determine your characteristic behaviours and thoughts. Generally, personality does not change fundamentally from day to day. It is formed through a combination of biological mechanisms and environmental factors ('nature and nurture'), and each personality is unique.

Understanding how personality traits might affect behaviour is important, more so in safety-critical roles. There are two important aspects or dimensions of personality. One considers where a person is on the scale from introvert to extrovert; the other considers how stable or unstable they are. An introverted person may not be comfortable asking questions in a group situation, whereas an extroverted person may be more likely to speak up and ask for help. Stable people tend to be organised, plan ahead, are reliable and do not act on impulse, whereas unstable people like taking chances and act on impulse.

Personality traits can therefore be important in health and safety. For example, take someone whose personality is such that they do not like following rules or procedures, or they continually think that they know best. Such a person could be at risk of injury to themselves and others if given a task that required a high degree of compliance with safety procedures.

Attitude

Attitude is a facet of personality; it is the way that a given person thinks or feels about a particular topic. We are aware of our own attitudes to certain things. Attitude is a mixture of beliefs and values. Our beliefs reflect our knowledge about a topic and it is the values that we put on our beliefs that define our attitude to the topic. Values give us standards against which we make judgements, and they motivate or drive certain behaviours. Attitudes can be influenced by our background or experience and by the influence of our peers, as we discussed in 3.1: Health and safety culture. Attitudes derived from long-held beliefs and ascribed high value are hard to change.

Attitudes are not directly observable. What we can see is how attitude manifests itself in people's expression of thought and feelings and through their behaviour. An individual's behaviour is therefore an indicator of their attitude. An individual's attitudes can be influenced by those of people they align themselves with.

In health and safety management, we are trying to create an attitude that the management of health and safety risks is not only necessary but desirable. Where negative attitudes, demonstrated through poor health and safety behaviour, have built up over time, these will be hard to change, and attempts to do so are likely to be resisted. Techniques that can be considered to influence attitudes include:

- strong leadership with positive attitude;
- training to reinforce safe behaviours;
- involvement in safety issues;
- education to reform attitudes;
- working with peers who have a positive attitude; and
- recognising those people who do demonstrate positive behaviours.

Perception of risk

Perception has to do with the way that people interpret the world around them. What one person sees, or perceives, as being relatively safe, another might think is quite dangerous – take parachuting or bungee jumping as examples. Our perception of risk is underpinned by our beliefs, attitudes, judgements and feelings as well as social or cultural dispositions that we have towards hazards and their benefits.

Our perceptions can be based on sensory information, which we take in from our five senses (hearing, taste, touch, smell and sight), or they can be based on experience or familiarity with a situation. The issue of perception is therefore a complex one, which needs to be thought about when trying to decide how to reduce accidents at work.

Imagine that a worker has a sensory impairment of some sort – we will say to start with that this is a natural impairment. That worker will not be receiving all the information needed to interpret their surroundings and will thus be more likely to have a distorted picture of what is going on, in which case they are liable to have an incorrect perception. An example would be a worker with colour vision deficiency; it will be difficult, if not impossible, for them to tell one coloured cable from another; therefore people who want to become electricians are routinely given colour vision assessments before they start their training. Another example would be a deaf person or someone with hearing loss; they may not be able to hear a fire alarm and so would be at increased risk.

Perceptions can also be affected by alcohol and drugs, which may affect the brain's ability to process information. It is worth remembering that 'drugs' include prescription medication, so there should be a policy in place to ensure that anyone on such medication makes themselves known to their line manager so that they can be given alternative work if necessary.

Carrying out dull, tedious or repetitive work, tiredness and intense concentration are all factors that can lower people's ability to perceive a hazard and the risk associated with it. Hazards can be hidden, such as radiation, or masked by environmental issues such as background noise or bad lighting.

Inadequate training will affect perceptions because the worker will have gaps in their knowledge or will not appreciate that something is hazardous when in fact it is hazardous. For example, their perception may be that the substance they are working with is harmless when in fact, due to insufficient training, they are unaware that it is a skin irritant.

Experiences also shape people's perceptions. You may occasionally hear someone say, 'this is safe – we have done it this way hundreds of times and never had an accident'. Their perception is clearly that the job is safe but it may not be; they might just have been lucky. As evidence of this, we often hear comments during accident investigations along the lines of, 'I don't know how that happened – we have always done it that way'.

Another issue is how the brain tries to make sense of unfamiliar patterns. If you see something that you have not seen before, you may find yourself saying that it 'looks like X'. That is a verbal expression of what your brain is trying to do – match the abstract or unfamiliar with something that is familiar.

Consider this picture as an example. At first sight, what do you see?

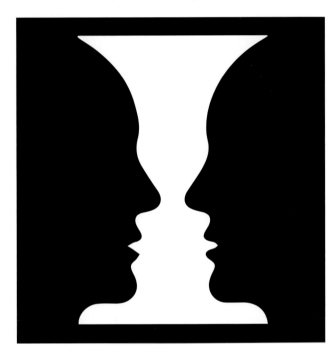

Some will see a candlestick while others will see two people face to face. It is all a matter of perception and how your brain makes sense of this unfamiliar drawing.

From a health and safety perspective, our subconscious perception process can mean that we may not see hazards in the workplace. In fact, we often see what we expect to see and do not see what we do not expect to see. Think about a delivery driver repeatedly driving the same route; they get used to the features on the route (such as junctions and traffic lights) so that they no longer perceive them. If there is a change, such as a different speed limit, they may not even register it.

In summary, different workers may perceive the same thing in entirely different ways, either because of their prior experience or because of a sensory impairment. The employer will need to try to make sure that all workers perceive things accurately and that they do not mistake a serious risk for something that is trivial.

Increasing visibility of hazard markings, providing information and training to improve hazard recognition and providing experiences through drills and scenarios can all reinforce the desired perception process and action.

Managing risk – understanding people and processes

> **TIP**
>
> What organisations need to do to manage individual human factors:
>
> - Use job and person specifications in recruitment, including consideration of individual factors where relevant to health and safety.
> - Apply ergonomic principles to match individual and jobs.
> - Support individual development though effective training.
> - Monitor performance.
> - Carry out pre-employment medicals and health surveillance when required.
> - Review individual factors after periods of absence.
> - Provide counselling and support for ill-health and stress.

3.3.4 Link between individual, job and organisational factors

From our consideration of the three types of human factors we can appreciate that an individual failing can be shaped by the circumstances in which the individual was placed; for example, the way the job was designed and inadequate management commitment in the organisation creating the job. Multiple human factors can combine, with the result being undesirable individual health and safety behaviour and even harm.

Individual factors such as attitude can be shaped by job factors. So inadequate equipment to do the job or unsatisfactory environmental conditions could lead to an individual developing the attitude that health and safety is not important and management do not care how the work is done. This in turn reflects a negative health and safety culture.

While this simple example shows how the link between factors leads to and reinforces negative attitudes and behaviours and increases the risk of accidents occurring, the linkage can be used to identify ways to improve health and safety behaviours. For example, initiatives that target improving attitudes through enforcement are unlikely to lead to sustained improvement without also addressing the job and organisational factors – in this case that might mean providing the correct tools and equipment, comfortable working conditions and a caring management. If the three human factors work together, they can reinforce and sustain safe and healthy behaviours.

> **APPLICATION**
>
> Consider how your organisation applies human factors in its management of health and safety. What could be improved?

KEY POINTS

- Human factors comprise organisational, job and individual factors. Singularly and collectively, they can exert influence on behaviour positively or negatively.
- Organisational factors include the organisational culture, leadership commitment, availability of resources, work patterns and how health and safety is communicated.

3.3 How human factors influence behaviour positively or negatively

- Job factors are concerned with how the work is designed and how workers must carry it out. They include the task, the amount of work, the environment in which it is done and the information provided through displays, controls and procedures.
- Job factors are strongly influenced by organisational factors.
- Individual factors include the individual's competence, skills, personality, attitude and perception of risk.
- There are strong links between individual, job and organisational factors.
- This linkage is useful in understanding what drives individual behaviours and can also assist in identifying changes required in all three areas to sustain safe and healthy behaviours.
- When all three factors work together positively, they help to sustain safe and healthy behaviours.

References

[1] HSE, *Reducing error and influencing behaviour* (HSG48, 2nd edition, 1999) (www.hse.gov.uk)

[2] Adapted from HSG48, see note 1

[3] Adapted from Rail Safety and Standards Board (RSSB), *Evaluating prevention and mitigation to manage cognitive underload for train drivers* (T1133, 2019) (www.rssb.co.uk) and RSSB, 'Human Factors Case Studies' (2021) (www.rssb.co.uk)

3.4: Assessing risk

Syllabus outline

In this section, you will develop an awareness of the following:
- Meaning of hazard, risk, risk profiling and risk assessment
- Risk profiling: What is involved? Who should be involved? The risk profiling process
- Purpose of risk assessment and the 'suitable and sufficient' standard it needs to reach (see HSG65: 'Managing for health and safety')
- A general approach to risk assessment (five steps):
  - identify hazards:
    - sources and form of harm; sources of information to consult; use of task analysis, legislation, manufacturers' information, incident data, guidance
  - identify people at risk:
    - including workers, operators, maintenance staff, cleaners, contractors, visitors, public
  - evaluate risk (taking account of what you already do) and decide if you need to do more:
    - likelihood of harm and probable severity
    - possible acute and chronic health effects
    - risk rating
    - principles to consider when controlling risk (regulation 4 and Schedule 1 of the Management of Health and Safety at Work Regulations 1999)
    - practical application of the principles – applying the general hierarchy of control (clause 8.1.2 of ISO 45001:2018)
    - application based on prioritisation of risk
    - use of guidance; sources and examples of legislation
    - applying controls to specified hazards
    - residual risk; acceptable/tolerable risk levels
    - distinction between priorities and timescales
  - record significant findings
  - reasons for review
- Application of risk assessment for specific types of risk and special cases:
  - examples of when they are required, including fire, DSE, manual handling, hazardous substances, noise
  - why specific risk assessment methods are used for certain risks – to enable proper, systematic consideration of all relevant issues that contribute to the risk
  - special case applications to young people, expectant and nursing mothers; also consideration of disabled workers and lone workers (see regulations 16, 18 and 19 of the Management of Health and Safety at Work Regulations 1999)

3.4 Assessing risk

Risk assessment is one of the most important skills that anyone managing or advising on health and safety matters needs to possess. If risks cannot be properly evaluated then appropriate control measures cannot be recommended. This could lead to situations where hazards are undercontrolled, which could result in an accident or incident.

Alternatively, the level of control may not be proportionate or inappropriate controls are imposed; this may lead to unworkable situations and health and safety can be viewed as an inhibitor to effective production or service delivery.

We start by defining some key terms before looking at a five-step process to conduct and record risk assessments. We also look at principles and practice in selecting risk control measures and the priority with which this should be done.

Risk assessment is required for many specific hazards, such as display screen equipment (DSE), manual handling, chemicals, fire and machinery. Although the principle of the assessment will be the same, these hazards will require specific factors to be considered. The workplace itself can present other hazards and risks that will require consideration, for example, access and egress, work patterns and workloads.

> **TIP**
>
> Legislation relating to risk assessment in the UK is contained in:
> - the Management of Health and Safety at Work Regulations 1999 (MHSWR).

3.4.1 Meaning of hazard, risk, risk profiling and risk assessment

> **DEFINITIONS**
>
> *Hazard: something with the potential to cause harm, which can include articles, substances, plant or machinery, methods of work, the working environment and other aspects of work organisation.*
>
> *Risk: the likelihood of potential harm from that hazard being realised, resulting in consequences.*
>
> *Consequence: the outcome of the hazard being realised.*
>
> *Risk profiling: a structured approach to shape risk management that requires a range of data and information. The output of the process should provide an organisation with an overview of the risk of its operations and the effectiveness of the controls in place to mitigate the risks.*
>
> *Risk assessment: the process of identifying preventative and protective measures by evaluating the risk(s) arising from a hazard or hazards, considering the adequacy of any existing controls and deciding whether the risk is acceptable. The level of risk associated with a hazard is assessed in the risk assessment process to determine which controls, if any, might be appropriate.*

It is important not to confuse hazards and risks – they are different. Hazards give rise to risks. Examples of hazards and their associated risks are given in Table 1.

Hazard	Risk
Trailing cable	Tripping and falling on the level
Unfenced edge	Fall from height
Sulphuric acid	Corrosive burns
Heavy loads	Back injuries, ligament or tendon damage, hernia etc
Unguarded machine	Contact with working parts leading to cuts, amputation etc

Table 1: Hazard examples and their risks

3.4.2 Risk profiling

What is involved?

Risk profiling is usually a high-level activity, carried out at the business unit, division or branch level. Its output is intended to identify high level areas of high risk or undesirable exposure to risk that require management attention. These threats are considered in terms of the harm to people, disruption to the organisation and the financial impacts.

The output of the risk profiling process will be a risk prioritisation, enabling organisations to focus on the important issues and not to give unnecessary priority and resource to minor risks. It also informs decisions about what risk control measures are needed.

Who should be involved?

The resource requirements to carry out a risk profiling exercise will depend on:

- the scope of the profiling exercise;
- the nature of the risks involved;
- the information available; and
- the depth required to give a meaningful profile that can inform management decision-making.

Any risk profile will only be as good as the information gathered. It is important to consider this at the outset. It is essential that those involved in creating the risk profile are competent to do so and can provide the necessary depth and breadth of information on the organisation's operations and associated risks and controls.

The risk profiling process

In terms of health and safety risks, the Health and Safety Executive (HSE) advocates that the risk profile is "the starting point for determining the greatest health and safety issues for the organisation".[1]

A risk profile examines:

- the nature and level of the threats faced by an organisation;
- the likelihood of adverse effects occurring;
- the level of disruption and costs associated with each type of risk; and
- the effectiveness of controls in place to manage those risks.

The way in which risk profiling is carried out varies across sectors. It is good practice to make it a collective exercise involving relevant people from across the organisation, unit or discipline for which the risk profile is being produced. Risk profiling usually involves the following key stages/elements:

- Agree definitions and descriptors for loss categories, likelihood, hazard categories, mitigation assessment and risk prioritisation.
- Gather information about operations and processes, loss data and scenario assessments using existing resources (eg documents, people).
- Analyse scenarios and situations and allocate risk ratings/levels.
- Produce a profile of the risks to inform further actions on risk management, which will be affected by risk appetite and the desired risk profile.

For example, in the major hazard industries, such as oil and gas, a risk profile can take about two weeks to complete by a team of perhaps 10 people. The process provides for transparency when all organisational risks are presented using the same units of measurement for all activities, all areas and all consequences. In turn, this assists the organisation-wide decision-making process for the continuous reduction of risk.

A risk register is a tool an organisation can use to record and monitor risks identified through the profiling process. It provides an overview and prioritisation of risk for management attention. It usually describes each risk, an assessment of the likelihood and consequences, a ranking, a risk owner and mitigation information. When populated with information on each risk, the register can be analysed to give risk profiles for different aspects of the organisation. Over time, it can be used to present trends in the risk profile and to focus management attention on the areas of greatest risk.

Risk registers are widely used across a range of industries, such as oil and gas, construction and utilities, to facilitate efficient operations. Risk profiles are not static and require review at intervals and especially when significant change occurs (for example, the introduction of new processes or an acquisition), and should be incorporated into a management of change process. In some situations, the risk profile can change throughout the lifetime of a project, for example, during a major construction project.

3.4.3 Purpose of risk assessment and the 'suitable and sufficient' standard it needs to reach

Put simply, the objective of risk assessment is to prevent workplace accidents and incidents that might give rise to injuries and/or occupational ill-health. The risk assessment process allows organisations to identify hazards and evaluate the priority and resources needed to control them. Preventative controls and precautions can be deployed to help to reduce the risk to an acceptable level or eliminate it.

Most of us carry out risk assessments every day, such as when we drive or cross a road. So most of us have an ability to recognise hazards and take any actions. We all have varying perceptions of risk, as we have discussed in 3.3: How human factors influence behaviour positively or negatively, but applying our experience to formal workplace risk assessments can be challenging. Risk assessment needs to be suitable and sufficient and cover the issues identified in the organisational risk profiles. Risk assessment should generally remain valid for a reasonable period and be reviewed periodically unless something happens to suggest they are reviewed sooner, such as a change in legislation or an accident.

As discussed in 1.3: The most important legal duties for employers and workers, carrying out suitable and sufficient risk assessments is a legal requirement under reg 3 of the MHSWR.

Risk assessments should be carried out in any of the following circumstances:

- when they have never taken place before;
- there are changes to the workplace eg new equipment is about to be introduced or significant structural alterations have been made;
- there are significant personnel changes;
- changes have been made to applicable legal requirements or codes of practice;
- external changes have occurred that might affect the site eg access to the premises; and
- on a regular basis to review the efficiency of present systems.

Suitable and sufficient

Risk assessments need to be fit for purpose. The detail they contain should be proportionate to the risks involved. You would not expect a hazardous chemical-transfer operation that might result in a serious explosion to be assessed on a single sheet of paper; neither would you expect an assessment on access to an office to be 10 pages long.

'Suitable' relates to the means or method used to assess the risk and 'sufficient' relates to the extent or depth of that methodology.

To be 'suitable and sufficient', an assessment must:

- identify risks arising from, or in connection with, the work being assessed;
- provide a level of detail that is proportional to the risk;
- ignore insignificant risks that will only result in a very minor outcome, eg the risk of getting a paper cut when loading a copier when there are much more significant issues in that task;
- be completed by a competent person;
- consider all those who might be affected eg workers, visitors, customers;
- make use of appropriate sources of information, such as relevant law, guidance and supplier information;
- be appropriate to the nature of the work; and
- remain valid for a reasonable period.

A risk assessment will not need to anticipate those risks that are not reasonably foreseeable.

> **ADDITIONAL INFORMATION**
>
> Consider a simple tripping risk in an office. In most cases, the person who trips will stumble and quickly regain their balance. Sometimes they may fall to the ground, but the injury they sustain, if any, will most probably be minor. While the trip hazard should be dealt with, the risk assessor would not need to think about the possibility that the person who trips could strike their head on the edge of a desk and be killed – that is such a remote probability that it can safely be ignored for the purposes of the assessment.

3.4.4 A general approach to risk assessment (five steps)

Carrying out a risk assessment in a systematic way helps to ensure that it is suitable and sufficient. A simple five-step approach to risk assessment is shown in Figure 1.

Figure 1: Five-step risk assessment process[2]

The process from hazard identification to reviewing the assessment can be broken down into slightly different steps but the overall process is the same. In the UK, the HSE adopts the following steps: identify hazards, assess the risks, control the risks, record the findings and review the controls. These different approaches lead to the same result – a suitable and sufficient risk assessment.

Step 1: Identify hazards

Hazards fall into five different categories:

- physical, eg trip hazards, machinery, noise etc;
- chemical, eg acids, solvents etc;
- biological, eg living organisms such as *Legionella* bacteria;
- ergonomic, eg poorly designed workstations; and
- psychosocial, eg stressors.

There are several ways to identify hazards to choose from, according to circumstances. These include the following:

- During workplace inspections: this is an ideal opportunity to look at unsafe conditions or hazards in the workplace and to discuss them with the workforce, which promotes worker involvement in safety matters.

- During a task analysis: this is a structured approach to analysing a task by breaking it down into its component parts and considering the hazards at each stage of the task.
- Accident/incident data: this provides valuable information. Not only is it possible to see what the hazard is, but this type of data will also give valuable indicators as to the likelihood and severity of injuries that the hazard might cause, which are important later in the risk assessment process. We should also point out that this type of data is not limited to internal information – it can be from external sources too, such as national statistics.
- Ill-health records: these provide valuable information on instances of ill-health that may be associated with a breakdown in control measures. For example, a higher than average frequency of ill-health in a particular area may signify a problem, especially if all those who go off sick do so with the same or similar occupational ill-health condition.
- Absence records: the safety manager may be unaware when workers are absent from the workplace for reasons of ill-health or injury. An analysis of workplace sickness absence records may therefore help to pinpoint common causes of absence that can then be tackled.
- Audit reports: these provide a more detailed analysis of the effectiveness of parts of the safety management system.
- Incident investigation reports: these help to focus on weaknesses in safety management that may not previously have been identified or considered as important. They are useful for identification of hazards and can also yield information on human behaviour that can then be used as the basis for behavioural change and training programmes.
- Workers: the workforce can be a valuable source of information on the reality of working conditions, problems encountered when performing tasks, hazards, risks, suggestions and so on. As pointed out earlier, workers can and should contribute to the risk assessment process since they often have a lot of useful information to give.
- Using legislation to guide hazard identification: for example, a knowledge of relevant law on machinery safety may provide inspiration for thinking about hazards.
- Manufacturer's information: the manual that accompanies new work equipment will often identify specific hazards.
- International bodies such as the International Labour Organization (ILO), World Health Organization (WHO) and European Agency for Safety and Health at Work, national and international standards bodies and national regulatory agencies provide information on hazards, risks and controls in a wide range of settings, which can inform hazard identification and risk assessment.

> **APPLICATION**
> Reflect on your own workplace – what could cause harm?

Step 2: Identify people at risk

Once the hazards have been identified, it is time to consider who might be harmed and how that harm might come about. One common fault in many risk assessments is to only consider the person doing the job – it is equally important to consider other people who might be affected by the work.

3.4 Assessing risk

The following groups (or categories) of people are frequently considered in risk assessments:

- **Workers:** this covers a wide range of people present in the workplace and may include those who are directly engaged in a task (eg a worker using a power saw to cut sheet metal) as well as those working in the immediate vicinity who might also be affected. In the example given here, one of the main hazards is high noise levels – the person doing the job might have the correct ear protection, but what about their nearby co-workers? If they are not adequately protected, they could suffer noise-induced hearing loss.
- **Operators:** this category relates more directly to those operating a piece of plant or equipment. For instance, the driver of a tipper truck is likely to be seriously injured/crushed underneath if the vehicle overturns but this is less likely to happen to other nearby workers. In workplaces where vehicles are regularly used, pedestrians and vehicles should be separated by designated walkways for pedestrian use; we will look at this in more detail in 8.6: Safe movement of people and vehicles in the workplace.
- **Maintenance staff:** these workers are often overlooked in assessments, although they are potentially at greater risk as they may have to deactivate safety mechanisms to get to the parts of equipment that need to be maintained. Many serious accidents occur each year because of poorly planned and conducted maintenance activities.
- **Cleaners:** this group will be exposed to hazards associated with their cleaning activities but may also be exposed to hazards associated with the equipment they are cleaning or the area in which they are working. Cleaners often work in small groups or alone outside normal working hours and so may be at greater risk for that reason. Furthermore, cleaners may not be aware of the hazards present in some equipment, which increases the chances of them encountering something that may cause them harm.
- **Contractors:** a contractor's focus is going to be on the job that they must do; they may not pay quite so much attention to what is going on around them. A lack of familiarity with the workplace or work equipment being operated in the area may mean that contractors will be at greater risk of injury.
- **Visitors:** they are at greater risk because of their lack of familiarity with the workplace. They will also have less knowledge of what to do in an emergency and so special precautions should be taken to help ensure their safety.
- **Members of the public:** those who enter the premises or site by invitation are classified as visitors. However, members of the public may simply be walking past an area where a task is being conducted and so may be at risk because of their proximity to hazards. For example, work at height being conducted in a busy street poses a risk to passers-by, who may be struck by falling objects if they are allowed to approach too close to the point of work. Conversely, members of the public might inadvertently pose a risk to the workers themselves (eg by knocking into a ladder that has been positioned on a public path, causing the worker to fall).

Consideration also needs to be given to more vulnerable people who may be at risk from particular hazards or who may be affected by hazards because of a lack of understanding or knowledge. The range of people who can be considered as vulnerable includes young workers, those with disabilities or on medication, pregnant workers or nursing mothers, lone workers and people from other countries where language and work practices may differ from those in the country of work.

Managing risk – understanding people and processes

> **APPLICATION**
>
> Consider your own workplace – which groups of people might be harmed?

Step 3: Evaluate risk

This stage of the assessment process allows the risk assessor(s) to decide whether the risks to which the various people are currently exposed are adequately controlled, or whether more needs to be done. This involves a judgement about each hazard and how likely it is that the hazard will be realised (likelihood) and, if it is, what the most probable outcome will be (severity).

There are two extremes – and variations between them – in how this assessment can be carried out. One is a simple, subjective judgement and the other involves complex techniques using quantitative data.

Likelihood of harm and probable severity

Likelihood is considered in the context of circumstances in which the hazard may be encountered and the control measures in place, as these affect the likelihood of someone being harmed. Circumstances can include the environmental conditions, the competence of the people involved, attitudes, supervision levels, frequency and duration of exposure and workload pressures.

Probable severity considers the probable outcome of contact with the hazard; this could be minor injury, major injury, death or damage to equipment. The focus should be on the probable outcome not just a possible outcome, which could involve some extremely unrealistic outcomes. Circumstances such as weather conditions play a part here too.

Possible acute and chronic health effects

Exposure to some substances might have an immediate (or acute) effect on the worker; however, if that exposure continues the worker may develop chronic effects. An example of this would be exposure of a construction worker to the sun without protection for a short period, causing sunburn. However, continued and repeated exposures could result in more permanent damage and chronic effects such as skin cancer.

Risk rating and prioritisation of risk

When evaluating risk the likelihood and severity are considered. Likelihood and severity can be assigned values on a scale:

- Likelihood of an undesired event occurring – eg on a scale of 1 to 5, 1 being highly unlikely and 5 being certain, how likely is it that you will trip over that trailing lead?
- The probable severity of the outcome – eg on a scale of 1 to 5, 1 being very minor and 5 being a fatality, how bad is your injury likely to be?

The assessor looks at the situation and considers the control measures that are already in place to help reduce the risk arising from a hazard. They will then choose the point on the scale (number) that they feel represents the reality of the situation. The chosen numbers are then combined as follows to give a figure for residual risk:

$$\text{risk} = \text{likelihood} \times \text{severity}$$

Table 2 shows examples of 5-point scales with numbers and descriptors for both likelihood and severity.

Likelihood scale

1	Very unlikely	Less than 10%
2	Unlikely	10% to 39%
3	Likely	40% to 59%
4	Very likely	60% to 80%
5	Almost certain	More than 80%

Severity scale

1	Very minor	Injuries such as cuts or scratches; no lost time other than first-aid or small repair
2	Minor	Injuries such as sprains or bruising
3	Moderate	Injuries such as fractures or burns
4	Major	Permanent disability
5	Severe	One or more fatalities and/or loss of/damage to plant that would cause a serious disruption to business

Table 2: Examples of 5-point likelihood and severity scales

This approach provides a 'semi-quantitative' evaluation of the risk. The use of figures (with meanings defined in the organisation's risk assessment system) helps to make this approach less subjective, but an element of subjectivity remains. However, properly trained and competent assessors should be able to arrive at reasonable conclusions using a system like this and so this approach is very widely used in all kinds of industries and workplaces. Some organisations use more complex '10 x 10' matrices with more gradations on the likelihood and severity scales.

Some organisations that wish to gain an even more accurate evaluation will use a quantitative assessment system that employs probability calculations based on researched data to come up with a more accurate, more objective assessment of residual risk. This approach is likely to be used in situations where accurate and dependable results are needed, such as in the petrochemical industry.

APPLICATION

Imagine a trailing lead across an office floor. Twenty people work in the office and the lead is across a main walkway; it is not covered. We start with the 'likelihood' that someone will trip.

Thinking about likelihood, in this situation it would seem almost certain that someone is going to trip over the cable – we will give this a '5'.

In terms of severity, there is a remote possibility that a very serious injury could occur, but the reality of the situation is that this probability is very small. The injury is probably going to be minor – we will give this a '2'.

Now, you may have thought differently. This highlights the point we made earlier about subjectivity. Using the scales helps to narrow the focus and reduce the subjectivity.

Therefore, if all the assessors have had the same training and use the system in the same way, they are more than likely going to reach similar conclusions.

> With a '5' for likelihood' and a '2' for severity, multiplying these together gives us a risk rating of 10 out of a possible 25. This is a moderate level of risk and is something that we should probably do something about.
>
> Imagine the cable is not in an office but across the top of a flight of heavily used stairs. The likelihood of a trip is still a '5' but falling down a flight of stairs would probably cause significant injury, so we are going to give it a '4' for severity. That will give us a residual risk figure of 20 – a much more serious proposition.

Prioritising risks

When it comes to prioritising risks, those with the highest rating are the priority. In the Application example, the more serious issue is the cable across the stairs, so it becomes our highest priority.

The figures that can be obtained from a risk assessment can be set out on a grid, as shown in Figure 2. You will see that the grid has been divided into colour-coded zones, representing different levels of risk.

		Likelihood				
		Almost certain	Very likely	Likely	Unlikely	Very unlikely
Severity	Severe	25	20	15	10	5
	Major	20	16	12	8	4
	Moderate	15	12	9	6	3
	Minor	10	8	6	4	2
	Very minor	5	4	3	2	1

Figure 2: 5 × 5 risk matrix

A useful phrase when thinking about how to react to a given level of risk is 'effort and urgency should be proportional to the risk'. Thus, the higher the number, the more effort should be put into dealing with the risk. This approach is encapsulated in Figure 3, which makes use of the coloured risk bands in the matrix in Figure 2 to define the nature of the response to a given level of risk.

Risk band	Risk level	Action
1–3	Trivial	No action needs to be taken. No requirement to record this risk.
4–6	Tolerable	An acceptable level of risk. Record and monitor. Action can be taken to reduce the risk provided no cost is incurred.
8–12	Moderate	Action should be taken within three months to reduce the risk into the tolerable zone.
15–16	Substantial	A higher level of risk, which must be reduced to 'moderate' before commencing work.
20–25	Intolerable	Work must not be allowed to continue until action has been taken to reduce the risk to an acceptable level.

Figure 3: Risk bands and actions

A note of caution should be applied to the simplistic approach of prioritising risks with the highest rating; based on this approach, hazardous situations, or events with high severity but low likelihood of happening, can be overlooked. If a situation is so serious that it would have catastrophic consequences, then this situation is best considered using other risk assessment techniques and included in an organisation's business continuity plan.

Principles to consider when controlling risk

Regulation 4 of the MHSWR directs duty holders to a good practice philosophy of risk prevention set out in Schedule 1 of the Regulations.

The general principles are as follows:

- Avoid risks where possible: clearly, if there is no risk, then no harm can occur, but this option may not always be possible.
- Evaluate unavoidable risks: meaning that a risk assessment will be needed for risks that cannot be avoided.
- Combat risks at source: dealing with the risk at source significantly reduces the chance of being harmed by it. An example is designing a power tool to have low vibration characteristics.
- Adapt work to the individual: this is about taking an ergonomic approach to the design of the job, tools and workstation.
- Adapt to technical progress: keeping up to date with technological developments that may make a task safer.
- Replace the dangerous with the non-dangerous/less dangerous: this really means substitution, such as substituting a water-based product for a solvent-based one.
- Develop a coherent overall prevention policy: this refers to having in place a consistent strategy for the control of risk.
- Give priority to collective protective measures over individual protective measures: 'collective protective measures' include engineering controls, barriers etc. 'Individual protective measures' include personal protection eg gloves, hearing protection and equipment such as power guillotines where the worker will have to simultaneously press two separated controls (left hand, right hand) for operation. An example of this approach is to totally enclose a chemical process rather than relying on everyone wearing respiratory protective equipment. Giving priority to collective over individual protective measures is about creating a 'safe place' rather than relying on a 'safe person' strategy.
- Give appropriate instructions to workers: information, instruction and training are important, but are at the bottom of this list because instructions can be easily misunderstood or forgotten.

There are circumstances when workers cannot be provided with a 'safe place of work' eg a firefighter or other emergency worker or a worker dealing with a hazardous spillage – every time they go into a fire or have to deal with a chemical spillage, they are at risk. In such situations, the only way that the risk can be reduced is to concentrate on a safe person strategy, considering personal protective equipment (PPE), information, instruction and training and the safe behaviour of the individual.

Practical application of the principles – applying the general hierarchy of control (clause 8.1.2 of ISO 45001:2018)

The most effective way of dealing with any risk is to eliminate the hazard that causes it. In some cases, this may be possible and reasonable, but in many situations it will not. When it is not appropriate or possible to eliminate the hazard, controls will need to be used to help reduce the risk.

The best option for control of risks when elimination is not possible is to create a safe workplace. In theory, therefore, the workplace can be made sufficiently low risk that all who enter it will be adequately safeguarded. This is normally achieved by using controls that apply directly to the hazard, such as guarding on machines or total enclosure of a process; if the hazard is enclosed then everyone in the area is protected regardless of what they are doing or where they are.

The less effective (though still necessary and widely used) option is to have a 'safe person' approach. This involves ensuring that individuals act safely so that their safety is not compromised. This can be achieved by having the worker follow a safe system of work, by providing training to increase skills or by using PPE. The problem with this approach is that it only protects the person who is using the 'safe person' control, and so it relies heavily on compliance with procedures and on human reliability.

What we have essentially described is a hierarchy of control options. ISO 45001:2018 clause 8.1.2 sets out a five-level hierarchy to provide a systematic approach to improve health and safety, eliminate hazards, and reduce or control risks. Each level in the hierarchy is less effective than the one above it. In practice, a combination of controls will be needed to reduce risks to a level that is as low as reasonably practicable.

The hierarchy is demonstrated in Figure 4.

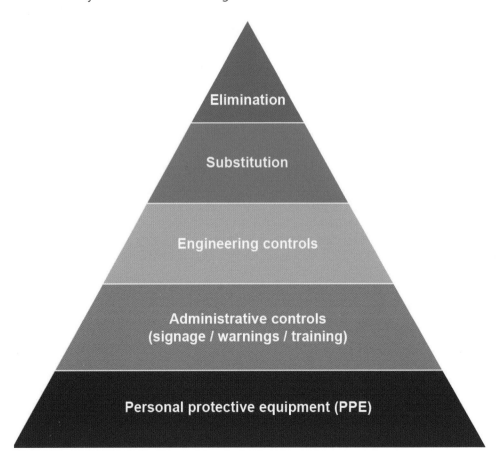

Figure 4: The hierarchy of control

3.4 Assessing risk

Table 3 provides examples of the hierarchy of control.

Eliminate	• Remove lead additives from petroleum products. • Remove monotonous repetitive work through automation.
Substitute	• The problems associated with the use of solvent-based paints and varnishes (drowsiness and fire, particularly in confined spaces) are eliminated through substitution with water-based products. • Use laser cutting in a ship-building shop, which dramatically reduces noise problems. • Control pedestrian/traffic conflicts by measures such as banning one or other from certain areas, separation of routes, use of an underpass or footbridge etc.
Engineering controls	• Use a power sander with a built-in dust extractor. • In a bottling plant, rubber pads are used to separate the bottles as they move along a conveyor belt, thus significantly reducing noise levels. • Components to be soldered are repositioned and fed mechanically to avoid the worker having to turn around to reach for them. • Partially enclose processes by use of, for example, a paint-spraying booth that is open on one side. • 'Ingredients' for processes are delivered pre-weighed and packaged to reduce exposure.
Administrative controls	• Safe systems of work • Permit-to-work • Safety signs
Personal protective equipment	• Chain mail gloves in a meat processing factory • Thigh-length boots for wading in water • Safety spectacles in a chemical laboratory • Goggles when using a grinding machine

Table 3: General hierarchy of control examples

Signs and warnings are often used to support administrative controls and raise awareness of hazards and the behaviours required. It is important that when signs are used, they are relevant and maintained, and, if circumstances change so that a sign is no longer appropriate, it is removed and replaced with the correct one. As Table 4 illustrates, signs fall into five main categories – prohibition, mandatory, warning, Emergency first aid or escape (sometimes referred to as safe condition) and fire safety – all with recognised conventions for shape and colour.[3]

Type of sign	Shape	Design	Pictogram
Prohibition Actions/equipment that are prohibited	Circular Red and white	Black pictogram on white background with red border and diagonal line	
Warning These signs give warning of potential risks	Triangular Black on yellow	Black pictogram on yellow background with black border	
Mandatory A course of action which must be taken	Circular Blue and white	White pictogram on blue background	
Emergency first aid or escape	Rectangular or square Green and white	White pictogram on green background	
Fire safety Location of firefighting equipment	Square Red and white	White pictogram on red background	Fire extinguisher

Table 4: Sign categories

Application of control measures based on prioritisation of risk

In circumstances where the risk priority determines that the current level of control is inadequate, additional controls need to be introduced. Essentially, the higher the risk the greater the priority to apply control measures. If it is going to take some time to implement the control measures, temporary measures may have to be put in place. If, however, the risks are high and temporary measures are not deemed adequate, then suspending the work might have to be considered. Judgements about adequacy of risk controls can be informed by the use of guidance and legislation.

Use of guidance, sources and examples of legislation

Specific legal requirements are in place for certain control measures. For example, it is a requirement under the Confined Spaces Regulations 1997 that confined space entry will be carried out under a safe system of work. In such cases, the need to consider risk levels can be less pressing than understanding what the law requires as a minimum. It is possible that a risk could be calculated to be tolerable, leading the assessor to conclude, not unreasonably, that no further controls are necessary. However, if the controls required by law are not in place, then the organisation will still be breaking that law, even though the actual risk is low. Therefore, the legal requirements should be the starting point when considering control measures (a topic we will come to shortly). Specific legal requirements pertaining to control measures for noise, vibration, radiation and chemicals are covered in 5: Physical and psychological health and 7: Chemical and biological agents.

3.4 Assessing risk

There is also a lot of guidance available from a range of national and international sources that provides details of controls that can usefully be employed to deal with specific hazards. Such guidance should always be consulted by the assessor as it will save a lot of time and effort in deciding what is and what is not needed to control the risk.

Applying controls to specified hazards

The means by which risks associated with specific hazards are reduced will vary according to the hazard. If you are looking to help reduce risks associated with chemicals, this may involve a change to a less hazardous substance. If the issue is reducing risk associated with manual handling, the use of smaller loads or mechanical aids should be considered. For all hazards, the controls need to follow the general hierarchy of control.

Residual risk: acceptable/tolerable risk levels

To recap, at this stage we have identified a hazard and considered who it might harm. We have then evaluated the risk by considering what control measures (if any) are already in place and then taking a view as to how likely it is that an incident might occur and, if it does occur, how serious the outcome is likely to be. We have assigned figures to likelihood and severity, by which means we have arrived at a figure representing the 'residual risk', which is simply the amount of risk that remains once existing control measures have been considered.

Now that we have a figure for 'residual risk', we can take one final step. This involves making a judgement as to whether the level of residual risk we have calculated can be accepted, in which case we need do nothing more, or whether the risk is still too high, in which case more controls need to be put in place.

Think about crossing a road – you need to get to the other side. Now, if the road was extremely busy, with fast-moving traffic, you would probably conclude that the risk was too great (intolerable) and so, under normal circumstances, you would either not cross the road at all or you would seek an alternative way across – say, via the footbridge half a mile away, which would reduce the risk to more acceptable levels. But what if there was some significant benefit to you being able to get across to the other side of the road immediately, by the quickest possible route? In such a situation, you might consider the risk to be worth taking – in other words you tolerate it. It is still the same risk as it was before, but now the circumstances have changed. The risk could still be managed so that it becomes lower, but in the circumstances, you are willing to accept it. This is the concept of 'tolerability' or 'acceptability' of risk.

Tolerability criteria will vary from organisation to organisation – what one organisation considers intolerable, another may accept because of the nature of its undertakings or other things, such as the benefit to be gained from taking the risk, the level of training and expertise within the workforce, the consequences of not taking the risk and so on. In essence, the decision as to what constitutes an acceptable level of risk is related to the context. This will show itself in the risk assessment matrix illustrated in Figure 3, where a company that has a low tolerance to risk may put the 'intolerable' range from 15 to 25. Alternatively, an organisation that is more accepting of risk will probably move its 'tolerable' range up to 8 or even higher. It really is dependent on the organisation's attitude to risk.

Distinction between priorities and timescales

Ideally, a high-priority action would be completed very quickly, with a low-priority action perhaps being left for some time. However, depending on the complexity of what has been recommended, it may take quite a while to action recommendations, even if they are ostensibly high priority. For example, training the workforce may be a high priority, but the timescale for delivery will need to account for training development time as well as the time taken to train all staff, which could take several months.

Conversely, an assessment might identify a low-risk issue, which because it is a low risk will would therefore be a low priority for action. However, if the actions can be done quickly and cheaply, the timescale for implementation of the recommendation would be measured in days rather than months, even though the action itself is not a high priority.

Step 4: Record significant findings

It is essential to record all the significant findings so that an accurate picture of the risks associated with a task or location can be built up and to enable appropriate controlling action to be taken. Insignificant or 'trivial' findings need not be recorded. If everything that was found during an assessment was to be recorded in writing, then assessments would often be a lot longer and more complicated than they need to be.

There is no standard way in which risk assessment findings must be recorded. In addition, the variety of different things to be assessed tends to work against having a 'one size fits all' approach that can be used to assess access and egress, chemical usage, DSE, manual handling, fire, noise, vibration and so on. Each of these requires its own format, the exact nature of which is left to the organisation to decide. There are, however, common contents that all risk assessments should share. These include:

- recognition of hazards;
- identification of persons at risk;
- consideration of existing control measures;
- further controls; and
- review date.

HSE produces risk assessment templates that can be downloaded and completed by any low to medium-risk organisation. This is in a simple format (see Figure 5).

Risk assessment template

Company name: **Assessment carried out by:**

Date of next review: **Date assessment was carried out:**

What are the hazards?	Who might be harmed and how?	What are you already doing to control the risks?	What further action do you need to take to control the risks?	Who needs to carry out the action?	When is the action needed by?	Done

More information on managing risk: www.hse.gov.uk/simple-health-safety/risk/
Published by the Health and Safety Executive 09/20

Figure 5: HSE risk assessment template[4]

Step 5: Reason for review

Risk assessment is not a one-off process. Recommendations for additional controls may have been made, or it may have been decided that the current situation is safe enough; either way, it will be important to keep the assessment under review.

When an assessment is completed, it is important to set a sensible review date. This is normally going to be in line with the highest level of residual risk calculated in the assessment. Imagine that you have completed a risk assessment and that the highest residual risk figure you have identified is a 12. If you look at Figures 2 and 3, you will see that, according to our system, this is a moderate risk, which should be actioned within three months. You would therefore set a review date for three months' time.

There are several other reasons why it would be appropriate to review a risk assessment.

Following an incident
Remember that the risk assessment does not have to be completely perfect; it merely needs to consider what is likely to happen. Unfortunately, despite our best efforts, things sometimes get missed or the unexpected happens. In such cases, you need to investigate to find out what went wrong and then take steps to stop it from happening again. This may involve revisiting the risk assessment to add further controls.

Changes in processes or equipment
This introduces obvious technological changes that may not have been considered in the original assessment. An opportunity to update the assessment to reflect the changes should therefore be taken.

Changes in personnel
This may mean a change in the level of skills or competencies that are present in the workplace, which can influence the potential for accidents. For example, imagine that a highly experienced worker retires and is replaced by a much younger, far less experienced worker. This would obviously call for a reassessment that considers their different levels of skill and knowledge.

Changes in the law
As already mentioned, the organisation will need a system for keeping up to date with changes in the law. When these occur, such as changes to hazardous agent exposure limits, a review of the assessment will be needed to check compliance with the new requirements.

Passage of time
Over time, control measures can break down and workers can fall into a more relaxed way of working, which may increase risks. Also, as time passes, the relevance of the assessment may reduce due to changes in organisational priorities etc. For these reasons, and in the absence of any other reason to review the assessment, it is important that the assessment is reviewed, say annually, just to check that it is still relevant and applicable.

3.4.5 Application of risk assessment for specific types of risk and special cases

Examples of when they are required

The general approach to risk assessment we have discussed is applicable in a wide range of work situations. However, some risks and circumstances require an enhanced approach to include specific factors in assessing the risks; for example, when considering risks arising from fire, noise and chemicals. It is usual to consider specific factors to assess these risks. In the case of noise, noise exposure levels are measured and compared with action levels. Similarly, occupational exposure limits must be considered for certain chemicals.

Manual handling activities also have a requirement to look at specific risks relating to the task, the person, the load and the environment in which the handling is done. This is discussed in further detail in 6.2: Manual handling.

Why specific risk assessment methods are used for certain risks

It is important that when certain risks do require specific methods to be used to assess the risk, such as at a chemical or nuclear plant, these methods are used to ensure a proper, systematic consideration of all relevant issues that contribute to the risk. Using a generic approach when specific risks should be considered might result in an inadequate assessment and therefore inadequate or inappropriate controls.

Special case applications to young people, expectant and nursing mothers; also consideration of disabled workers and lone workers

Young workers
Young workers are those who have not yet reached the age of 18. They will be at additional risk due to their physical and/or mental immaturity and because of their absence of awareness of existing or potential hazards. Young workers will also want to fit in with the rest of the work group and may therefore try too hard to impress, for example by doing their work too quickly, which may create safety issues. They may also be more easily influenced by peer pressure (which we discussed in 3.3: How human factors influence behaviour positively or negatively).

The risk assessment should take young workers into account; in fact, a special risk assessment may be needed just for the tasks to be done by the young persons. Regulation 19 of the MHSWR places specific requirements on employers to consider the protection of young people at work from any risks to their health or safety that are a consequence of their lack of experience, absence of awareness of existing or potential risks, or the fact that they have not yet fully matured. They will generally need a lot more supervision than their older and more experienced counterparts and it may be that the organisation decides not to allow them to do certain jobs due to the risks involved. They will, of course, require additional training and instruction to bring them up to the same standards as the rest of the work group.

New and expectant mothers

Pregnant workers, and especially their unborn child, may be at risk from exposure to certain substances, such as lead, or from exposure to harmful levels of physical agents such as ionising radiation. There may also be problems when trying to lift and carry heavy or awkward loads, plus they are more prone to becoming fatigued during the working day. Regulation 18 of the MHSWR requires a worker to notify the employer in writing that they are pregnant, have given birth within the previous six months or are breastfeeding.

Regulation 16 also requires that a risk assessment should be done on a case-by-case basis for work done by a woman who is pregnant so that appropriate measures can be taken to safeguard her health and that of her unborn child. It may be necessary to restrict certain activities, such as jobs that involve heavy lifting, and it may be appropriate to allow more frequent rest breaks. In some cases, particularly where night work is concerned, the employer may need to adjust the worker's working pattern to accommodate the pregnancy, such as by moving the worker to day work. In all cases, it will be important to involve the worker in the discussion to find the best solution.

Disabled workers

Much depends on the nature of the disability. Workers with hearing impairments may struggle to hear alarms or verbal instructions, whereas those with mobility problems may encounter difficulties in evacuating the premises in the event of an emergency – clearly, disability can take many forms, of which these are but two examples. Importantly, disability may not always be visible or obvious to co-workers. This is especially so with cognitive issues such as dyslexia or autism. The employer will need to consider the needs of each disabled individual and may need to make reasonable adjustments to the workplace or to the way in which the task is done so that the disabled worker is not placed at additional risk.

It is also quite proper to have a recruitment policy for some tasks, whereby employers specify the key physical attributes that the worker should have. For example, a person with a lower back and mobility disability should not be engaged in a task involving lots of strenuous lifting and carrying since that will only exacerbate their condition. Although such a policy may have the effect of excluding some disabled people from a task such as this, it should not be seen as an absolute barrier to their employment within the organisation; employers should be careful in circumstances such as these not to fall foul of anti-discrimination laws.

Lone workers

The nature of lone working is such that there is unlikely to be someone else around to render assistance if the worker is injured or falls ill while at work. In some occupations, such as sales executive or domiciliary care worker, lone working is a natural part of the job and cannot be avoided. In others, especially those where higher-risk activities such as machinery maintenance are being conducted outside core working hours, the need for lone working should be by exception. As the risk control measures provided by a co-worker being able to help with a task or to raise an alarm if needed are removed in this situation, the risk assessment needs to identify what can reasonably be done to control the risk and whether this is adequate.

Managing risk – understanding people and processes

KEY POINTS

- A hazard is something with the potential to cause harm.
- Risk is the likelihood of potential harm from that hazard being realised.
- Risk profiling is a structured approach to shape risk management that requires a range of data and information. The output of the process should provide an organisation with an overview of the risk of its operations and the effectiveness of the controls in place to mitigate the risks.
- Risk assessment is the process of identifying preventative and protective measures by evaluating the risk(s) arising from hazards, considering the adequacy of any existing controls and deciding if the risk is acceptable.
- The purpose of risk assessment is to prevent workplace accidents and incidents that might give rise to occupational ill-health.
- A general approach to risk assessment is to use five steps:
  - identify hazards;
  - identify people at risk;
  - evaluate risk (taking account of what you already do) and decide if you need to do more;
  - record significant findings; and
  - review.
- Considering the likelihood of harm occurring, and probable severity if it did, allows a risk rating to be determined.
- Risk control should be based on the broad 'principles of prevention' that encourage risk to be avoided or dealt with at source.
- The general hierarchy of control provides a practical framework for controlling risks in order of importance and effectiveness.
- Control measures should be applied based on risk prioritisation and informed by legislative and other good practice guidance.
- Residual risk is the amount of risk that remains once existing control measures have been considered. Organisations have differing standards for what they will consider an acceptable/tolerable risk level.
- Some risk areas, such as fire, DSE, manual handling, hazardous substances and noise, require specific factors to be considered to carry out the risk assessment.
- Certain risks, such as those associated with a chemical or nuclear plant, require specific risk assessment methods to enable proper, systematic consideration of all relevant issues that contribute to the risk.
- There are certain types of worker who require special consideration when carrying out a risk assessment; these are young people, expectant and nursing mothers, disabled workers and lone workers.

References

[1] HSE, *Managing for health and safety* (HSG65, 2013) (www.hse.gov.uk)

[2] Adapted from ILO, 'OSH Management System: A tool for continuous improvement' (2011) (www.ilo.org)

[3] International Organization for Standardization (ISO), ISO 7010:2019 *Graphical symbols — Safety colours and safety signs — Registered safety signs*

[4] Source: HSE (www.hse.gov.uk)

3.5: Management of change

> **Syllabus outline**
>
> In this section, you will develop an awareness of the following:
> - Typical types of change faced in the workplace and the possible impact of such change, including: construction works, change of process, change of equipment, change in working practices
> - Managing the impact of change:
>   - communication and co-operation
>   - risk assessment
>   - appointment of competent people
>   - segregation of work areas
>   - amendment of emergency procedures
>   - welfare provision
> - Review of change (during and after)

Change is bound to happen in any organisation, no matter how big or small. Events such as a change in leadership, growth, product changes or technical equipment upgrades can have a dramatic impact. The impact of a change can sometimes be difficult to appreciate. This is why organisations need a logical, systematic way to manage changes.

Health and safety risks to workers are potentially many times higher during change than during routine operations. Workforce health and safety should be an integral part of managing organisational change. Sometimes health and safety aspects of changes are seen to be a related yet separate set of issues within the organisational change process. However, to ensure the effective implementation of the change process, they need to form an integral part of the overall change process planning.

A change to core processes, even temporary ones, can endanger the health and safety of the workers. It is critical to recognise, define and risk assess the change before introducing it to the process in a planned way.

Management of change

Changes in the environment in which an organisation operates may bring both challenges and opportunities. Changes are often driven by external influences, such as legislation and standards, new technologies or materials or major global events, such as a pandemic. Change may also be driven by internal influences, such as a change in top management.

It is important that the organisation has an effective process that allows it to identify and to minimise the impact of any hazards or risks that arise from the change. As with any risk assessment process, the scope, complexity and level of detail required should be proportionate to the risk. In other words, the extent of the management of change process should be balanced against the scale of the challenge or opportunity and potential impact on the organisation. By exploring potential future development and anticipating associated risks and opportunities, organisations can be prepared for change.

However, there are instances when developments occur rapidly and there is little time to carry out a thorough management of change process, for example, as occurs with a pandemic or major natural disaster. It is advantageous to consider potential scenarios and work out some strategies that could be drawn on in such emergency situations. You can then rely on dynamic risk assessment to add the detail at the time. This is usually the case in emergencies – like those faced by emergency services – when the exact details are not known until the time comes.

3.5.1 Typical types of change faced in the workplace and possible impact of such change

Changes in the workplace are many and varied and are driven by a wide range of issues.

Common examples are:

- Organisational changes, including:
  - restructuring, with changes to key personnel, roles, responsibilities, teams or departments; and
  - downsizing, accompanied by increased outsourcing, flexible working, reduced/ flattened hierarchy, self-management and increased automation.
- Process or technical changes, including:
  - changes to administrative arrangements and working practices, such as working hours, staff relocation, new equipment, new materials or changes to material packaging or composition of material (eg using pellets rather than powder).

Failing to manage change effectively is a major cause of serious incidents and especially so in high-hazard installations. Potential outcomes of ineffectively managing change include:

- **Construction works**, such as renovations and extensions introduce new activities to the workplace and with those come additional hazards such as obstruction, changes in access and egress and hazardous substances. The nature of construction work means that the workplace can be subject to constant change over the lifetime of the project. As a result, the hazards on site can change daily and escape routes may also be adjusted as a project progresses.
- **Changes in process**, such as increasing concentrations of a reagent to speed up a reaction, which could generate an uncontrolled thermal reaction that the equipment cannot withstand, and fire occurs.
- **Changes in equipment**, such as the speed of a conveyor on a packing line, may result in worker fatigue and additional ergonomic hazards, or increased automation may lead to worker concerns about lack of skills and job security.
- **Changes in working practice**, such as introducing mechanical aids (eg a hoist) to replace manual handling, could introduce the hazards of moving equipment, falling objects and mechanical failure.
- **Changes in workforce structure**, such as a smaller workforce, smaller teams doing the same work or fewer layers of management, may lead to overload, the need for multi-skilled workers or self-managed teams, unclear reporting lines and systems that do not work well with a smaller workforce.
- **Changes in reliance on contractors** may result in situations where contractors lack the workers' skills and experience and workers need to develop skills in supervising contractors, increasing their own workload.

3.5 Management of change

> **APPLICATION**
>
> Consider the potential changes that may take place in your organisation. How would they be managed? What are the potential outcomes of ineffectively managing them?

CASE STUDY 1
Flixborough explosion

On 1 June 1974 the Nypro (UK) chemical plant at Flixborough was severely damaged by a large explosion. Twenty-eight workers were killed and a further 36 suffered injuries. Off site, there were 53 reported injuries and property in the surrounding area was damaged to a varying degree.

Prior to the explosion, on 27 March 1974, it was discovered that a vertical crack in a reactor was leaking cyclohexane. The plant was subsequently shut down for an investigation. The investigation that followed identified a serious problem with the reactor and the decision was taken to remove it and install a bypass assembly to connect two other reactors so that the plant could continue production.

During the late afternoon of 1 June 1974, a 20-inch bypass system ruptured. This resulted in the escape of a large quantity of cyclohexane, which formed a flammable mixture and found a source of ignition, leading to a massive vapour cloud explosion.

One of the key failings identified was that a plant modification occurred without a full assessment of the potential consequences. Only limited calculations were carried out on the integrity of the bypass assembly. No drawing of the proposed modification was produced. No pressure testing was carried out on the installed pipework modification. Those concerned with the design, construction and layout of the plant did not consider the potential for a sudden failure rapidly escalating to a major disaster.

This highlights the consequences of not effectively managing a temporary change.[1]

CASE STUDY 2
Chemical manufacturer

The purchasing department in a chemical manufacturer sourced a cheaper supplier of a hazardous liquid raw material.

The liquid could be delivered to the dedicated delivery point on site using the same type of road tanker with the same type of connections as the existing supplier. Based on this and the reduced costs, an order was placed.

The purchasing department did not consult with the plant staff on the implications of this change. When the delivery arrived, it demolished the delivery gantry, resulting in a moderate loss of hazardous material but a significant loss of production. Although the new supplier's tanker and connections were compatible with the delivery point, the trailer carrying the tanker was higher than had been previously used.

This illustrates the potential benefits of identifying what has actually changed – some changes may have an importance that is not immediately obvious.

3.5.2 Managing the impact of change

The impact of change can be managed effectively by adopting a systematic and structured process. ISO 45001:2018 sets out requirements for the management of change process. This involves developing a process for temporary and permanent changes that affect health and safety. Changes to be considered include: new or changed products, services and processes; changes in legal requirements and in knowledge about health and safety risks; and knowledge and technology developments. The consequences of unintended changes should be reviewed, and action taken to mitigate any adverse effects. It is worth noting that while change may present risk to the organisation, it may also present opportunities, for example, greater efficiencies.

So how might an organisation go about setting up a management of change process? Using the following three-step framework (Figure 1) is a useful starting point.[2] Although this was developed initially in the process industries, it has wide applicability and relevance.

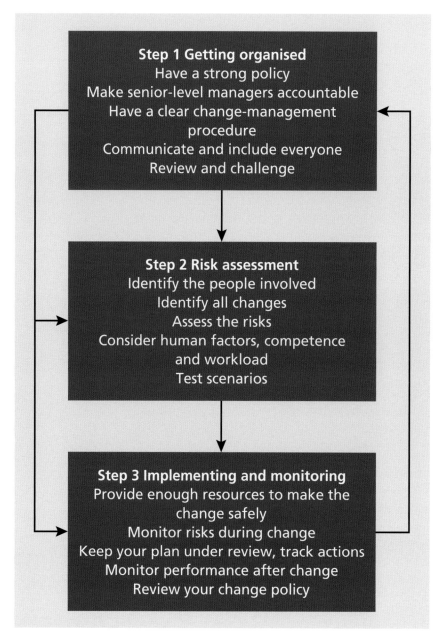

Figure 1: Framework for managing organisational change

Step 1: Getting organised

- **Policy:** having a clear policy for the management of organisational change will clarify principles, commitments and accountabilities in relation to their impact on health and safety. Changes vary significantly in scale and in their impact on health and safety. Not all changes require a formal management of change process; some can be managed within the existing management system. It is important to identify what constitutes a significant change. The policy should commit to considering change in a proportionate way. Changes that do not seem obviously connected to health and safety need to be considered. The effort and resource invested in the change management process must be proportionate to the complexity of the change, the scale of the hazards concerned and the extent to which the change may impact on the management of major hazards.
- **Senior management commitment:** top management commitment is essential. It helps if an influential senior manager can sponsor or champion for the elements of the change process that relate to health and safety. They should ensure the safety aspects of the change receive an appropriate level of resource and attention.
- **Effective formal change process procedures, consultation and communication arrangements:** these are important elements of success that will enable the early identification of potential impacts as well as potential improvements.
- **Change-management procedure:** the change should be planned in a thorough, systematic and realistic way. The procedure for each element of organisational change management should be documented and structured. It should be clear what is to be done, by whom and in what way, what the potential risks are and who reviews the changes, when and how. The stages in the process should be adequately recorded to ensure transparency, auditability and accountability.
- **Competence:** the change process should be supported by competent individuals. For some detailed technical changes to a process or plant, specialist skills may be required, such as in quantitative risk assessment.
- **Communication:** this is key to the success of the process. All stakeholders should be involved from an early stage and kept involved throughout the process. This helps build acceptance of the change and provides the best opportunity to use the vast knowledge of workers and contractors.
- **Review and challenge:** the process should be subject to regular progress reviews by senior management, for which they need sufficient information.

Step 2: Risk assessment

The risk assessment should ensure that, following the introduction of the change, the organisation will have the resources and competence to maintain adequate standards of health and safety.

Two aspects of the change need consideration:

- risks arising from the process of change; and
- risks and opportunities resulting from the outcome of change.

The assessment can be based on the mapping of the existing arrangements to the new arrangements, taking into account individuals and tasks. By considering various foreseeable scenarios that might arise, the adequacy of proposed new arrangements can be assessed.

The assessment needs to identify non-routine situations, such as infrequent tasks (eg maintenance) and cover for sickness and leave, work priorities and new skills, knowledge and experience required to undertake the work activities. Other factors to consider are workloads to avoid the consequences of overloading and the potential for human failure. If contractors will be used during and after changes, the risks associated with this should be considered; for example, a contractor being unwilling or unable to meet requirements.

During and after the change, amendments to emergency arrangements and welfare provision may be necessary and work areas may need to be segregated. See 3.8: Emergency procedures and 8.1: Health, welfare and work environment for more details.

> **CASE STUDY 3**
> **Construction work**
>
> In the case of construction work to refurbish an industrial unit, additional welfare facilities for the site staff and contractors were needed, as the existing facilities were in the area to be refurbished.
>
> The construction site was segregated from the operational part of the unit to limit and control access. The evacuation routes, muster point and first-aid provision were relocated during the work.

Step 3: Implementing and monitoring

- Implementation: resources should be sufficient to ensure that exposure to risks is not significantly increased during the change. Plans should account for an increase in workload during the transition.
- Monitoring: risk assessments and plans for both the transition and progress should be regularly reviewed to determine progress against set objectives and key performance indicators.

Health and safety performance during and after the change should be monitored through both proactive and reactive measures. Where there is evidence of significant risk, decisions may need to be changed or reversed.

3.5.3 Review of change

There will always be a degree of uncertainty about the impact of change. Sometimes the effects of a change can be quite subtle or delayed. Periodic, planned reviews should assess whether the changes have been effectively implemented or whether additional action is required. Identifying what went well and what could have gone better helps you learn from the process to benefit the management of future changes.

> **APPLICATION**
>
> Consider how change is managed in your organisation. How could the process be improved?

3.5 Management of change

KEY POINTS

- Changes in the workplace such as construction works, change of process, change of equipment and change in working practices may introduce new hazards and risks.
- To manage the impact of change it is important to adopt a systematic approach.
- Control measures to mitigate the impact of change include: communication and co-operation of all parties involved, risk assessment, appointment of competent people, segregation of work areas and amendment of emergency procedures and welfare provision.
- Periodic reviews of the progress of the change process and the effectiveness of the change should be carried out.

References

[1] Adapted from HSE, Flixborough (Nypro UK) Explosion 1 June 1974

[2] HSE, 'Organisational change and major accident hazards: Chemical Information Sheet No CHIS7' (2003) (www.hse.gov.uk)

3.6: Safe systems of work for general work activities

Syllabus outline

In this section, you will develop an awareness of the following:
- Why workers should be involved when developing safe systems of work
- Why procedures should be recorded/written down
- The differences between technical, procedural and behavioural controls
- Developing a safe system of work:
  - analysing tasks, identifying hazards and assessing risks
  - introducing controls and formulating procedures
  - instruction and training in how to use the system
- Monitoring the system

Thinking back to our discussion on the content of the health and safety policy, you will recall that the policy document should include general and specific arrangements for the implementation of the policy. These arrangements are usually set out in procedures, rules, handbooks etc.

A safe system of work is a formal procedure that results from a systematic examination of a job or task to identify the hazards associated with it. The system of work sets out the method to be used, including the controls required. There are broadly three categories of control – technical, procedural and behavioural.

Once a safe system of work has been developed through a structured process with worker involvement, it must be implemented. Communication with, and training of, those affected by the system are two important requirements for ensuring effective implementation of the system. Ongoing monitoring of the system supports its continued implementation and identifies potential modifications that may be required over time.

TIP

Legislation relating to safe systems of work is contained in:
- Section 2(2) of the Health and Safety at Work etc Act 1974 (HASAWA).

3.6.1 Why workers should be involved when developing safe systems of work

It is of vital importance that those involved in carrying out the task are involved in the development of the safe system. There are two main reasons for this:

- They will be a valuable source of information regarding hazards and risks, and they will be able to describe in detail what needs to be done so that an appropriate procedure can be devised.

- Their involvement will help to ensure that they will accept and follow the safe system of work once it is introduced, since they will understand how it came about and will feel a sense of ownership because they have contributed to its development.

CASE STUDY 1
Failure to provide a safe system of work

The idea of safe systems of work is not confined to the UK. Though this case study is from Australia, the same legal duty to provide a safe system of work applies. In this case, a worker employed as a slaughterer at an abattoir suffered a severe laceration of his left wrist and hand when an animal struck his right arm, leading to a knife injury on his left arm. The injured worker had worked in other abattoirs for 21 years.

In court, the injured worker argued that his employer failed to provide him with cut-resistant gloves, instruct him to wear the gloves and provide him with training on how to perform his duties while wearing the gloves.

Cut-resistant gloves were available at the time of the worker's injury and, had the worker been wearing a glove on his left hand, the injury could have been prevented.

Further evidence proved that the employer did not direct or insist that certain workers (those employed in the same area as the injured worker) wear cut-resistant gloves. Instead, the employer informed these workers that they did not need to wear the gloves while performing their duties.

The employer explained to the court that they did not insist that these particular workers wear cut-resistant gloves as the workers were resistant to change. The employer claimed it was a 'militant' workforce and so was reluctant to carry out any action that might provoke industrial conflict.

The court accepted the injured worker's evidence that the employer had failed to provide a safe system of work and, although the worker was supplied with cut-resistant gloves, he was not instructed or mandated to wear them by the employer.

The court ruled there was an unsafe system of work that was reasonably foreseeable and preventable by the employer. Damages were awarded to the injured worker.[1]

CASE STUDY 2
Failure to follow a safe system of work

A qualified train driver was driving a train on a main line connecting several major towns. Two crucial safety devices – the Driver Safety Device (DSD) and National Radio Network (NRN) radio – stopped working shortly after they set off. Going by their training and the recognised rail industry rule book (on which procedures were based), the driver should have notified the signaller immediately so that they could get instructions that would have allowed the train to safely continue the journey or to move to a safer place where it could be assessed. The driver did not do that. Further into the journey, the train's speedometer also stopped working and the driver chose to ignore safety procedures for the second time by failing to stop the locomotive and notify the signaller.

This case study illustrates that even when workers are trained and qualified, adherence to safe systems of work, particularly for high-risk activities, needs monitoring through periodic assessment.[2]

3.6.2 Why procedures should be recorded/written down

Some tasks can be very simple and straightforward, so would not strictly require a written safe system of work – a verbal instruction would be adequate. However, many tasks are more complex, with many steps and potentially serious consequences if steps are missed or taken out of sequence. Human beings are generally prone to be rather forgetful, so a procedure, backed up with the right training, will be an ideal way of helping to ensure that the task is carried out the same way every time it is done, by whoever does it. In this way the chances of human error can be minimised.

Written procedures can relate to specific tasks to be performed or to general aspects of health and safety such as reporting incidents or auditing. Procedures need to be realistic and reflect how the work should be carried out safely in practice. A procedure that is unrealistic and theoretical is unlikely to be followed. Worker involvement in developing procedures helps ensure reality is factored in. Procedures that are clear and well-presented will be more likely to be understood and implemented than those that are not. Procedures set standards of performance for the task and risk mitigations.

In summary, the written safe system of work provides a formal procedure and consistency of approach, which not only reduces the potential for human error but also offers a record that can be examined in the event of an accident or audited to ensure continued compliance with good working practice.

3.6.3 The differences between technical, procedural and behavioural controls

There are a range of controls available in a safe system of work to mitigate the risk of harm. These fall into three categories: technical, procedural and behavioural.

> **DEFINITIONS**
>
> *Technical control* *deals with risks from the perspective of physical controls or hardware related to the workplace or the activity being carried out.*
>
> *Procedural control* *is about 'how things should be done in the organisation', ie encompassing all aspects of work on particular equipment or in a particular environment: normal operation, maintenance, breakdown/spillage, emergency, use of personal protective equipment (PPE).*
>
> *Behavioural control* *is about shaping how the worker performs. Individual competence, attitudes, perception and motivation will have an impact on performance, which may be affected by training, development, coaching etc.*

Effective systems of work will often use a blend of controls as illustrated in the following example.

Consider the controls in a relatively simple situation in which ingredients such as sacks of flour must be cut open and then poured into a food mixer in a food processing factory.

3.6 Safe systems of work for general work activities

- Technical controls: these would include provision of equipment to move and lift the sacks, shrouded knife for slashing open bags and ventilation system to reduce dust levels.
- Procedural controls: system of work that considers, for example, the number of sacks that should be 'waiting ready', whether they should be left on a trolley, when the ventilation system should be switched on to reduce dust levels as the bags are slashed open and what happens to the 'empty' bags. The bags are not quite empty and over the months a build-up of flour can cause a very significant explosion hazard. The dust extraction system and disposal of the waste also need to be checked. We discuss risks from dusts in 7.1: Hazardous substances and explosion risks in 10: Fire.
- Behavioural controls: in this case, there will be considerable overlap with procedural controls. Behavioural controls would encompass training in manual handling techniques, use of respiratory equipment, elementary checking of ventilation systems, food hygiene etc. Behavioural controls would also encompass informing workers about the problems associated with dermatitis.

In Figure 1 we summarise the technical, procedural and behavioural controls that should be considered.

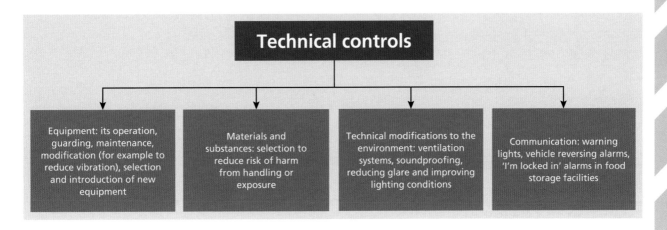

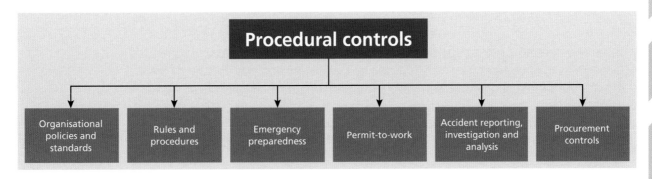

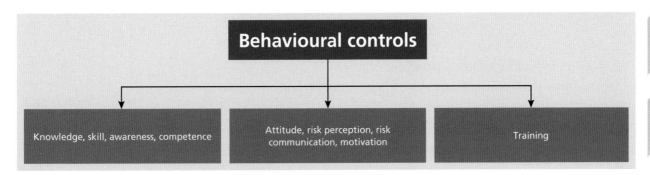

Figure 1: Technical, procedural and behavioural controls to consider

> **CASE STUDY 3**
> **Benefits of improving technical and procedural controls**
>
> A small printing company employing 20 workers received some free advice from its trade body on how to improve health and safety. This included rewriting procedures for working safely with chemicals to make them specific to the activities in the workplace and making adaptations to equipment to ensure a safer working environment.
>
> The company identified several benefits from implementing the changes to its systems of work, including reduced operating costs because of fully documented and more efficient procedures. It also saw improvements in worker morale as workers were confident that management cared about their wellbeing. The company's insurance premiums remained static.
>
> When assessed against its trade body's health and safety checklist, the company's performance scores had almost doubled.[3]

3.6.4 Developing a safe system of work

The development of a safe system is a systematic process involving a number of people. It has the following key stages:

- Stage 1: Analysing tasks, identifying hazards and assessing risks;
- Stage 2: Introducing controls and formulating procedures; and
- Stage 3: Instruction and training in the operation of the system.

Stage 1: Analysing tasks, identifying hazards and assessing risks

Analysing the task

All aspects of the task and the risks to safety and health that it presents must be assessed. This means considering the environment in which the task is carried out as well as the task itself. Simpler, lower-risk tasks may adopt a structured process but without the need for use of supporting documentation. More complex or higher-risk activities may involve a process known as job or task safety analysis. Although the terms are used interchangeably, generally a job is made up of a series of tasks and so the steps in the task analysis will be smaller than in a job analysis. The steps in a job safety analysis are the tasks that make up the job.

Job/task safety analysis involves:

- identifying the job/task to be analysed and its purpose;
- breaking the job/task down into the sequence of actions that need to be carried out and how they need to be done to achieve the task objectives;
- identifying the potential hazards associated with each action;
- identifying the current control measures in place;
- considering the effectiveness of the control measures and what might impact their effectiveness; and
- determining what additional control measures might be needed.

Job/task safety analysis is usually carried out by observing the process but could be done without this by drawing on the experience of those carrying out the analysis.

3.6 Safe systems of work for general work activities

When developing any safe system of work, typical questions or prompts to consider are:

Where and when are the various tasks carried out? For example:

- how they interact with one another; and
- how they affect others in the vicinity.

What is used? For example:

- plant and substances;
- potential failures of machinery;
- toxic hazards;
- electrical hazards;
- design limitations; and
- risk of inadvertent operation of automatic controls.

Who does what? For example:

- delegation;
- training;
- foreseeable human errors; and
- ability to cope in an emergency.

How are the tasks carried out? For example:

- procedures;
- potential failures in work methods;
- shortcuts; and
- insufficient foresight of infrequent events.

Why are the tasks done this way? For example:

- particular problems with the process; and
- alternative methods (possibly safer).

Table 1 shows an example of part of a job safety analysis for using a hand truck.

Figure 2: A hand truck in use

Step	Description of job/task steps	Potential incidents or hazards	Risk control measures	Actions/additional control measures
1	Pre-operation safety check	• Untrained operator	• Training on hand truck design and controls, stability, the appropriate way to transport, load and stack on the hand truck	
2	Assembling the load	• Rolling the wheels off the edge of ramps	• Stay well back from the edge • Never turn round on the ramp/slope • Keep the truck ahead when going down the ramp • Pull the truck behind when going up the ramp/slope • Make sure the bottom plate of the truck is all the way under the load	
3	Operating the two-wheeled hand truck	• Slip/trip/fall	• Slow down for turns • Ensure enough clearance	
4	Transporting the load	• Pinching hands between the truck and other objects	• Be alert • Wear gloves to protect hands • Secure bulky or hazardous loads to the frame • Stack heavier objects at the bottom	
5	Storing the hand truck	• Tripping	• Store in designated area	

Table 1: Excerpt from job safety analysis

Defining the hazards and risks

When a task has been analysed, its hazards should be clearly identified and the risks weighed up. Whenever possible, the aim should be to eliminate the hazards and reduce the risks before you rely on a safe system of work – if the hazards can be eliminated altogether, there is no need for the safe system of work.

The analysis stage provides information on the steps that make up the work, the hazards related to these steps and the control measures currently in place. It provides an opportunity to consider if further controls are required. This will be determined by the consequences of the potential hazards being realised through a hazardous event and then the likelihood of that happening, or put another way, the risk. If a significant risk is identified, there might be a need to carry out a detailed analysis of a specific aspect of the work.

If a job analysis has identified that a hazard is controlled in a certain way, for example, through the wearing of hearing protection in a noisy area, this could be taken as acceptable and included in the system of work because it is the current practice. However, reflecting on the risk, and the adequacy of the control measure to manage that risk to an acceptable level, would lead to challenging the reliance on PPE for a serious risk. This illustrates the importance of reflection and consideration at the risk assessment stage to ensure the appropriateness of control measures within the system of work.

Further and more detailed consideration of hazard identification and risk assessment can be found in 3.4: Assessing risk.

Stage 2: Introducing controls and formulating procedures

Defining safe and healthy methods

Your safe system of work may be defined verbally, by a simple written procedure or in exceptional cases by a formal permit-to-work scheme (see 3.7: Permit-to-work systems).

> **TIP**
>
> When deciding on methods, you should consider:
>
> - the previous analysis;
> - adequate control of the hazards where they cannot be eliminated at source and any protective equipment requirements;
> - meeting legal, best practice and organisational standards;
> - incorporating manufacturer or supplier instructions;
> - defining the method as simply as possible to make it understandable to workers;
> - defining the person in control of the work and responsibilities at various stages;
> - defining who does what, when, where and how; and
> - dealing with any likely emergencies.

If the method is new or significantly different from the previous method, then testing or piloting it might be a good idea.

Implementing the system

To help ensure that the safe system of work is effectively implemented by workers, it needs to be communicated properly so that it is understood by workers and applied correctly. Gaining the support and commitment of supervisors and line management to the safe system of work helps to encourage workers to follow the system. If the safe working method requires use of specific tools or materials, it is essential that management ensure they are available to the workforce, otherwise this could lead to inadequate compliance with the procedure and workers developing shortcuts as 'work-arounds'.

Stage 3: Instruction and training in operating the system

Simply issuing a procedure through the organisation's intranet or similar and expecting compliance is unlikely to be effective. When a safe system of work presents a change to a well-established way of working, it is essential that adequate communication and training are put in place to support adoption of the new system of work. As discussed in 3.2: Improving health and safety culture, using a blend of communication methods would be most effective, such as issuing the procedure in writing, discussing it at team briefings or during toolbox talks and displaying information on noticeboards or as a presentation on the organisation's intranet.

Workers, supervisors and managers should be trained in the necessary skills and made fully aware of potential risks and the precautions they need to adopt. Supervisors need to be able to recognise that the system of work is being followed and notice any deviations from it. They need to understand the consequences of deviations. The task analysis stage discussed earlier can be incorporated into the training required.

For high-risk tasks, training programmes based on a job/safety analysis followed by practical application and assessment should be used. When formal training is required, it should be documented; once completed, records should be maintained.

3.6.5 Monitoring the system

All systems of work should be monitored to check that:

- employees continue to find the system workable;
- the procedures laid down in the system of work are being carried out and are effective; and
- any changes in circumstances that require alterations to the system of work are considered.

Monitoring can be carried out through direct observation or discussion in team meetings. Feedback from workers is a valuable resource in helping to ensure that systems of work remain effective. Audits and investigations following accidents/incidents or ill-health occurrences can also provide insights into the effectiveness of safe systems of work. Records of monitoring activities should be kept and reviewed by management as part of the health and safety management system.

APPLICATION

Compare your organisation's approach to developing safe systems of work with the approach we have described. Can you identify any improvements?

KEY POINTS

- Involving workers in the development of safe systems of work provides a valuable resource of practical information that can only be gained from actually doing the tasks that are part of the system of work being developed. It also increases the chances that workers will implement the system.
- Recording/writing down all but the simplest systems of work helps to ensure the system is consistently followed. It also provides a useful reference for training.
- Controls fall into three categories:
    - technical controls – physical things/hardware to give protection;
    - procedural controls – 'how things should be done in the organisation'; and
    - behavioural controls – drivers affecting how the person performs.
- Developing a safe system of work is a structured process involving:
    - analysing tasks, identifying hazards and assessing risks – this can involve questioning who does what, where, when, why and how or a job or task safety analysis;

- introducing controls and formulating procedures – this needs to take account of elements such as legal and organisational standards and best practice, be easily understood and deal with likely emergency situations; and
- instruction and training in how to use the system.

■ Safe systems of work require periodic monitoring so they remain workable and are implemented correctly, and to check if there have been any changes in circumstances that could affect the effectiveness of the system.

References

[1] Adapted from Worksafe Queensland Government, 'Employers must provide safe work system' (2004) (www.worksafe.qld.gov.au)

[2] Adapted from Office for Rail Regulation, 'Health and safety prosecutions' (www.orr.gov.uk)

[3] Adapted from HSE, Dolphin Printers (www.hse.gov.uk)

3.7: Permit-to-work systems

> **Syllabus outline**
>
> In this section, you will develop an awareness of the following:
> - Meaning of a permit-to-work system
> - Why permit-to-work systems are used
> - How permit-to-work systems work and are used
> - When to use a permit-to-work system, including: hot work, work on non-live (isolated) electrical systems, machinery maintenance, confined spaces, work at height

In 3.6: Safe systems of work for general work activities we mentioned that some work activities might require a formal permit.

> **DEFINITIONS**
>
> A *permit-to-work system* is a formal written system used to control work that is potentially hazardous.
>
> A *permit-to-work* is a document (printed or digital) that is used as part of the permit system – it specifies the work to be done and the precautions to be taken to ensure the worker's safety while doing it.

Permit-to-work (PTW) systems can form a critical element of a safe system of work for many maintenance activities. They allow work to start only after safe procedures have been defined and they provide a clear record that all foreseeable hazards have been considered. In situations where permits are required, effective functioning of the PTW system and training of all parties involved in, and affected by, the work are key. The Piper Alpha oil rig fire and explosion is a powerful reminder of the consequences of failures in PTW systems.

3.7.1 Meaning of a permit-to-work system

A PTW system is a formal written system covering all aspects of the work to be carried out, with signed authority required for the various activities to take place. Every aspect of the work is planned, overseen, checked, recorded and, when the time comes, confirmed as having been completed satisfactorily. The system is used to control certain types of work with high hazard potential. It also provides a way of communicating between managers, supervisors and workers and those carrying out the hazardous work.

3.7 Permit-to-work systems

Figure 1: A PTW system provides a means of communication with people doing hazardous work

> **TIP**
>
> A PTW system should not be used for every task, except in some permanently high-risk work such as a maintenance shutdown on a complex plant, as this detracts from its importance.

A permit-to-work refers to the permit (paper or electronic) issued as part of the overall system of work.

3.7.2 Why permit-to-work systems are used

A PTW is part of a system of work and can help manage specific work activities, usually where there is a high risk of hazards. In these situations, the use of a PTW ensures that the risks associated with the work and the controls needed to ensure safe working are fully considered.

A PTW system aims to ensure:

- awareness of the hazards between all relevant parties;
- awareness of the extent, nature, limitations and timescales of the hazardous work activity;
- formal checks, such as plant isolations and gas tests, are completed before the work starts;
- activities are co-ordinated if multiple people or groups are involved, eg contractors; and
- the work is authorised before it starts.

Its objectives are to:

- identify the scope of the work and the hazards associated with that work;
- set out the standards of protection required;
- identify the responsibilities for requesting and being involved in the work;
- provide a formal written transfer of information from those requesting the work to those carrying out the work; and
- provide a mechanism for returning plant and equipment to an operational state after completion of the work.

3.7.3 How permit-to-work systems work and are used

System requirements

For a PTW system to be effective there are several requirements. It should:

- be documented;
- be simple to use;
- have the commitment of system users – this is more likely if managers and workers have been consulted about the system;
- provide concise, clear and precise information;
- clearly identify areas to which the PTW applies;
- provide for the display of permits to show that the nature of the work and the precautions needed have been checked by appropriate persons;
- provide instances when the work has to be suspended, ie stopped for a period before it is complete;
- provide for a formal handover between shifts when the permit duration is longer than one shift;
- provide for a formal hand-back of the plant or equipment that has been worked on to help ensure it is in a safe condition and ready for reinstatement;
- allow for a process for change, such as when hazards need to be reassessed, and for controlled communication of change;
- allow for liaison with those in charge of other parts of the site or facility who might be affected by the activities taking place under the PTW;
- provide training in the system for all those who work under the control of the PTW or might be affected by it. Training helps to achieve quality and consistency in the use of the PTW system and should be periodically refreshed; and
- be monitored and audited to ensure the system works as intended.

Permit-to-work document

A PTW document (or certificate) has several important features, including:

- a description of the work or the task to be carried out and where this will be done;
- the duration of the permit, including the start and end time and date;
- identification of all foreseeable hazards associated with the work, including those arising because of the work;

- identification of the control measures required, including personal protective equipment (PPE);
- a record of the isolations made, such as electrical tests carried out and gas tests in a confined space;
- signature of an authorised permit 'issuer', who authorises that the job has been inspected, the controls implemented and those carrying out the work briefed;
- signature of an authorised permit 'acceptor', indicating acceptance of the task and that the controls indicated have been implemented;
- indication by the permit acceptor of whether the work has been completed or not, and any extension required to complete it; and
- indication by the permit issuer that the work has been completed and the equipment or area recommissioned or returned to service.

Once a permit is issued it overrides all other work instructions until cancelled. If there are changes to the planned work, the permit must be amended or cancelled and a revised or new permit issued. Amendments or cancellations should only be carried out by the authorised permit issuer.

Figure 2: A PTW should outline foreseeable hazards associated with the work and the control measures required

Figure 3 shows a generalised outline of a PTW certificate. In practice, organisations will hold standard colour-coded copies of different PTWs, such as red for hot work permits and green for confined space permits. PTWs may be printed or digital.

Type of permit: For example, hot work, confined space	Reference no
Job location: information to be supplied on • Location • Identification of equipment • Diagrams	

Withdrawal of the plant/equipment from service: Confirmation of when the specified plant has been taken out of service and duration of the permit Available from: Time/date Available until: Time/date Signed: ISSUER	
Hazard identification: List hazards related to the work and any arising because of the work being done under the permit Authorised signature: ISSUER Date/time	**Control measures:** List all control measures, including personal protective equipment Authorised signature: ISSUER Date/time
Isolations/other permits associate with the work: Information on isolations etc Electrical: Ref no: Signature: ISSUER Date: Other: Ref no: Signature: ISSUER Date:	
Permit issue: Authorisation that job has been inspected, precautions implemented and workers briefed Authorised signature: ISSUER Date/time	**Permit acceptance:** Acceptance by workers doing job that they have seen the job and are satisfied the precautions have been implemented Authorised signature: ACCEPTOR Date/time
Completion/extension of the work: Statement that the work has been: • Completed • Not completed, including time of extension Authorised signature: ACCEPTOR	**Cancellation of job/return to service:** Conformation that work has been done and plant recommissioned Authorised signature: ISSUER

Figure 3: Outline of a PTW certificate

Issuers and acceptors of permits must be competent persons, trained, assessed and approved in the issue or acceptance of PTWs. After their initial training, their competence will need to be assessed and reassessed periodically. Authorised issuers will often have their permits countersigned by an experienced issuer for a time after completing their training so that they can demonstrate they have reached the required level of competence.

3.7.4 When to use a permit-to-work system

A permit should not normally be used for carrying out routine operational activities. PTW systems are usually suited to non-production work (such as maintenance, repair, inspections, testing and cleaning) and to situations involving two or more parties needing to co-ordinate their activities to carry out work safely or when responsibility for work is transferred from one party to another.

In some non-production situations, such as maintenance shutdowns on complex plant, there can be a need to use multiple permits or combined permits covering more than one activity. Co-ordination of permits is particularly important in these situations.

The following are examples of situations in which a PTW will be required.

Hot work

Hot work PTWs are used in cases where there is a risk of fire or explosion, or emission of toxic fumes from the application of heat. This might involve ignition sources or heat being used in an environment that may contain flammable or combustible substances. Ignition sources or heat could be provided by activities such as welding and flame cutting, tools that may produce sparks and any electrical equipment that is not intrinsically safe. Protection of the flammable/combustible substances (or elimination of them if possible) and the provision of firefighting equipment on standby (and training in its use) are common control measures.

Work on non-live electrical systems

High risks associated with working on or near electrical systems often necessitate a PTW. Key considerations will be isolation of the equipment, access and egress, work at height and lifting. The use of the permit helps ensure the appropriate isolations and precautions and the competence of the workers involved. The issue of an electrical PTW indicates that an electrical circuit or piece of equipment is safe and that work can be carried out on it. Permits should not be issued to justify working on live electrical equipment. Instead, live working requires a different approach to help ensure the safety of workers involved. Control measures for working with electricity, including electrical permits, are covered in more detail in 11.2: Control measures.

Machinery and maintenance work

This type of work can require different disciplines or parties to work together at the same time at a complex plant or site, for example, during a planned maintenance programme on a power station or chemical plant. These situations present a wide range of risks, from access to dangerous parts of the equipment and falls from height to the release of stored energy. A PTW helps to ensure that the work is planned, power sources are isolated and there is communication across the disciplines, particularly when different work activities are being carried out at the same time.

Confined spaces

A confined space is any substantially enclosed space or enclosure (such as a vat, pit or tank) where there is a risk of death or serious injury arising from hazardous substances or dangerous conditions, such as a lack of oxygen or the presence of explosive gases. In this type of hazardous situation there is increased risk of injury from fire or explosion, losing consciousness from asphyxiation from gases and fumes, lack of oxygen and drowning from a rise in liquid levels. Confined space work has resulted in fatalities among both those working in the space and those attempting to rescue them. A confined space PTW should specify the precautions to help eliminate exposure to dangerous fumes or to an oxygen-depleted environment before workers are allowed to enter the space, confirm that the space is free from such fumes and asphyxiating gases and specify the measures to protect the space from becoming contaminated. These measures could include ventilation or breathing apparatus. We consider this further in 8.3: Safe working in confined spaces and 8.2: Working at height. Working in a confined space could involve working on elevated platforms or ladders but also near an excavation where a fall below ground level could occur. If the work at height cannot be eliminated, then the PTW should help to ensure the potential fall distance is minimised and due consideration is given to weather conditions that could affect the safety of the task.

Work in environments that present considerable health hazards

Examples of this type of work include:

- radiation work;
- repair work in kilns or food freezer storage systems where there is severe thermal stress; and
- work involving toxic dusts (eg asbestos), gases and vapours, often in confined spaces.

> **TIP**
>
> Considerations when developing an appropriate PTW structure for a planned maintenance programme in a food cold storage system might include:
>
> - systems of communication;
> - insulated clothing, gloves and other PPE;
> - emergency procedures;
> - monitoring of the environmental conditions (in this case temperature);
> - control of working times and rest breaks (the required balance being determined by the conditions being encountered);
> - clear identification of all components, including pipework, power supplies and pumps;
> - isolation of services (gas, electricity, water);
> - tools and equipment to be used; and
> - lines of authority.
>
> All these elements, and perhaps others, would need to be considered in developing the PTW.

When developing a PTW system, some considerations will be common to different hazardous environments, such as emergency procedures, environmental monitoring, communications, PPE and isolation of power supplies.

> **APPLICATION**
>
> Consider a worker in a deep sewage system. What would need to be considered when developing a PTW system for work in the confined space presented by this sewage system?

3.7 Permit-to-work systems

One of the greatest challenges faced in a hazardous working environment is ensuring that different workers or different groups of workers know what the others are doing. Failure to do this has led to the loss of many lives and major industrial catastrophes such as the Piper Alpha oil rig disaster in the North Sea and the Bhopal disaster in India. The PTW system must encompass the co-ordination between these different groups of workers.

CASE STUDY
Piper Alpha oil rig fire and explosion, 1988

Failure of the permit-to-work system was at the heart of the Piper Alpha oil rig accident.

A back-up condensate pump was taken out of service for a routine check on the pressure safety valve (PSV) under a permit-to-work. Either an open connection was left in the pump where the valve had been fitted or a blanking plate secured by only finger-tight bolts was put in its place.

The work could not be completed before the end of the shift as there was no lifting equipment available to hoist the PSV into position. The fitters received permission to complete the work the following day. The PTW was suspended. Later, during the evening shift, the primary condensate pump failed. The control room operators, who did not know the back-up pump had been removed for maintenance, tried to restart the pump. A large volume of gas escaped from the hole in the pump where the PSV had been fitted. The gases ignited and the fire escalated to an inferno.

The PTW was criticised for not showing a clear enough distinction between suspension and cancellation of the permit. The fitters are believed to have left the suspended PTW in the office without pointing that out to the operators, who were, in any case, involved in a shift changeover. Information about the status of the back-up pump failed to get through to the control room operators. They therefore thought the PSV had been replaced as planned.

The operation of the PTW system had become extremely poor, with operators relying on informal communication during handovers between shifts. Lock-out procedures were also not implemented. If the PTW system had been working effectively, the accident could have been averted, avoiding the loss of 167 lives.

APPLICATION
Consider the permit-to-work systems in your organisation. How do they compare against the requirements we have outlined?

KEY POINTS

- A permit-to-work (PTW) system is a formal written system covering all aspects of the work to be carried out, with signed authority required for the various activities to take place.
- A permit-to-work refers to the permit (paper or electronic) issued as part of the overall system of work.

- PTW systems are used to manage specific work activities, usually where there is a high risk of hazards. This ensures that the risks associated with the work and the controls needed to ensure safe working are fully considered.
- PTW systems are used in situations such as:
  - maintenance, repair, inspections, testing and cleaning; and
  - those situations involving two or more parties where there is a need to co-ordinate their activities to carry out work safely or where responsibility for work is transferred from one party to another.
- Specific examples of activities in which PTWs are used are hot work, work on non-live (isolated) electrical systems, machinery maintenance, confined spaces and work at height.

3.8: Emergency procedures

Syllabus outline

In this section, you will develop an awareness of the following:

- Why emergency procedures need to be developed
- What to include in an emergency procedure (see HSG268: *The health and safety toolbox*)
- Why people need training in emergency procedures
- Why emergency procedures need to be tested
- What to consider when deciding on first aid needs in a workplace:
  - shift patterns
  - location of site
  - activities carried out
  - number of workers
  - location relative to hospitals/emergency services

Introduction

We have seen that the risk of incidents occurring can be minimised, or at least reduced, using such techniques as risk assessments, safe systems or work, permits-to-work and the provision of health and safety information, training and supervision. However, despite such systems being in place, incidents do occur.

Some incidents may occur fairly frequently, but with minor injuries. However, other incidents may have an extremely low probability of occurrence but can result in serious consequences. It is therefore necessary to have procedures in place to cope with such incidents and we refer to these as emergency procedures.

Firstly, it will be useful to review some situations where emergency procedures would be appropriate.

TIP

The main pieces of legislation relating to emergency procedures and first aid at work are the:

- Management of Health and Safety at Work Regulations 1999 or Management of Health and Safety at Work Regulations (Northern Ireland) 2000 (MHSWR); and
- Health and Safety (First-Aid) Regulations 1981 or Health and Safety (First-Aid) Regulations (Northern Ireland) 1982 (hereafter referred to as the First-Aid Regulations).

Regulation 9 of the MHSWR specifically looks at emergency arrangements. There is no official Approved Code of Practice for the MHSWR. However, the Health and Safety Executive (HSE) has produced the following guidance to explain what should be considered in an organisation's emergency procedures:

> - *The health and safety toolbox: How to control risks at work. Part 1: How to manage health and safety* (HSG268).[1]
>
> The First-Aid Regulations also do not have an Approved Code of Practice. However, the HSE has produced the following guidance documents to help employers understand their duties:
>
> - *First aid at work. The Health and Safety (First-Aid) Regulations 1981. Guidance on regulations* (L74);[2] and
> - *First aid at work: Your questions answered* (INDG214).[3]
>
> Both sets of guidance have also been approved for use in Northern Ireland by the Health and Safety Executive for Northern Ireland (HSENI).
>
> We will draw on these guides here.

Situations where emergency procedures would be appropriate include:

- a large fire;
- an explosion: perhaps due to a fractured gas main, with the escaping gas finding a source of ignition; an example of this was the Texas city disaster in 1947. An explosion on a ship started a chain reaction of fires and explosions and more than 500 people died;
- a natural disaster: for example, an earthquake caused a tsunami that disabled the power supply to the Fukushima nuclear plant in Japan; three of the reactor cores melted within three days;
- a bomb threat: for example, an unexploded bomb from a past conflict or terrorist action from a current conflict; and
- a train crash.

Figure 1: A train crash would require emergency procedures

3.8 Emergency procedures

> **APPLICATION**
>
> Have you experienced an emergency event in your organisation? If so, what was it and how was it investigated?
>
> If you have not experienced any emergency event, what do you think is the greatest hazard at your workplace? Fire? Chemical spill? Traffic collision? Something else?
>
> Why do you think your selected hazard is the greatest hazard? From incident records? Near misses? Another reason?

3.8.1 Why emergency procedures need to be developed

Emergency measures can do the following:

- They help to protect workers and any others who may be affected by a serious incident. For example, an explosion on an organisation's site may not only injure workers but also those near to its premises.
- They help to reduce the severity and consequences of an incident. For example, in the case of a fire, quick action such as shutting off the gas supply may prevent a subsequent explosion. This would limit injuries and damage to plant and equipment.
- They help to ensure that workers in an organisation are aware of the actions they need to take. Some workers may simply need to proceed to a known place of safety whereas others may have specific duties under the emergency procedure. For example, there may be a person nominated to provide the emergency services with information such as workers not accounted for or locations where hazardous substances are stored.
- They help to ensure there is compliance with legal requirements.

3.8.2 What to include in an emergency procedure

Workplaces need a plan for emergencies that may occur. Special procedures will be required for emergencies that could cause serious injury, for example, those involving fire, chemical spills or electrocution.

The HSE HSG268 health and safety toolbox guidance directs that people involved in emergency procedures must be competent and trained (meaning they need appropriate skills, knowledge and experience) and they must take part in regular and realistic practice of emergency procedures (that is, emergency drills).[4] To be realistic, workers should not have advance warning that an emergency drill is about to take place.

An actual emergency is likely to occur without warning, so such drills should be unannounced and there must be agreed, recorded (in writing) and rehearsed plans, actions and responsibilities.

Potential emergencies

Firstly, the type of emergency that might occur needs to be identified in the procedure. Given that emergencies tend be uncommon but can have devastating consequences, there needs to be an in-depth study of how an emergency might occur. This can be done by using techniques that consider situations where the possibility of accidents is likely to have more than one cause, as is often the case. You do not need to know about these techniques (outlined in the Tip box), but you may be interested in studying them further.

> **TIP**
>
> Two examples of techniques for such an in-depth study are:
> - failure modes and effects analysis (FMEA); and
> - fault tree analysis (FTA).
>
> A simplistic way of considering FTA is that this technique estimates the probability of the worst possible outcome (known as the 'top event') that is likely to lead to an emergency situation. This is done by estimating the probabilities of the various system failures that could lead to the top event.

The procedure should include the mechanism by which the alarm is raised – a manual procedure is unlikely to take account of the fact that an emergency could occur outside normal working hours when no workers are present or even when the organisation is temporarily shut down. The procedure would need to examine whether an alarm is transmitted directly to the emergency services or if it is transmitted to a person who then needs to contact the emergency services.

Producing a plan

A plan should be produced showing the locations of hazardous items, particularly hazardous substances. The plan might also include arrangements for regular visits by the emergency services to give them opportunities to familiarise themselves with the layout of the organisation.

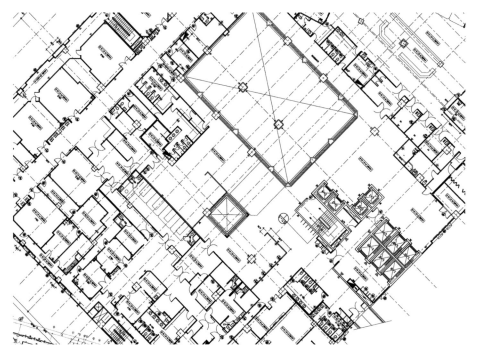

Figure 2: A layout plan for a hospital

Naturally, any significant change to the layout should be communicated to the emergency services.

There needs to be a sufficient number of emergency exits for all persons to escape quickly and escape routes must be kept unobstructed and clearly marked. There may be regular inspections to keep such exits clear.

Worker involvement

The procedure should include the nomination of competent workers to take control in the event of an emergency. These workers would need to know who to contact in the event of an emergency, when to get assistance, who to contact for help and at what point to contact the emergency services.

The workers may be able to control the emergency themselves, for example, putting out a fire with a fire extinguisher. On the other hand, if the workers decided they were unable to control the situation (the fire may be too big), they would need to know who they should contact to manage the emergency; this might mean calling the emergency services. The emergency services would have the appropriate skills and equipment to address the situation safely. For example, if an emergency (such as a fire) occurred on the upper floor of a tower block, the emergency services would be aware of the length of ladders that might be needed during a rescue operation.

All workers must be included in the emergency plan. The plan will specify the duties of all workers and the actions they need to take in the event of an emergency, such as the following examples.

- Technical duties: competent persons are selected to carry out specific technical operations, such as electrical isolation, shutting off the gas supply and closing valves to stop substances entering a particular part of the plant.
- Administrative duties: other workers may have various duties. These may include:
  - acting as an incident controller, who may liaise with the emergency services when they arrive on site;
  - providing technical or site-specific information, for example, where hazardous substances are stored;
  - acting as first-aiders to provide initial treatment for injured persons if needed;
  - providing disabled or other vulnerable workers with suitably placed workstations to minimise their difficulties if evacuation from the workplace is necessary; for example, a worker who uses a walking stick could be located on the ground floor of the building near an emergency exit;
  - being ready to assist disabled or other vulnerable workers in the event of an emergency;
  - checking that all areas of the building are clear of everybody (such as workers and any visitors). For this to work well, all workers leaving the building (for example, visiting clients, going out to lunch) should enter this fact in a register that would be collected by a specified worker during an emergency event or a practice drill; and
  - checking that all persons had reached an assembly point (a place of safety) and reporting this to the incident controller.

All other workers not allocated special duties in the plan would have instructions to make their way to a place of safety in an emergency.

The workers involved in the plan must receive appropriate training. This would include practice drills.

A practice drill would examine whether features of the plan were appropriate and working well. In addition, a specified worker would note the time taken for the building to be evacuated and make a record of this.

There are specific reporting requirements if there are large quantities of dangerous substances on site. For example, if there are more than 25 tonnes of dangerous substances on site, the organisation must notify the fire and rescue services and put up warning signs.

Finally, the procedure must make it clear that work may not resume if danger remains. In practice, when there has been a serious incident the decision to enable work to restart is likely to be made by the emergency services.

> **APPLICATION**
>
> What are the emergency plans in your organisation?
>
> What else (if anything) do you think needs to be added?
>
> How is your emergency plan tested for its efficiency?

3.8.3 Why people need training in emergency procedures

There are three general situations at work: work proceeding normally, work interrupted by predictable situations (such as maintenance) and work interrupted by unpredictable situations such as emergencies. People need training in emergency situations as they may need to take appropriate competent, non-routine decisions in a short space of time.

There will be a team of people with specific duties to act in the event of an emergency. Members of this team need to know what actions to take should an emergency occur; the training could take the form of emergency drills so that the people involved become familiar with the procedure.

The training should include:

- methods of communication between team members, for example, a team member may observe a leaking or ruptured container of highly flammable liquid and will then inform the incident controller, who in turn can advise external emergency services; in this way, important information is communicated;
- training for certain designated people in the use of emergency equipment such as firefighting equipment;
- how the alarm should be raised;
- familiarising team members with specific hazards such as the locations of highly flammable liquids and explosive chemicals; and
- information on how (and when) to contact the emergency services – the team will need to decide quickly whether any emergencies are outside their control.

3.8.4 Why emergency procedures need to be tested

Testing of procedures is necessary to check they work as intended. For example, you should make sure:

- people know what to do (see worker involvement);
- people reach appropriate assembly points (places of safety);

3.8 Emergency procedures

- everyone can hear the alarm; and
- enough assembly points have been created.

When testing is taking place, any weak points in the system can be identified so that improved procedures can be drawn up.

Testing should improve communications between team members or perhaps identify areas where communication signals are not sufficiently clear.

> **TIP**
>
> - The testing should include the evacuation of people with disabilities during the test procedure, particularly any workers who require wheelchairs, as this part of the procedure will need especially competent assistance.
> - The testing should also be carried out using back-up people in case any of the main members of the emergency team are absent, for example, due to holiday or sickness.
> - A written record of any testing of procedures should be kept, including the evacuation time needed for everybody to reach a place of safety, and all those with specific duties can record any problems encountered during the procedure.

Despite carefully designed plans and procedures being put in place, emergency situations can still occur. Then careful investigation needs to take place to attempt to prevent a recurrence – see the following Case study.

> **CASE STUDY**
>
> A manufacturing company employed 35 people. Work took place from 08.00 to 18.00 hours. Nobody was on the premises (kept securely locked) overnight. No problems involving vandalism, break-ins etc had occurred in the nine years that the company had been trading.
>
> The manufacturing process required chemicals and components that were delivered in cardboard boxes. Highly flammable (HF) solvents were used in the process. A fire risk assessment was carried out, the result of which was to ensure that all cardboard boxes were transferred to a specific room (the 'waste room') immediately after they had been unpacked. Also, no containers containing more than 50 ml of HF solvents were permitted to be left in the production building overnight. Any containers with more than 50 ml were transferred to an HF store situated well apart from the production building. A manual fire alarm system was installed.
>
> Overnight a fire occurred – the resulting investigation showed that it had started in the waste room. It burned slowly at first until it reached the production area, where some of the bottles of HF solvent caught fire. Severe fire damage resulted. Analysis indicated that the cause of the fire was thought to be a rat chewing through an electrical cable, causing a short circuit and subsequent fire (the remains of a dead rat were found in the waste room). It was suggested that an automatic fire detection system might have alerted authorities to the fire earlier and reduced the damage, but it was difficult to predict the presence of the rat near that electrical cable as no rats had previously been reported.

3.8.5 What to consider when deciding on first-aid needs in a workplace

There are many things that need to be considered when deciding on first-aid provision in workplaces. These include:

- Shift patterns: steps should be taken to help ensure that first-aid cover is available at all times. For example, an organisation may work night shifts and there needs to be first-aid cover for these as well as during the day.
- Location of site: you need to know whether the organisation operates from just one site or whether there are many sites. For example, a water company may have a head office but many other small sites, such as remote pumping stations and sewage treatment works. Such sites are likely to be visited from time to time by maintenance workers and workers needing to take samples for subsequent analysis. Such facilities are likely to have very limited (if any) first-aid supplies and, due to their locations, workers may need to carry their own first-aid kits when visiting these sites.
- Activities carried out: this will have a bearing on the number of first-aiders required. The greater the number of hazards and risks at the workplace, the more first-aiders (and other first-aid facilities) that will be required. For example, an engineering company with workshops will need more first-aid facilities than an office.
- Number of workers: the more workers on a site, the greater the first-aid needs. The types of worker should also be considered. For example, if there are vulnerable workers (such as pregnant workers or workers with disabilities) on the site, more first-aid facilities may be needed.
- Location relative to hospitals/emergency services: if a workplace is close to a hospital (or other emergency services), then only basic first-aid facilities would be needed, as professional medical help should be accessible quite quickly. However, if a workplace is a considerable distance from a hospital, then greater first-aid facilities are likely to be required. For example, in a remote location it might be good practice to have a defibrillator on site and, of course, people trained to use it. This might save a life while waiting for the arrival of emergency services.

> **TIP**
>
> Other things that may be worth considering include the following.
>
> - The history of any accidents, injuries or ill-health at the organisation's site: the greater the number of recorded incidents, the more first-aid facilities that should be available, although some sites that have few recorded results may not be efficient at recording them.
> - If the site is open to the public, the organisation may need to possess sufficient first-aid facilities, for example, at a shopping centre.
> - Where there are specific hazards there may be the need for a first-aider to be specially trained in providing oxygen or, for example, administering a cyanide antidote if the organisation uses cyanide at its premises.
> - Lone workers driving to remote workplaces or to several workplaces in the course of their work should carry an appropriately stocked first-aid box with them.

3.8 Emergency procedures

KEY POINTS

- Situations that are likely to require emergency procedures include large fires, explosions, natural disasters, bomb threats and train crashes.

- Emergency procedures should be developed to protect workers and other people close to the site, limit the severity and consequences of an incident, make workers aware of what they need to do in an emergency and comply with legal requirements.

- The content of emergency procedures will probably cover the likely nature of an emergency, mechanisms for raising the alarm, a plan of the site and the duties of workers generally and in specific roles in case of emergency.

- People need training in emergency procedures so they are clear about what they need to do. The procedures need to be tested to check that they work properly and to identify anything that needs to be improved.

- Decisions on what first-aid facilities might be needed at a workplace will be based on factors such as its shift patterns and location, the hazards and risks of the work activities and the number of workers and where they are based, especially those whose duties include driving to remote or multiple workplaces, especially on their own.

References

[1] HSE, *The health and safety toolbox: How to control risks at work* (HSG268, 2014) (www.hse.gov.uk)

[2] HSE, *First aid at work. The Health and Safety (First-Aid) Regulations 1981. Guidance on regulations* (L74, 3rd edition, 2013) (www.hse.gov.uk)

[3] HSE, *First aid at work: Your questions answered* (INDG214, 2nd edition, 2014) (www.hse.gov.uk)

[4] See note 1

ELEMENT 4

HEALTH AND SAFETY MONITORING AND MEASURING

4.1: Active and reactive monitoring

> **Syllabus outline**
>
> In this section, you will develop an awareness of the following:
> - The differences between active and reactive monitoring
> - Active monitoring methods (health and safety inspections, sampling and tours) and their usefulness:
>   - differences between the methods; frequency; competence and objectivity of people doing them; use of checklists; allocation of responsibilities and priorities for action
> - Reactive monitoring measures and their usefulness:
>   - data on accidents, dangerous occurrences, near misses, ill-health, complaints by workforce, and enforcement action and incident investigations
> - Why lessons need to be learnt from beneficial and adverse events
> - The difference between leading and lagging indicators

Introduction

You have seen that, in a health and safety management system, initial steps consist of planning, for example, by setting policies and then putting plans in place to implement these policies, for example, by introducing safe systems of work. You need to know how well you are doing, or to express this more formally, you need to measure health and safety performance. This section considers a number of techniques available to you, starting by examining the differences between active and reactive monitoring.

4.1.1 The differences between active and reactive monitoring

There are two basic types of monitoring.

1. Active monitoring: this takes place **before** any loss-making event occurs, such as personal injuries or damage to plant. It is usually in the form of some type of inspection of workers' actions, plant and premises. Such an inspection will observe whether standards are being maintained and, in some cases, whether control measures are working effectively. Active monitoring is useful in helping to

prevent loss. It is sometimes described as 'proactive monitoring', but we will only use the term 'active monitoring' here.

2 Reactive monitoring: this takes place **after** a loss-making event occurs. Its purpose (and usefulness) is not only to learn lessons from such an event, for example, by means of accident investigation, but also to recommend actions to help prevent a recurrence of the loss-making event.

4.1.2 Active monitoring methods

There are several of these methods and we will discuss each of them.

Health and safety inspections

Health and safety inspections consist of a formal examination of part, or all, of a workplace to identify hazards, unsafe working practices and unsafe working conditions. Much of the inspection will be based on observations, although an inspector may ask workers questions, perhaps to find out if they had always carried out a particular operation in the same way or whether they had received training on how to carry out the operation.

Figure 1: Inspections are mainly based on observation and asking questions

An inspection may be carried out by internal or external people. Whoever carries out inspections must be competent to do so. There is no specific definition of competence for safety inspectors, but they should possess the necessary observation skills, communication skills, health and safety knowledge and inspection experience to carry out this type of work.

Internal inspections might be carried out by just a competent health and safety adviser or by a team consisting of the health and safety adviser, a manager responsible for the area of inspection and a safety representative who represents the workers in the area to be inspected. A team inspection has the advantage that the manager and representative can make observations simultaneously and agree on what has been observed. If the whole plant is to be inspected, different managers and safety representatives might be involved at appropriate times.

External inspectors are likely to be enforcement officers. Examples could be Health and Safety Executive (HSE) inspectors, environmental health officers or fire officers. Insurance officers may also carry out inspections.

External inspectors will also make observations and perhaps wish to speak with employees, but they may also ask to see documentation, for example, relating to statutory inspections such as for boilers or lifting equipment.

A report will be created after the end of the inspection.

The report should include:

- observations of unsafe acts and conditions;
- records of any complaints from workers;
- identification of any breaches of health and safety legislation;
- recommendations for remedial actions, subdivided into actions for high, medium and low priority.

The next thing to consider is whether an inspection should be preannounced or not.

An internal inspection can be either. If managers and safety representatives are involved, the inspection will almost certainly take place on a prearranged date and the workforce will know about it. A health and safety adviser working alone may wish to carry out an inspection without advising anyone. This would have the advantage that the inspection would be more realistic; that is, it tends to avoid the possibility that workers make 'special arrangements' just before the inspection is due.

Frequency of inspections

An internal inspection would usually take place approximately every three months. However, certain things can influence this frequency, including:

- results of previous inspections – the more unsatisfactory observations from the previous inspection were, the shorter the time interval before the next inspection;
- the level of risk – the higher the level of risk, the shorter the time interval before the next inspection;
- introduction of new work equipment or processes – such circumstances are likely to require an inspection irrespective of when the previous inspection was carried out;
- complaints from the workforce;
- a significant increase in workplace deaths, non-fatal injuries, building or equipment damage, dangerous occurrences or near miss reports;
- specific recommendations from manufacturers of work equipment;
- recommendations following risk assessments;
- requirements of the organisation's safety management system as a result of any enforcement actions; and
- any legal requirements.

On the other hand, an external inspection is usually unannounced.

There is no set frequency for this type of inspection. As a general rule, if an external inspection is deemed to be largely satisfactory, there may be a significant period before a subsequent inspection takes place. However, if many bad practices are observed during an inspection, and particularly if there have also been any reportable injuries, a follow-up inspection is likely after a short time period.

Having determined the frequency of an inspection, it will then be appropriate to examine the factors that decide how the inspection is to be planned – such factors are shown in the Tip box.

> **TIP**
>
> **Planning an internal inspection**
>
> If you carry out an internal inspection, you need to adopt a plan before starting the physical inspection itself. This plan might include:
>
> - Persons involved: decide on the composition of the inspecting team or whether the inspection is to be carried out by an individual such as a safety adviser.
> - Studying the results of the most recent inspection: note any deficiencies identified during that inspection and observe whether any improvements have been made. Also decide what additional aspects you would observe.
> - Identifying the scope of the inspection: whether the inspection should cover only some areas of the premises – and if so, which ones – or all locations.
> - How to record the results: whether notes are taken as observations made or whether checklists are used.
> - Time of the inspection: this may be significant if the organisation operates day and night shifts. For example, there may be less supervision at night, resulting in more unsafe practices taking place. It is likely, therefore, that you would need to carry out two separate inspections, one for the day shift and a further one for the night shift.
> - Whether any special requirements need to be in place for the inspection: for example, the inspection may take place in an area where the wearing of eye protection is mandatory, so eye protection must be available to (and worn by) inspectors in that area.
> - Who will need to see the inspection report: clearly, this will include the managers or supervisors of the areas inspected. There may, however, be others who will need to see the report. For example, in a manufacturing organisation, if a machine guard is found to be defective, the maintenance manager should receive a copy. In a shopping centre, if the examination of fire extinguishers is found to be out of date, the centre manager and/or the landlord of the centre should receive a copy of the report.
> - How recommendations in the report will be actioned: in a large organisation, there may be regular health and safety meetings at which members of management, worker representatives and competent health and safety persons are present. These meetings will ideally be chaired by the managing director/ top management. The findings of the inspection report should be discussed, recommendations and the priorities for actions agreed, persons responsible for carrying out the actions identified and timescales for action specified. For leased premises, landlords would have responsibility for actioning some of the items.

Health and safety monitoring and measuring

Use of checklists

There are both advantages (strengths) and disadvantages (weaknesses) associated with the use of checklists.

Strengths

- The inspection is more structured and a checklist can be used as part of the inspection planning process.
- The checklist provides an immediate record of the findings of the inspection.
- It helps to ensure that important issues will not be overlooked.
- It assists consistency of inspections, especially when using different competent inspectors.
- It provides documentation for health and safety auditors that can be useful for analysing trends.

Figure 2: Checklists can be used in inspections

Weaknesses

- Following a checklist may be too restrictive; important issues not on the checklist may be missed.
- There may be a temptation to assume that, because there is a checklist, any untrained person can be designated to carry out the inspection.
- Due to the habitual nature of using a checklist, there may be a lowered perception of risk.
- The checklist may not be regularly reviewed; therefore, new hazards and risks from, for example, new processes or new equipment, may be missed.

Allocation of responsibilities and priorities for action

Following the inspection, during which observations are made, a report should be issued that includes recommended actions. It should also identify the people who will pursue these actions and the date by which the actions are expected to be completed.

CASE STUDY 1

This form is an example of part of a report from an inspection carried out in a manufacturing area.

Observation	Recommended actions	Person(s) responsible for action	Action completion date	Comments
Pothole on the road along forklift truck vehicle route Risk of truck overturning	**Short term:** Place warning notice and diversion sign **Medium term:** Repair hole **Long term:** Review traffic route, with a view to directing traffic on a more resilient surface	Area supervisor Maintenance manager Transport manager	Immediately Within 1 week Presentation to be made within 4 months	Done
High level of noise in machine area – complaints from workers; risk of hearing damage	**Short term:** Provide earmuffs and instruct workers to wear them	Supervisor	Within 2 days	Delivery of earmuffs expected in 24 hours

4.1 Active and reactive monitoring

Observation	Recommended actions	Person(s) responsible for action	Action completion date	Comments
	Medium term: Arrange for noise assessment to be carried out by competent consultant	Safety adviser	Within 2 months	Meeting has been arranged with competent consultant in 10 days
	Long term: Investigate the possibility of providing a noise enclosure	Safety adviser/ engineering manager/external supplier	Viability study to be completed within 6 months	
Examination of machine guarding – all guarding found to be properly in place. Maintenance records to check guarding found to be up to date	**Short term:** None **Medium term:** Continue maintenance activities **Long term:** Review at next inspection	Maintenance engineers Inspectors	Ongoing 2–3 months	Praise communicated to maintenance department Observation of good practice
Worker noted to be using portable electric drill with badly damaged insulation Risk of electrocution	**Short term:** Take drill out of service **Medium term:** Provide training on electrical hazards to all workers, in particular the importance of a visual check of portable electrical equipment before every use **Long term:** Review the system of portable appliance testing (PAT) within the organisation	Supervisor Safety adviser/ engineering manager Engineering manager	Immediately Within 2 weeks Within 2 months	Done – safe, checked drill provided for work to continue. Worker advised on visual checking immediately

Notes

1 Short, medium and long-term measures need to be identified, not just a short-term measure. Short-term measures can usually be put in place quickly at fairly low cost, whereas long-term measures require more resources in terms of cost and time though they tend to be far more effective over time. A good action programme will include all three of these measures.
2 Machine guarding was satisfactory at the time of the inspection and those responsible were praised for this. It is helpful to identify good practice and to praise those responsible. It increases morale and motivation.

Different organisations will contain different hazards. Case study 2 shows an example of part of a report that could apply to an office.

Health and safety monitoring and measuring

> **CASE STUDY 2**
>
> The following inspection was carried out in one of the offices of a publishing company.
>
> | Observation | Recommended actions | Person(s) responsible for action | Action completion date | Comments |
> |---|---|---|---|---|
> | Trailing cables found across floor – risk of tripping | **Short term:** Place warning signs by cables | Supervisor | Immediate | Done |
> | | **Medium term:** Place rubber covers over cables | Maintenance manager | Within 1 week | |
> | | **Long term:** Reroute cables | Safety adviser/ maintenance manager | Within 3 months | |
> | Fire extinguisher used to prop a door open – incorrect use of fire extinguisher

Extinguisher is not in the correct place if needed to fight fire | **Short term:** Remove fire extinguisher and relocate it to its proper position | Supervisor | Immediate | Done |
> | | **Medium term:** Provide training on use of fire extinguishers | Safety adviser | Within 1 month | |
> | | **Long term:** None | None | | |
> | Computer work; all workers noted to have good posture at desk; chairs in good condition; no reports of any health problems; display screen equipment (DSE) assessments carried out | **Short term:** None | | | Good practice |
> | | **Medium term:** Review risk assessments at appropriate time | Safety adviser | Ongoing | Praise DSE assessor |
> | | **Long term:** Review at next inspection | Inspectors | 3 months | |

In summary, an inspection consists of a physical inspection of the workplace looking for unsafe acts and conditions. It is carried out by a competent person who could use a checklist. A report will be issued after the inspection that will include opportunities for improvement, such as remedial measures for hazards identified.

Safety sampling

This technique is similar to safety inspections, but involves making observations in a small, specific area.

- These observations are carried out on a prearranged regular basis (perhaps once a week) and take approximately 30 minutes to complete.
- The same work equipment, environment and working procedures are viewed on each occasion.

- The sampling is normally carried out by workers who are familiar with the area.
- They will have a checklist containing questions that can be answered 'yes' or 'no'. For example, are all workers who need to wear safety shoes wearing them?
- At the end of the sampling exercise, the worker conducting the safety sampling gives the completed checklist to the supervisor, who will take any necessary action.

This technique can be useful for studying trends.

Safety tours

This technique is useful to demonstrate management commitment to health and safety. The tour is carried out by a senior manager.

- The observing manager must wear appropriate personal protective equipment (PPE) while carrying out the inspection; for example, if eye protection is required in an area, the manager must wear it.
- The manager should have some awareness of health and safety but need not be an expert. However, the manager should be a competent observer and note any obvious non-compliance with health and safety requirements; for example, unguarded moving parts of machinery or failure of workers to wear appropriate PPE.
- A safety tour will typically take 30–40 minutes.
- The manager should also be prepared to briefly discuss health and safety matters with workers or safety representatives if these workers want to.
- The tour will be carried out at irregular intervals and should be unannounced to give a realistic impression of normal working.

TIP

Some other techniques that may be used in active monitoring are outlined here.

Safety surveys

- A safety survey is an in-depth study of a particular aspect of health and safety, for example, a noise survey. Another example could be the study of fire extinguishers in the organisation, including where they are placed, the types used, the arrangements for inspecting them, for replacing them and for providing training on their use.
- The survey would be carried out by a person with appropriate expertise in the aspect of health and safety being studied.

Safety audits

- This is also an active monitoring technique and is considered in depth in 4.3: Health and safety auditing.
- Other active monitoring techniques include risk assessment and environmental monitoring. These are covered in 3.4: Assessing risk.

APPLICATION

What active monitoring methods are used in your organisation?

4.1.3 Reactive monitoring measures

Data on accidents

One of the most widely used reactive measures is to analyse accident statistics. There are many ways of expressing accident data from simple numbers of accidents.

Such statistics can be useful in observing trends in accidents, but simple numbers will not provide a complete picture; for example, consider Case study 3.

> **CASE STUDY 3**
>
> An organisation has 200 workers in a given year, during which 10 accidents take place (0.05 accidents per worker in a year).
>
> The following year the organisation sells part of its operation so that only 80 workers remain. During this year, 8 accidents take place (0.1 accidents per worker in a year). The number of accidents has reduced, but you can see that there are now more accidents per worker, so the health and safety performance appears to have got worse.

The commonly used statistics in benchmarking are accident frequency rate (AFR) and accident incidence rate (AIR). However, care must be exercised in using these; for example, there may be different definitions of an accident in different organisations. Also, one organisation might include contractors working on the site whereas another might not. Different organisations may arrive at their figures in a different way, so like-for-like comparison may not always be advisable.

Dangerous occurrences

> **DEFINITION**
>
> *The definition of a **dangerous occurrence** may be regarded as an incident that had a high potential to cause death or serious injury but did not result in any injury.*

Dangerous occurrences that are required to be reported are covered by the Reporting of Injuries, Diseases and Dangerous Occurrences Regulations 2013 (RIDDOR). An example would be a scaffold, the substantial part of which was more than five metres in height, that collapsed but did not fall on anyone. Specific dangerous occurrences that must be reported are found in Schedule 2 of RIDDOR.

Recording of dangerous occurrences is essential so that actions can be put in place to prevent a recurrence.

Near misses

A near miss is an incident that could have resulted in personal injury or damage to property or equipment but did not do so. Many studies have been carried out that show that the number of near misses at work far exceeds the number of serious injuries, which in turn far exceeds the number of fatalities. The usefulness of a near miss is that it can be regarded as an early warning. Statistically speaking, the greater the number of near misses, the greater the potential for serious injury or death.

Therefore, near misses should be recorded and actions taken to try to reduce the risk of recurrence of incidents causing such near misses.

Ill-health reports

These can be useful to discover instances of exposure to hazardous work activities (including harmful agents). When ill-health conditions resulting from exposure occur quickly, remedial action must be actioned immediately. For example, if several computer workers are complaining of neckache, backache or headaches, this may indicate that they are working for a long time without suitable breaks or are using unsuitable equipment. These complaints should therefore be followed up with display screen equipment (DSE) risk assessments (this will be covered in detail in 6.1: Work-related upper limb disorders.) However, exposure to some chemicals may only produce a long-term effect. A subsequent ill-health report may only be available once damage to health has occurred and is perhaps irreversible.

An example of this is exposure to asbestos in various industries. In the past, asbestos was used in many applications due to its excellent heat and fire-resisting properties. It was only years later, after the ill-health effects and catastrophic disease death toll associated with its use became widely known, that asbestos was banned for almost all applications.

Complaints by the workforce

This can be useful data, since it will identify issues that the workforce is particularly concerned about and may help to apportion resources where they are most needed, as demonstrated in Case study 4. If prompt action is taken to address legitimate complaints, then this can be seen as being an effective way of helping to build a positive safety culture.

CASE STUDY 4

A school cook developed breathing problems after working with flour in the school kitchen. The room was small. There were no controls to minimise exposure to the flour dust and the ventilation was particularly poor.

The cook's daily job included making dough in a large mixer.

She had complained to the school council about her breathing problems but no action was taken to address this.

Eventually, her breathing problems became so bad that she could hardly walk. She had to sit up to be able to sleep.

The cook had become severely asthmatic and retired from the job on health grounds, subsequently leading a restricted lifestyle.

She was awarded substantial compensation in court.

If there was a good safety culture, management would have responded quickly to the cook's initial complaints. However, a failure to respond to her complaints resulted in the permanent ill-health of the cook and consequences for the school council, including the payment of substantial compensation and payment of court costs. The council was also served with an enforcement notice to carry out health and safety improvements to the school kitchen.

> **APPLICATION**
>
> How is data from reactive monitoring used to analyse your organisation's health and safety performance?

> **ADDITIONAL INFORMATION**
>
> A history of compensation claims by workers can be regarded as the civil equivalent of enforcement action. The number of claims, for example, due to injury at work or ill-health resulting from work, is again a good indication of the level of health and safety performance.
>
> Appropriate responses to active monitoring methods may reduce the risk of civil proceedings against the organisation.

Enforcement action

The number of enforcement actions received by an organisation can indicate its level of health and safety performance. For example, if organisation A reports a larger number of notifiable accidents, incidents and dangerous occurrences than organisation B, which has similar numbers of workers and similar numbers and types of hazard, then organisation A is likely to attract more visits from enforcing officers than organisation B.

This may be useful in that it could be a good indication of organisation A's level of compliance with statutory requirements. It may also be useful for organisation A, because once the required enforcement action has been carried out, the workplace should be safer.

A large number of required actions demonstrates shortcomings in the organisation. It may seem that organisation B has a superior health and safety performance to organisation A.

Such conclusions should be viewed cautiously as other aspects may need to be considered. For example, the reporting efficiency of organisation A could be far better than that of organisation B.

Incident investigations

These can reveal the root causes of incidents, which is the key reason for investigating them and in turn may prevent other incidents. For example, an incident may have occurred due to insufficient training, and giving attention to improved training could prevent other incidents.

Other reasons to investigate incidents include to:

- determine any non-compliance with legislation;
- monitor trends for that type of incident;
- determine the cost of the incident, for example, in terms of lost worker time, damaged equipment and investigation time;
- process any claims from workers; and
- determine whether a negative safety culture contributed to the incident; for example, if there was not enough management commitment to health and safety.

4.1.4 Why lessons need to be learnt from beneficial and adverse events

The main reasons to learn lessons from events are to build on knowledge from beneficial events and prevent recurrences of adverse events.

This should help to reduce the risk of prosecution due to criminal offences and the risk of civil actions from injured workers. It is likely there will be financial benefits due to less sickness absence, less cost to replace damaged plant, fewer first-aid costs, improved productivity and fewer increases in insurance costs. There may be other benefits as well, such as an improved image for the organisation and higher morale in the workforce due to fewer incidents.

Benefits to workers will include less pain and suffering due to reduced numbers of accidents. Also there is likely to be increased morale as workers perceive that management have learned lessons from previous incidents, for example, by installing more effective control measures and responding more quickly to worker complaints.

Further benefits to an organisation could include lower turnover of workers, as they will have more confidence in management aiming to protect their health and safety. Perhaps management have noticed better worker participation in safety discussions as a result of them holding more meetings with safety representatives (that is, improved consultation).

4.1.5 The difference between leading and lagging indicators

Leading indicators are active measures that can be observed and provide current information about the effectiveness of the health and safety management performance of an organisation prior to any incidents causing loss.

Lagging indicators are reactive measures that can only track negative outcomes, such as personal injury, after they have occurred.

Here are two specific examples of these indicators in practice, one for health and one for safety:

- **Health:** A working area is thought to be noisy. The noise level is measured during a noise survey and found to be at a level that might lead to hearing damage – this is a leading indicator. Hearing protection can now be provided. If no noise survey had been carried out and no hearing protection provided, workers in the area might subsequently have had hearing tests that showed damage to their hearing. The results of these hearing tests would be a lagging indicator.
- **Safety:** During a safety inspection, a machine guard was found to be missing – this is a leading indicator. The guard may now be replaced. If the safety inspection had not taken place and the guard not been replaced, a worker might subsequently have sustained a serious hand injury. The record of an injury of this type would be a lagging indicator.

General examples of leading indicators are safety inspections and safety audits.

General examples of lagging indicators are accident statistics and ill-health reports.

Health and safety monitoring and measuring

KEY POINTS

- The difference between active and reactive monitoring is that active monitoring takes place before a loss-causing event and reactive monitoring occurs after a loss-causing event.

- Types of active monitoring include formal safety inspections, which can be internal (by a competent adviser or a team) or external (by enforcement officers) and will physically look at a whole workplace or just part of it to identify any hazards, unsafe working practices and unsafe working conditions.

- Other active monitoring techniques include safety sampling, safety tours and safety surveys, which usually focus on a particular area and are usually carried out by workers and managers.

- How frequently inspections are carried out can be influenced by things like unsatisfactory results from previous ones, the introduction of new equipment and processes, workforce complaints, a big increase in incidents and near misses, risk assessment recommendations and legal requirements.

- Inspection reports should give details of the types of hazard observed and recommended actions and control measures for each, including the person(s) responsible for completing actions and timescales for their completion.

- There are both strengths and weaknesses in using checklists to carry out inspections. For example, they are structured and can help ensure consistency and that important issues are not overlooked, but they may be too restrictive, out of date and used by untrained people.

- Reactive monitoring measures include analysis of accident data, dangerous occurrences (that may need to be reported), near misses, ill-health reports, worker complaints, enforcement actions and incident investigations.

- The difference between leading and lagging indicators is that leading indicators provide information about an organisation's current health and safety performance before any loss-causing incident and lagging indicators can only track negative outcomes from incidents that have already occurred.

4.2: Investigating incidents

> **Syllabus outline**
>
> In this section, you will develop an awareness of the following:
>
> - The different levels of investigations: minimal, low, medium and high (see HSG245)
> - Basic incident investigation steps:
>   - step 1: gathering the information
>   - step 2: analysing the information
>   - step 3: identifying risk control measures
>   - step 4: the action plan and its implementation
> - How fatalities, specified injuries, 'over 3- or 7-day injuries', diseases and dangerous occurrences must be recorded and reported

Introduction

The objective of health and safety management is to prevent, or at least minimise, incidents that cause occupational injury or ill-health to workers. This is what you must strive to achieve. However, accidents happen and ill-health occurs, so you need to have systems in place to learn lessons and introduce measures to prevent such events occurring again.

We will start by providing some important definitions, then move on to examining the four basic incident investigation stages. Case studies have been used to illustrate these four stages.

Finally, there is an overview of how to record and notify the regulatory authority about occupational accidents and diseases.

> **TIP**
>
> We will draw on the following legislation:
>
> - the Reporting of Injuries, Diseases and Dangerous Occurrences Regulations 2013 (RIDDOR); or
> - the Reporting of Injuries, Diseases and Dangerous Occurrences Regulations (Northern Ireland) 1997.
>
> We will also use information from the Health and Safety Executive (HSE) and Health and Safety Executive for Northern Ireland (HSENI) guidance:
>
> - *Investigating accidents and incidents* (HSG245);[1] and
> - *RIDDOR (NI) 97: Reporting of Injuries, Diseases and Dangerous Occurrences Regulations (Northern Ireland)*.[2]

Here are some terms you may come across in incident investigation.

Health and safety monitoring and measuring

> **DEFINITIONS**
>
> An **accident** is an unplanned event that results in personal injury or ill-health.
>
> A **dangerous occurrence** is a specified adverse event identified in Schedule 2 of RIDDOR that will need to be reported to the enforcing authority. There is the potential for personal injury, although none occurred.
>
> A **damage-only incident** is an event that damages equipment, property or materials but does not result in personal injury.
>
> An **incident** is an unplanned event that caused neither injury nor damage but had the potential to do so – this is often referred to as a **near miss**.

(Note: you may occasionally see the term 'dangerous incident'. This is similar to a dangerous occurrence, where no injury occurred, although a potentially serious injury outcome was possible. The term 'accident' implies injury and/or ill-health has resulted.)

4.2.1 The different levels of incident investigation

Now we examine the level of investigation required for incidents presenting different levels of risk and their likelihood of recurrence.

Incidents with minor consequences occur much more frequently than those with more serious consequences. Table 1 demonstrates different combinations of frequency, level of risk and level of investigation.

Likelihood of recurrence	Minor consequences	Serious consequences	Major consequences	Fatal consequences
Certain	Low	Medium	High	High
Likely	Low	Medium	High	High
Possible	Low	Medium	High	High
Unlikely	Minimal	Low	Medium	High
Rare	Minimal	Low	Medium	High

Table 1: Different levels of risk and likelihood

In a minimal-level investigation, the supervisor will review the circumstances of the incident and try to put in measures to prevent a recurrence.

During a **low-level investigation**, the relevant supervisor will carry out a short investigation with the worker and their safety representative to investigate the circumstances of the incident and try to agree on measures to prevent a recurrence.

A **medium-level investigation** will involve the supervisor, worker, safety representative, safety adviser and possibly a senior manager in identifying all causes of the incident and drawing up a formal plan to prevent a recurrence.

A **high-level investigation** includes everyone involved with a medium-level investigation, is supervised by senior management and is a thorough, formal investigation. A timetable of immediate and longer-term actions will be agreed and a formal record made. It may well be that an external enforcing officer will become involved at an early stage and the officer will probably identify certain actions to be carried out. In an extreme case, for example a fatality, the enforcing officer may require work that led to the fatality to stop until certain actions are carried out.

Any investigation will need to answer six basic questions, sometimes known as '5 Ws and a how', as shown in Figure 1.

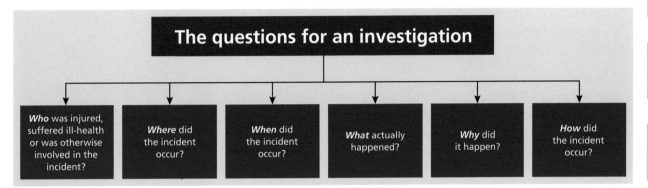

Figure 1: Questions for an investigation

4.2.2 Basic incident investigation steps

The HSE guidance document HSG245 identifies the four stages of an incident investigation as:[3]

1 Gathering information;
2 Analysing this information;
3 Identifying risk control measures; and
4 Implementing an action plan.

Step 1: Gathering information

There are basically three ways in which information (also referred to as evidence) can be obtained.

1 **Information obtained from people:** this will include any workers involved in an incident, also those who were not involved but were witnesses to the incident. In our discussions we will refer to both as witnesses. When conducting interviews, the interviewer will need to use a certain method of behaviour to make the interview effective. In particular, the interviewer should do the following:

- Try to ensure the interview is conducted as soon as possible after the incident, so that the events of the incident are still fresh in the mind of the interviewee.
- Put the interviewee at ease from the start and assure them that the purpose of the interview is simply to establish facts and not to apportion blame.
- Allow for the state of mind of the interviewee, who may be upset following the incident.
- Ask 'open' questions inviting some detail in answers rather than 'closed' questions that could be answered by 'yes' or 'no' responses.
- Ensure that the interview is conducted in private, with no interruptions, for example, mobile phones to be switched off.
- Arrange that only one person is interviewed at a time.
- Take account of any disabilities the interviewee has; for example, if the interviewee has hearing loss, questions should be asked clearly and slowly.
- Permit the interviewee to be accompanied if they so wish, for example, by a worker representative.

- Review the answers to questions at the end and agree their accuracy with the interviewee.
- Thank the interviewee for the information they have provided.

Personal information obtained from witnesses should include their name, address, job title and, for any injured persons (that is, workers or others who may have been involved, such as visitors), the name of the supervisor or manager to whom they report. In the case of a fatality or major accident, details of next of kin should be available.

2 **Physical information:** this will include determining details such as what machinery was in use, what substances were being used and what personal protective equipment (PPE) was being worn. The general condition of any machinery might be observed and whether all necessary guards were in place. Environmental aspects, such as whether the floor was damaged or slippery, the adequacy of the lighting, whether there were any spillages visible and the temperature and humidity of the workplace, may also be relevant.

3 **Documentary information:** the investigator will wish to see certain documentary information. This will assist the investigator to determine the quality of the organisation's health and safety management system. Documents to be examined would include:

- operating instructions for any machinery;
- safe systems of work in place;
- safety data sheets for any substances used;
- results of any safety inspections carried out in the incident area;
- results of any safety audits;
- any photographs taken or sketches made of the incident area;
- footage from any security videos covering the area;
- any training records of persons involved in the incident;
- maintenance records of any relevant machinery;
- risk assessments for the area;
- any permit-to-work system relating to the incident; and
- any previous accident or ill-health records associated with the relevant workers in the incident area (relating to the particular incident) ie previous relevant accident history.

Step 2: Analysing the information

You should now know who was involved in an incident, where and when it took place and what happened. Now you need to find out how and why the incident occurred.

There are different types of accident causes:

- immediate;
- underlying; and
- root causes.

Immediate causes

These are usually fairly easy to determine, although you should be careful not to jump to conclusions as there may be more than one cause. Immediate causes result directly from unsafe acts (for example, worker behaviour, such as ignoring safety instructions) or unsafe conditions (for example, controls on a saw not working properly).

4.2 Investigating incidents

> **APPLICATION**
>
> Consider a scenario where a worker was injured when their hand came into contact with the blade of a circular saw. Immediate causes could be that the saw blade was not guarded and the worker's hand came into contact with it when reaching for a piece of wood. However, another possibility is that the worker slipped and caught their hand on the saw.

Underlying causes

These are less obvious causes. In the example with the saw, perhaps the hazard of an unguarded saw had not been recognised or perhaps the guard had been removed for maintenance purposes and not been replaced. Perhaps the worker had not received any safety training in using the saw.

Root causes

These are associated with an initial failure leading to other causes. In the saw example, perhaps there had not been resources provided for safety training, so the worker was not aware of the hazard of an unguarded blade, or maybe there was insufficient supervision or risk assessment. Root causes are often described as 'management failures'.

> **APPLICATION**
>
> The technique used to establish causes of incidents is to start with immediate causes and ask, why? This will lead to underlying causes and again you ask, why? This will lead to root causes, and you ask why until it is difficult to go any further.
>
> We need to investigate immediate, underlying and root causes of incidents.[4]
>
> One possible chain of events is as follows:
>
> - In the saw example, the worker cut their hand on the saw blade – why?
> - Because there was no guard on the blade – why?
> - Because the worker did not think it was necessary to use it – why?
> - Because the worker had not received safety training – why?
> - Because the management system did not include it in its policies – why?
> - Because there was a lack of resources.
>
> It is now difficult to take this analysis further.

The difference between underlying and root causes can be a little blurred at times and the two types of cause are sometimes lumped together as root causes. However, the difference between root and immediate causes is distinct.

Step 3: Identifying risk control measures

At this stage of the investigation, you know what happened and why. You now need to put in control measures to eliminate the risk or minimise it if elimination is not practicable.

The method for this is to use a hierarchy of measures. This means establishing a list of measures in order of priority, considering the top measure first, then proceeding down the list.

A well-established hierarchy is covered in 3.4: Assessing risk.

> **CASE STUDY**
>
> A worker arrived at work one morning, walked into the reception area and fell over onto the floor. They suffered a sprained ankle and bruised knee. During the information gathering part of the incident investigation, six questions were identified:
>
> 1 Was the worker hurrying?
> 2 Was the worker wearing inappropriate footwear?
> 3 Was the reception floor damaged?
> 4 Was the reception floor slippery?
> 5 Were there unseen obstacles on the reception floor?
> 6 Was the lighting in the area inadequate?
>
> It was found that the answer to question 4 was 'yes' and the answer to all the other questions was 'no'. Therefore, the immediate cause was falling on the slippery floor.
>
> This led to more detailed analysis:
>
> - Why was the floor slippery?
>   - Answer: there was too much polish on the floor and it was the wrong type.
> - Why was there too much polish of the wrong type on the floor when it was known that a competent cleaner had been cleaning that floor for several years?
>   - Answer: this cleaner was on holiday and a temporary cleaner (TC) was tasked with cleaning the floor.
> - Why did TC use the wrong polish and too much?
>   - Answer: TC had received no training and instruction on polishing the floor.
> - Why not?
>   - Answer: The job was thought to be too simple to require training – this was a management failure. Also, TC had tried to find someone to advise on the type of polish. The instructions for TC were to ensure the floor looked shiny because the managing director thought that this represented a good company image.
> - Why was there nobody for TC to ask?
>   - Answer: TC's working hours were 17.30 to 20.30 when most staff had gone home. The relevant supervisor's hours were 09.00 to 17.00.
>
> Several risk control measures were identified:
>
> - TC was immediately given training on polishing the floor.
> - The supervisor roster was changed to ensure that a supervisor was present for some time after 17.30.
> - The organisation's management system was reviewed to ensure that all jobs received necessary training.

The case study example shows that even simple cases can give rise to incidents. Now consider what analysis and control measures would have been necessary if, in addition to a slippery floor, the lighting had been inadequate and there had been a trailing cable across the floor. Also consider that the worker who fell might have hit their head and been admitted to hospital with severe concussion. This demonstrates that incidents can have many causes and consequences and investigations can be complex.

Step 4: The action plan and its implementation

You have reached the point in the investigation when you have decided what control measures are needed following an incident to prevent its recurrence.

Return to the Case study where the worker fell on a slippery floor. The template in Table 2 shows an action plan for this incident. It identifies the short-, medium- and long-term actions needed, the workers responsible for carrying them out and appropriate timescales.

Recommendation	Short-term action	Medium-term action	Long-term action	By whom	By when	Review date**
Ensure reception floor is not slippery	Train TC on type and quantity of polish to use on floor			Supervisor	Immediate	1 day after training
		Ensure supervisor present during working hours of TC		Department manager	Within 3 days	1 week
			Review management system to determine training necessary	Senior manager	Within 4 weeks	6 weeks

**Normally, actual dates would appear here; this table just gives an idea of timescales.

Table 2: Incident action plan example

Incident report forms should have a section titled 'Action to prevent a recurrence'. If you see an entry in this section that simply states, 'Worker must be more careful', this suggests the writer may not have investigated the incident properly.

4.2.3 How fatalities, specified injuries, 'over 3- or 7-day injuries', diseases and dangerous occurrences must be recorded and reported

Drawing on RIDDOR, we will look at four types of incident involving workers that should be reported:

- non-fatal injuries;
- additional requirements for fatalities;
- dangerous occurrences; and
- diseases.

The responsible person

Firstly, you need to identify who is required to report incidents (the 'responsible person').

Who is this responsible person?

- for workers, this will be the employer;
- for a self-employed person, it will be themselves; and
- where members of the public may be injured while using a building (for example, a shopping centre), it will be the person in control of the building.

Non-fatal injuries

The responsible person is required to follow reporting procedures for specified injuries that appear in reg 4 of RIDDOR. These are:

- the amputation of a finger, thumb, toe, hand, arm, foot or leg;
- a bone fracture other than to a finger, thumb or toe;
- an injury causing permanent loss or reduction of sight;
- any crush injury to the head or torso causing damage to the brain or to internal organs in the chest or abdomen;
- any burn or scalding injury that covers more than 10% of the whole body surface area or causes significant damage to the eyes, respiratory system or other vital organs;
- any degree of scalping requiring hospital treatment;
- loss of consciousness caused by head injury or asphyxia, or other injury arising from working in an enclosed space leading to hypothermia or heat-induced illness, requiring resuscitation or admittance to hospital for more than 24 hours; and
- injuries to workers or self-employed people must also be reported when a person cannot carry out their normal job for more than seven consecutive days following the day of the incident (including weekends and rest days).

The reporting procedure is set out in Schedule 1 of RIDDOR. The responsible person must:

1 Notify the enforcing authority (EA), such as the Incident Contact Centre of the HSE, by the quickest possible means without delay, for example by telephone or email. Note: the telephone service is available for fatal injuries and specified incidents. For non-fatal and non-specified incidents, the responsible person will normally report these online.
2 Send the EA a report in an approved manner (for example on HSE form F2508) within 10 days of the incident that caused the injury. If an injured person is away from work or unable to perform their normal duties for more than seven consecutive days, not including the day of injury, but including weekends or public holidays in this period (an 'over 7-day' injury), then the responsible person is required to submit a report to the EA within 15 days of the date of the incident.

Fatalities

Where a person dies within a year of the date of an accident, the responsible person must report the death even if the accident had been previously reported under a non-fatal specified injury. For example, a worker lost consciousness due to a head injury after hitting their head as a result of a fall. They appeared to have recovered and returned to work, only for the head injury to result in a chronic condition, for example, a slow bleed on the brain (ie different from the acute effect from the accident, but still due to that accident) that caused their death within a year. The intention of RIDDOR is that a delayed death occurrence report is required when the death results from a direct cause of the original injury.

In the event of a fatality, other parties should be advised as appropriate, including the police, next of kin, the coroner and the organisation's insurers.

Dangerous occurrences

There are many dangerous occurrences listed in Schedule 2, Part 1 of RIDDOR. We will look at examples that relate to most workplaces:

- malfunction of lifting equipment;
- failure of pressure systems;
- contact with uninsulated overhead electric lines where the voltage exceeds 200 volts or close proximity with an electric line, causing electrical discharge;
- electrical incidents that cause fire or explosion;
- malfunction of breathing apparatus;
- collapse of scaffolding where a substantial part of it is more than five metres in height;
- structural collapses arising from ongoing construction work;
- any fire or explosion that stops work in the premises for more than 24 hours;
- release of flammable liquids and gases in specified quantities, either inside a building or in the open air; and
- escape/release of hazardous substances and/or biological agents that could cause injury/ill-health.

Schedule 2 of RIDDOR also gives details of reportable dangerous occurrences relating to specific workplaces, for example, collision between passenger trains, and specific industries such as mining or quarrying.

The reporting procedure is as follows. The responsible person must:

- notify the enforcing authority (EA), for example, the Incident Contact Centre of the HSE, by the quickest possible means without delay, for example by telephone or email; and
- send the EA a report in an approved manner (for example, on form F2508) within 10 days of the dangerous occurrence.

Diseases

Examples of these are:

- carpal tunnel syndrome;
- hand or forearm cramp resulting from prolonged periods of repetitive movement;
- occupational dermatitis;
- occupational asthma;
- hand-arm vibration syndrome;
- tendonitis or tenosynovitis in the hand/forearm; and
- occupational cancers if the person's current job involves exposures to carcinogens or mutagens.

The reporting procedure is set out in Schedule 1 of RIDDOR. The responsible person must send a report of the diagnosis of the disease in the approved manner to the EA without delay.

When an employee dies, having suffered a non-fatal injury reportable under reg 4 (see non-fatal injuries) within one year of the date of the accident, the employer must notify the relevant enforcing authority of the death in an approved manner without delay, whether or not the injury has been previously reported under reg 4.

There are specific reporting requirements for mines and quarries. These are not covered here but can be found in reg 13 and Schedule 1 of RIDDOR.

Recording incidents

When the responsible person is required to report accidents, dangerous occurrences and diseases as we have described, they are also required to keep associated records.

If a person is injured resulting from a workplace accident and the person is unable to perform their normal duties for more than three consecutive days (an 'over 3-day' injury), the responsible person is required to keep appropriate records. These records will include information such as the injured person's full name, the nature of their injury, the location of the injury, the location of the accident, the date on which the accident was first reported to the EA (if applicable) and a brief description of the circumstances of the accident; such details can be found in Schedule 1, Part 2 of RIDDOR.

'Over 3-day' injuries do not need to be reported unless the worker is incapacitated for seven days or more following the date of the incident.

Such records must be kept for at least three years from the date they were made and must be kept at the usual place of business of the responsible person.

The responsible person will be required to provide extracts of the records when asked to do so by the EA.

KEY POINTS

- Terms used in incident investigation include accidents, dangerous occurrences, damage-only incidents and incidents, which have different definitions.
- Investigation levels for incidents vary depending on their seriousness and likelihood of occurrence.
- There are three sources for gathering information in the first stage of incident investigation: people, physical (including the environment) and documents.
- Analysis of the information obtained considers possible immediate, underlying and root causes.
- Risk control measures in response to the causes identified by analysis can be short, medium or long term.
- Action plans to prevent recurrence of an incident should set out the control measures needed, who is responsible for implementing them and appropriate timescales.
- Even apparently simple cases can give rise to incidents that have multiple causes and require complex investigation.
- Incidents need to be reported, recorded and notified under RIDDOR.

References

[1] HSE, *Investigating accidents and incidents: A workbook for employers, unions, safety representatives and safety professionals* (HSG245, 2004) (www.hse.gov.uk)

[2] HSENI, *RIDDOR (NI) 97: Reporting of Injuries, Diseases and Dangerous Occurrences Regulations (Northern Ireland)* (2018) (www.hseni.gov.uk)

[3] See note 1

[4] See note 1

4.3: Health and safety auditing

> **Syllabus outline**
>
> In this section, you will develop an awareness of the following:
> - Definition of the term 'audit' (clause 3.32, ISO 45001:2018)
> - Why health and safety management systems should be audited, including:
>   - negative: identifying failing of a management system
>   - positive: organisational learning and assurance
> - Difference between audits and inspections
> - Types of audit: product/services, process, system
> - Advantages and disadvantages of external and internal audits
> - The audit stages:
>   - notification of the audit and timetable for auditing
>   - pre-audit preparations, including competent audit team, time and resources required
>   - information gathering
>   - information analysis
>   - completion of audit report

Introduction

It is essential that the health and safety performance of organisations is monitored. There are many ways to do this; here we concentrate on the technique known as a health and safety audit.

We will look at the differences between audits and inspections. We shall see that audits often require more preparation than inspections and that people carrying out audits need a wider range of skills than those involved with inspections.

We shall discuss a system for analysing data to help evaluate the results of the audit and the importance of communicating the results of audits.

4.3.1 Definition of the term 'audit'

> **DEFINITION**
>
> *According to ISO 45001:2018, an **audit** is a systematic, independent and documented process for obtaining audit evidence and evaluating it objectively to determine the extent to which the audit criteria are fulfilled.*
>
> Note: ISO 45001:2018 is an international standard that has been widely adopted by national standards organisations. In the UK, the standard has been adopted and published by the British Standards Institute as BS ISO 45001:2018, but is otherwise identical to ISO 45001:2018.

'Audit criteria' is effectively the standard that the evidence is judged against. This will simply be all the policies, procedures or other requirements relevant to the audit in question. These document what an organisation says it does or is trying to achieve.

'Audit evidence' is what the person carrying out the audit looks at and gathers during the audit, for example records (such as of training, complaints or permits issued), notes from interviewing workers or from what has been observed.

An audit can, in principle, be any size or duration (there is no reference to either in the definition we have given). What we mean by an audit can therefore vary enormously in practice.

Types of audit

As we will see later, audits can be either internal or external. This is based on the purpose of the audit (rather than who does them):

- internal – mainly used for internal purposes, such as to feed into management reviews. These are usually done by the organisation itself, using its own, trained workers. But sometimes an organisation may use external people to do this.
- external – mainly used to give assurance to external parties. These are done by a separate organisation, typically a customer, independent certification body or regulator.

As an example, consider the external audits carried out by a certification body. These audits are designed to independently verify whether the audited organisation's management system conforms (or continues to conform) to a recognised standard (such as ISO 45001). If the auditor concludes it does, you are issued with a certificate confirming this.

These bodies generally conduct two basic types of audits – certification (or recertification) audits and surveillance audits. A certification audit would cover the entire management system and be carried out infrequently (every 2–3 years is common); more complex or larger sites might need the services of a team of auditors for several days. Surveillance audits, done in between the certification audits, might be carried out every six to twelve months, but would focus on specific aspects of the system. For external audits like these, you may have relatively little influence over the scope (in term of the areas covered).

For internal audits, it is entirely up to the organisation to determine the scope. In general, they should focus more attention on areas of significant risk for the organisation (this is called 'risk-based' auditing). This will depend on the risk profile of the organisation (the industry, nature of hazards, size, complexity etc). As a result, internal audits are typically much more frequent (eg monthly) and are more focused on aspects of the system, such as permit-to-work systems, training or accident reporting. Because of their focus, each audit might require only a single auditor for just a couple of hours.

4.3.2 Why health and safety management systems should be audited

The purpose of auditing is to see how effective your management system is. The output of audits is just one of many things that can help you to continually improve the system. Auditing a health and safety management system (or parts of it) enables the identification of weaknesses and failings in the system by evaluating the level of compliance of individual elements to the required standard. The extent to which elements are not meeting the required standards can also be identified. This enables the organisation to take action to address these failings to improve the system.

Audits are also used to identify strengths of the health and safety management system and where the organisation has performed well, in addition to areas offering opportunities for improvement. This in turn will assist management in directing resources to areas where improvement is advised. For example, the audit could show that most general risk assessments had been carried out, but few manual handling risk assessments had been completed. Thus, future resources in terms of time and finance would be directed more to carrying out manual handling risk assessments until most of these were completed.

4.3.3 Difference between audits and inspections

As we have seen, audits can vary enormously in scope. So, in some cases, there might be little practical difference between a focused internal audit and an inspection. However, in general, inspections tend to be based on observation of the physical workplace, whereas audits tend to be concerned with systems and processes. The differences between inspections and external audits are much greater; this is illustrated in Table 1.

Inspections	External audits (eg certification audits)
Consist of straightforward visual observations of the workplace and workplace activities at a particular point in time (eg identification of hazards)	Consist of a complete examination of the documented health and safety management system
Usually take a short time to complete	Lengthy processes that take much longer to complete than inspections
Tend to take place frequently; timescales vary, but they could occur every 1–3 months	Only take place occasionally; timescales vary, but they could perhaps take place once every 2–3 years
Usually carried out by an internal worker	Carried out by an external auditor
Require less skill to complete; they could be carried out by workers, managers, supervisors or worker representatives	Higher degree of skill required to complete; they need to be carried out by competent auditors
Result in the production of a short report/completed checklists, identifying specific actions that need to take place	Result in the production of a comprehensive report that identifies strengths and weaknesses of the management system
The resources needed for inspections are fairly limited; for example, they could be completed by one person	The resources needed for audits are significant and considerably greater than for inspections; for example, they may require input from multiple people

Table 1: The differences between inspections and external audits

4.3.4 Types of audit

There are many different types of audit. We will look at three specific types here.

Product or services audit

This type of audit looks at specific products and/or services to establish whether they are conforming to relevant requirements. The requirements could be set by performance standards (for example, car emissions must be below a certain level) or customer requirements.

Process audit

This type of audit will establish whether processes are working within expected limits. The process will be evaluated against a set standard to check its effectiveness. This type of audit will look at the process requirements, that is, whether the process is being done within a set time, level of accuracy, or component mixture (if relevant). The audit will also review the resources required to make sure that the process is working as expected. Finally, it will look at the effectiveness and adequacy of the process controls.

System audit

This type of audit looks at the performance of management systems, so is the one that we will look at in detail.

There is likely to be common ground among the types of health and safety audit. For example, all audits will include the examination of the health and safety policy, to check if the policy is still effective, whether identified persons have specific duties and if the practical arrangements for implementing the policy cover all activities. The topic of risk assessments will also be common to all audits.

Differences will start to appear for different types of work. For example, for product and process audits, major audit topics such as the systems for receiving raw materials and for processing the materials at minimum risk to health and safety, their packaging and their delivery to customers are likely to be key topics.

In a service industry, the major systems will probably include interface with customers and continuity of supply. There will need to be systems for emergencies, such as a gas leak in a main road or the repair of electric cables following a severe storm and perhaps in adverse weather.

It is vital that any audit system is relevant to the type of work concerned. For example, there will not generally be the need for a system to measure radiation in a normal office; the exception might be if there was a leak of radon into the office, however, it is gratifying to note that this would be the exception rather than the rule.

For this reason, it is wise to be careful before using an 'off-the-shelf' audit system for an organisation. The hazards relative to an organisation need to be understood before developing the specific system for that organisation.

We have already identified the need for health and safety audits to be carried out by competent persons. For large organisations, it is possible that there is such a competent worker in the organisation; this is unlikely to be the case for smaller organisations.

You need to be aware of the benefits and limitations of internal and external personnel carrying out audits and we consider this now.

4.3.5 Advantages and disadvantages of external and internal audits

External auditors

Advantages and disadvantages of using external auditors are shown in Table 2.

Advantages	Disadvantages
Likely to have more auditing skills and expertise than those in the organisation	May recommend unrealistic standards or targets
Not biased and not concerned about criticising management or other workers	Not familiar with the organisation's workplace and processes
Likely to be more up to date with legal requirements	More resources (in terms of cost) needed than for an internal auditor
Can see the organisation through a fresh pair of eyes so may see things missed by the organisation's own workers	Not familiar with the qualities, competences and attitudes of the workforce
May know of best practice and solutions to problems found in other organisations	Workforce may be less co-operative with an external auditor
Crucially, seen to be independent	Not able to see improvements since previous audit (unless the same external auditor is used)

Table 2: External audit advantages and disadvantages

Internal auditors

Advantages and disadvantage of using internal auditors are shown in Table 3.

Advantages	Disadvantages
More likely to be familiar with the organisation's workplace and processes than external auditors	Not likely to have the necessary auditing skills and expertise
Aware of what is practicable for this type of organisation	May not be up to date with current legislation
Familiar with the qualities, competences and attitudes of the workforce	Less aware of good practice and solutions to problems found in other organisations
The workforce may be more co-operative with an internal auditor	Likely to be unwilling to criticise management or other workers in their organisation

Advantages	Disadvantages
Less resources (in terms of cost) needed than for an external auditor	Less resources (in terms of time) may be available compared with an external auditor; may be under time pressure from management
Can see an improvement from previous audit more than external auditor (unless the same external auditor is used)	Crucially, not likely to be seen as independent

Table 3: Internal audit advantages and disadvantages

It may be that an organisation wishes for both internal and external audits to take place. In such cases, the scope of internal audits will be restricted to topics within the expertise of the internal auditor(s) while the external audit includes all relevant topics. The internal audits will probably take place more frequently than the external audits.

4.3.6 The audit stages

Depending on the scope of an audit and the organisation (size, complexity, risk profile etc), not all the stages outlined will be needed. For example, a focused internal audit can be handled informally. However, all stages are included to cover a range of audit types.

Notification of the audit and timetable for auditing

For an audit to be thorough, various methods are required to gauge the effectiveness of the system. These include:

- examination of documents;
- interviews with selected workers; and
- workplace observations.

Therefore, it is important to create an audit timetable with agreed timescales to include all these activities. There must also be an agreement with the organisation on a suitable date for the audit. This enables the organisation to make necessary arrangements for the day; for example, making documentation and relevant workers available to participate and enabling a workplace observation to be carried out.

Pre-audit preparations, including competent audit team, time and resources required

Firstly, a competent team of workers should be assembled to be involved in the audit and the decision made as to whether an internal auditor or an external auditor will be used. Typical workers included in the team might be a manager, a safety representative and a health and safety practitioner, although for a simple, low-risk workplace, a competent auditor may be the only person conducting the audit. The auditor should be independent, but this can be difficult to guarantee if they are internal.

4.3 Health and safety auditing

The team will need to agree, or the lone auditor needs to decide on the:

- scope – what are you looking at? For example, whether the audit covers just one site or, for a larger organisation, several sites, the size of and the number of topics to be considered;
- objectives – what are you looking for?
- criteria – what are you comparing it against?

There are two basic methods to determine how far audit criteria are fulfilled:

(a) a non-scoring system that obtains evidence to show whether compliance with specified criteria has been achieved; or

(b) a scoring system that determines the extent to which compliance with specified criteria has been achieved.

It is for each organisation to decide which system they find most helpful.

Audit questions are usually created to help guide the auditors as they gather evidence against the audit criteria.

The Application box shows some example questions that might be used to help audit the use of personal protective equipment (PPE).

APPLICATION

- Are sufficient steps taken to ensure that any PPE provided is suitable?
- Is suitable storage provided for PPE?
- Are areas/circumstances clearly specified for PPE to be used?
- Is instruction and training given on the use of PPE?
- Are records kept of PPE issued?
- Are reasonable steps taken to ensure PPE is properly used?

A scoring system may also be used. Success or failure is judged against whether compliance with a relevant standard has been achieved or not. Also, organisations need to be wary of off-the-shelf generic audit systems, as specific topics and questions need to be developed for specific organisations. Some topics/questions may be selected for virtually all organisations, such as questions regarding general risk assessments.

The documentation to be examined during an audit will vary from organisation to organisation. The nature of the documentation will vary according to the size and type of organisation, but a typical list might be as shown in Table 4.

List of documents to be examined during an audit	
Accident reporting procedures	Emergency procedures
Risk assessments	Environmental monitoring records
First-aid procedures	Health and safety training records
Maintenance records	Safety monitoring procedures
Minutes of health and safety meetings	Records of any enforcement action
Procedures for selecting contractors	Permit-to-work procedures
Documents for statutory inspection records	Health and safety policy

Table 4: Documents likely to be examined during an audit

It is important to decide on the standards against which the audit takes place so that a meaningful conclusion can be reached regarding the efficiency of the health and safety management system.

Finally, it needs to be agreed how feedback from the audit results is presented; for example, whether the feedback is provided to managers, who in turn provide information to all workers, or whether it is provided at meetings between managers and safety representatives.

If an external auditor is used, it is unlikely that an auditing team will be created because the auditor will be considered sufficiently competent to conduct the audit alone. Of course, the auditor will still need to agree times to meet interviewees and will advise the organisation in advance about which documents they want to analyse. Other features, such as agreeing the scope of the audit, the timescales and identification of resources, will follow a similar path. The audit questions will be created by the external auditor, but the auditor should make efforts to understand the business of the organisation beforehand.

Information gathering

Information will come from documents, interviews and workplace observations.

The external auditor or internal auditing team will review the chosen documents; a team will need to agree which documents are analysed by which member. Some audit questions may not be fully answered until interviews have taken place. With regards to workplace observation, there can be discrepancies between information from interviewees, documentary information and what has actually been observed in the workplace.

The information gathered must now be thoroughly reviewed.

Information analysis

The next step is to analyse all the information gathered from the audit questions, interviews and workplace observations.

One way to analyse the information is to score each of the questions. The scores will indicate the extent of compliance with the standards identified. As an approximate guide, a score of zero will mean that the particular topic has not been considered; a score in excess of about 75% will indicate that the topic has received significant attention, with some more work to be done; for example, if more than 75% of manual handling assessments have been satisfactorily completed.

As an alternative, for a non-scoring audit system there may be a 'standard' identifying the level of manual handling assessments expected to be completed and the result may be expressed as having achieved or not achieved that standard.

The contents of the audit report

After the information has been gathered and analysed, it is time to prepare a report.

If a scoring system has been used, a final score will indicate the overall level of compliance with standards. For non-scoring systems, it is usual to highlight areas of nonconformity (often classified as major versus minor) that need attention. Observations (areas that represent opportunities for improvement), as well as examples of good practice, may also be included.

A conclusion section will include the extent of compliance with standards. It is important to note that this section will include areas that have achieved a high level of compliance with standards as well as those areas that require attention to bring them up to the required standards.

The conclusion will be followed by a section giving recommendations. It is essential that the recommendations are subdivided into recommended actions of high, medium and low priority. Each recommendation should have a date by which action should be taken, or at least reviewed. The person responsible for taking the action should also be identified.

Communication of the findings of the audit report

In the first instance, the report should be sent to a senior manager. In the case of an external audit, the report will be sent to the manager with whom the audit was agreed. In the case of an internal audit, the identity of the manager receiving the report will have been established before the start of the audit.

As the audit is likely to have covered all aspects of the organisation, all senior managers (including top management, such as the managing director or chief executive) should receive a copy of the report.

Workers who have been allocated actions in the recommendations section of the report should also receive either the whole report or information regarding the relevant sections of the report.

Senior managers may then advise other relevant workers in the organisation, including other managers and supervisors. Other workers may also be advised of the contents of the report through meetings, such as those with safety representatives.

Finally, the main findings of the audit should be communicated to workers:

- In some organisations, the findings may appear in the annual report.
- In other organisations, the results might be discussed in meetings between management and safety representatives. The results may then be made known to all workers by supervisors and/or safety representatives.
- In small organisations (for example, a workforce of 6–8 workers), the managing director may simply call a meeting of all workers and communicate the audit results to them directly.

> **TIP**
>
> It is not recommended that the results are simply posted on a noticeboard. Workers may not read them.

Overall review of the audit system

We will now review some of the main features of the audit.

It was noted that the auditor, whether external or internal, needs to be competent and experienced.

Sufficient time needs to be allocated to the audit. In the case of an internal audit, it must be understood that the participants are not likely to be able to continue with their normal tasks as the audit will be time consuming. In the case of an external audit, much of the allocated time, such as pre-audit preparations, will be spent by the external auditor. There will still be some time allocated for specific internal workers to be available for interviews.

Health and safety monitoring and measuring

The scope of the audit will have been defined; for example, whether the audit covers a single site or the whole business.

The importance of the audit being independent has been noted.

A record of the audit must be retained. It is recommended that records of the previous three audits are retained to show the level of improvement from audit to audit.

It has been noted that a scoring system for the audit can be helpful to demonstrate the level of compliance with predetermined standards, but also note that some audit systems do not use scoring systems.

The organisation will need to allocate resources for carrying out recommendations identified in the audit.

It is essential that the results of the audit are considered reliable. Therefore, for consistency of approach, it would be advantageous for the same auditors to be used for each audit.

The audit system consists of pre-audit preparations, the audit itself and the report of the findings, which includes recommendations for improvements.

KEY POINTS

- The term 'audit' can be defined as an independent, systematic process to gather evidence and evaluate it objectively.
- The reasons for carrying out audits include identifying weaknesses and system failings, system strengths and opportunities for improvement.
- The differences between audits and inspections include what is involved (complete examination or simple visual observation), who carries them out, their frequency, how long they take, the resources required and their results.
- The three main types of audit focus on products or services, processes and systems.
- Internal and external audits have both advantages and disadvantages.
- Pre-audit preparations should consider the audit's scope, objectives and criteria.
- There are three sources of audit information: documents, interviews and workplace observations.
- Analysis of audit information may be done by scoring each audit question or by comparing findings against certain standards.
- The audit report should cover the extent of compliance with standards, areas that require attention, recommended actions in priority order, when these actions should be completed and who by.
- How the audit report findings should be communicated to managers and workers (and their safety representatives) will vary according to the size of the organisation and who was involved in the audit.
- The general features of a health and safety auditing system include allocating sufficient time to the process and adequate resources to carrying out recommendations.

4.4: Review of health and safety performance

Syllabus outline

In this section, you will develop an awareness of the following:

- Why health and safety performance should be reviewed
- What the review should consider:
    - level of compliance with relevant legal and organisational requirements
    - accident and incident data, corrective and preventive actions
    - inspections, tours and sampling
    - absences and sickness
    - quality assurance reports
    - audits
    - monitoring data/records/reports
    - external communications and complaints
    - results of participation and consultation
    - objectives met
    - actions from previous management reviews
    - legal/good practice developments
    - assessing opportunities for improvement and the need for change
- Reporting on health and safety performance
- Feeding review outputs into action and development plans as part of continuous improvement

Introduction

You have now read about many aspects of monitoring to determine the level of health and safety performance achieved.

This section presents the reasons why you should review this performance and explores what items you should consider in the review.

It also covers who the review findings should be reported to and suggests methods for achieving continual improvement.

4.4.1 Why health and safety performance should be reviewed

A review of health and safety performance should be carried out before any future accidents or incidents occur.

It should be carried out for the following reasons:

- To ensure compliance with legislation: this may require reviewing standards, especially in the event of new legislation being enacted prior to the review.
- To make a judgement on the adequacy of the health and safety performance: this will indicate how well the health and safety management system is working and whether it is working as intended.

Health and safety monitoring and measuring

- To identify areas where the health and safety performance may be improved.
- To be able to respond to change: for example, the health and safety audit would have identified areas where improvement was necessary, so a purpose of the review will be to identify both good and bad health and safety conditions and practices.
- To learn lessons from experiences such as accidents, incidents and near miss reports.
- To take action based on lessons learnt, such as reviewing the health and safety policy and risk assessments already carried out.
- To identify trends of accidents and incidents.
- To take account of human factors that may have contributed to accidents, incidents or instances of occupational ill-health that, for example, may have occurred due to failure to comply with systems (such as instructions or procedures) already in place. These systems may have been ignored by employees. The failure to comply may have been intentionally ignored or not perceived by management.

In summary, you should review your health and safety performance to ensure that it is effective in managing risk and protecting people.

CASE STUDY

A manufacturing organisation employed 60 people.

The managing director (MD) allocated overall responsibility for health and safety to the production manager. This overall responsibility was written into the health and safety policy.

This resulted in the production manager having two roles: responsibility for overall health and safety and responsibility for ensuring that product output was maximised. Sometimes these two roles were in conflict.

An accident occurred when a worker, under pressure to work more quickly, ignored a safety procedure and took a shortcut to speed up production. This was known to management, who allowed this shortcut to take place. The accident caused the worker to sever two fingers on one of their hands.

An enforcing officer (EO) visited the premises soon afterwards and was critical of the health and safety structure in the organisation as dictated by its health and safety policy.

In addition to other enforcing actions, the EO required the MD to assume overall responsibility for health and safety and improve the policy with immediate effect. The MD was shocked into rapid action and not only assumed this responsibility as required, but also allocated other responsibilities to workers, with necessary training provided. The health and safety policy was extensively rewritten.

As a result of these improvements, the following changes occurred:

- The accident rate in the organisation fell significantly.
- Any health and safety complaints were responded to and addressed more quickly and thoroughly.
- There was improved consultation between management and all workers.
- With workers perceiving that management were taking health and safety more seriously, the safety culture improved considerably.

Clearly the overall health and safety performance of the organisation had significantly improved. However, it was unfortunate that a worker had to suffer a serious accident before the improvements identified took place.

4.4.2 What the review should consider

There are many elements to be considered in a performance review.

Level of compliance with relevant legal and organisational requirements: the review should consider whether the organisation's health and safety management system complies with legal standards. Any information regarding enforcement action from enforcement agencies such as the Health and Safety Executive (HSE) or environmental health officers would be valuable. It should also consider whether the system complies with organisational requirements.

Accident and incident data, corrective and preventive actions: relevant information here would be the number of:

- fatalities;
- major injury accidents;
- less serious injury accidents; and
- near misses reported.

The review should consider the type of accidents reported, for example, if there have been several instances of crushed feet due to dropping loads during manual handling, and the location of accidents, for example, a considerable number in a particular part of the works premises. Accident trends are useful in showing improving or declining accident prevention performance.

Summarised results of inspections, tours and sampling: other monitoring data such as health surveillance information could also be reviewed. This will show progress with active monitoring initiatives, including the number of inspections, tours and surveys completed against the target number, plus information on key findings and the success of actions taken.

Absences and sickness: sickness absences may be due to:

- occupational accidents;
- occupational ill-health; and
- non-occupational causes.

Occupational accidents and the amount of resulting sickness absence must always be recorded. Such records need to be full and accurate. Without such accuracy, performance reviews based on this data will be of only limited use.

In addition, sickness absence due to occupational ill-health needs to be recorded. However, it may be a considerable time after exposure to a hazardous substance, such as a chemical, that a specific ill-health condition is identified (see 7.1: Hazardous substances). Mental ill-health conditions (see 5.4: Mental ill-health) must also be recorded.

Quality assurance reports (where relevant): these may identify non-compliances in a system that affect health and safety. For example, the wiring in an electric fire intended for sale may be faulty or simply not to the appropriate safety standard. The end user may subsequently suffer an electric shock or be injured from a resulting electrical fire.

Audit results: the information from audits is one of the principal indicators of health and safety performance. The audit is a systematic, critical examination of the health and safety management system and the resulting data will provide information on the level of compliance of many elements of the system.

Another method of evaluating performance is to compare audits carried out at different times and look for significant differences in the data. However, it is important to remember that data may change between audits. For example, new processes and/or equipment may have been introduced, making it difficult to compare.

Monitoring data/records/reports: there may be reports from workplace monitoring (for example, air monitoring for hazardous substances), dynamic personal monitoring (for example, noise monitoring) and medical personal monitoring. The results of any health surveillance may also be reviewed.

Trends in monitoring results can be useful in a performance review; for example, workplace and dynamic personal monitoring of dust levels may show a decrease in exposure. In a previous review, the system for controlling dust levels may have been identified as requiring improvement; the improvements may have taken place and the decreased dust levels provide evidence of an improved system performance.

External communications and complaints: there may be complaints from individuals, such as a neighbour, or from organisations, such as professional bodies and possibly even enforcement agencies. A neighbour may complain of a smell in the air. The smell may be due to emissions from the outlet of a local exhaust ventilation (LEV) system. This in turn may be due to an inadequate system of examination and testing of the LEV system. The testing regime may have improved as a result of the complaint.

Results of participation and consultation: the review should include outputs (such as actions) of regular health and safety meetings.

Whether objectives have been met: the review may question whether some targets were too challenging and others not challenging enough. If objectives have not been met, the reasons why need to be identified and action taken.

It is important to note that where objectives have been met, there should be positive acknowledgement of the workers concerned.

Actions from previous management reviews: if actions have not been carried out, it is important to find out why. For example, it may be that circumstances changed to the extent that the actions were no longer appropriate, or it could be that the workers responsible for such actions were overloaded with work and did not perceive the actions as high priority.

Whatever the reasons, it is important to ensure that follow-up actions are viewed and dealt with as a high priority. Otherwise, the effectiveness of the reviews may be undermined.

Assessing opportunities for improvement and the need for change: as time goes by, changes are inevitable. Some of the changes that could take place include changes in legislation, purchase of new equipment due to technological evolvement, change of premises, employing more or fewer workers, the use of temporary workers for the first time, use of different contractors and organisational change brought about by changes of key workers.

Here are some possible resulting changes in the health and safety management system.

- When new workers are employed, particularly at senior level, the safety policy and organisational roles and responsibilities probably need to be updated. Workers may find that they have different responsibilities, which may lead to more training being required.

- With new equipment, there will certainly be a need for more training for operational and maintenance workers.
- The safety monitoring system may need to change; perhaps different items will need to be added to inspection checklists. There may be different sets of questions that need to appear in audit documents. There will almost certainly be a change in risk assessments. The performance review will need to take account of anything new affecting the organisation or its workers.
- If changes involve recruiting new workers (perhaps due to the organisation's changing requirements), the review should consider new resulting risks and whether risk reassessment adequately ensures that the risks are as low as reasonably practicable to all workers.
- The review should also identify whether the organisation's health and safety policy is still valid and, if not, what changes should be made.
- Changes may require new plans to continue to minimise health and safety risks to workers. The review should consider whether such plans are sufficiently discussed with workers and/or their safety representatives, as adequate discussions will encourage more success in securing improvements if workers are fully involved.

Benchmarking: the performance of the health and safety management system may be compared with others in a similar business sector. The review will consider the organisation's own performance compared with the others.

4.4.3 Reporting on health and safety performance

Results of the performance review should be reported to senior people at board level. The impact of any changes, such as the purchase of new equipment or change of premises, needs to be reported to the board as soon as practicable. This will enable the board to allocate sufficient resources for such changes to be managed.

The board must then ensure that all appropriate workers in the organisation receive the necessary information so that the health and safety management system is kept up to date. For some organisations, the results of the health and safety performance review might be discussed in meetings between management and safety representatives. These results may then be made known to all workers by supervisors and/or worker representatives.

In small organisations, the managing director may simply call a meeting of all workers and communicate the results to them directly.

Many large organisations include a report on safety and health performance in the company annual report. This is becoming increasingly common as the importance of a safe and healthy workforce is realised. A positive report, showing real success in reducing loss, can be a useful tool to help attract potential investors, who will be evaluating the management of the organisation as a whole.

4.4.4 Feeding review outputs into action and development plans as part of continuous improvement

Many organisations have an efficient health and safety management system but nevertheless employ a policy of continuous improvement. This will involve continually reviewing and updating the system.

Continuous improvement requires organisations to keep reviewing and improving the quality of their health and safety management system. This could be done by considering the review outputs and implementing actions associated with audits (maybe by looking at the scope of audits and what happens to audit results).

In particular, this continuous improvement may be achieved by the following:

- Promoting a culture that supports a health and safety management system.
- Promoting the participation of workers in implementing actions aimed at continuous improvement of the health and safety management system.
- Communicating results to workers: there could be improved methods of communication – for example, instead of posting safety information on a noticeboard, all supervisors receive a written communication and then workers are told of the contents in briefings. Another improved communication method might be either implementing suggestions and recommendations from workers or explaining to them why their suggestions or recommendations are not applicable at that time. These actions might improve the motivation of workers because they feel more involved and encourage their participation in continuous improvement.
- Maintaining and retaining documented information for a reasonable time to demonstrate evidence of continuous improvement.
- Hazard identification could perhaps be continuously improved by continually creating and making use of improved checklists, which may then contain more relevant items.
- Reporting systems: these might be improved by being made more 'user-friendly' so that workers are not put off reporting hazards or dangerous occurrences.

KEY POINTS

- Reasons why health and safety performance should be reviewed include ensuring compliance with legislation, identifying areas for improvement, responding to change, learning lessons from accidents and incidents and identifying trends.
- There are a lot of things a review should consider, including data on accidents and incidents and actions taken in response, sickness absence records, audit results, quality assurance reports, monitoring information, external complaints, previous management reviews, targets and objectives, and the need for changes.
- The results of a health and safety performance review should be reported to senior management as soon as possible to ensure resources are allocated to manage any changes, and then shared more widely with the rest of the organisation.
- Aspects of health and safety management systems where continuous improvement might take place include involving all workers and communicating with these workers, promoting a positive culture and making reporting systems easier to use.

ELEMENT 5

PHYSICAL AND PSYCHOLOGICAL HEALTH

5.1: Noise

Syllabus outline

In this section, you will develop an awareness of the following:
- The physical and psychological effects of exposure to noise
- The meaning of commonly used terms: sound pressure, intensity, frequency, the decibel scale, dB(A) and dB(C)
- When exposure should be assessed; comparison of measurements to exposure limits established by recognised standards
- Basic noise control measures, including: isolation, absorption, insulation, damping and silencing; the purpose, use and limitations of personal hearing protection (types, selection, use, maintenance and attenuation factors)
- Role of health surveillance

Introduction

This section examines the ways in which noise can affect physical and psychological health; that is, we will look at how exposure to noise can cause harm to hearing (such as hearing loss) and harm to mental health (such as causing stress).

In this context, noise can be regarded as 'unwanted sound'.

Certain industry sectors are often associated with hearing damage, especially long term. Examples of these are:

- manufacturing (from noise sources such as rotating machinery and compressed air equipment);
- construction (from site equipment such as generators, the process of demolition and activities such as site transport); and
- entertainment (from frequent exposure to loud music and noise from large crowds).

Exposure to noise can result in an immediate temporary loss of hearing (an acute effect) or a gradual long-term effect causing some permanent loss of hearing (chronic effect).

Here you will learn some noise terminology and what you need to consider when carrying out a noise risk assessment.

You will learn about ways to minimise harm from noise exposure following such assessments and the role of health surveillance, which is important to identify the extent of damage to hearing over time.

> **TIP**
>
> The main pieces of legislation relating to noise at work in the UK are:
>
> - the Control of Noise at Work Regulations 2005 (NAW); and
> - the Control of Noise at Work Regulations (Northern Ireland) 2006.
>
> The Health and Safety Executive (HSE) has produced the following guidance to supplement the information in the Regulations:
>
> - *Controlling noise at work* (L108).[1]
>
> The Health and Safety Executive for Northern Ireland (HSENI) has approved L108 for use in Northern Ireland.
>
> We will also refer to the HSE guidance entitled *Sound Solutions*, which contains some examples of engineering techniques to reduce noise.[2]

5.1.1 The physical and psychological effects of exposure to noise

Structure of the ear

In trying to understand hearing damage, you may find it helpful to know about the structure of the ear (see Figure 1).

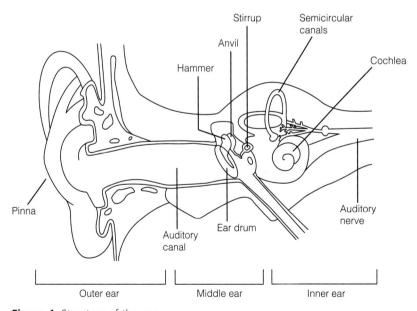

Figure 1: Structure of the ear

The ear has three basic parts: the outer ear, middle ear and inner ear. Sound waves enter the outer ear, then this sound energy proceeds to the middle ear, which contains three small bones called ossicles. The movement of the three ossicles causes the ear drum to vibrate, which, in turn, transmits this energy to the inner ear.

Within the inner ear is the **cochlea**, which contains many thousands of tiny cochlear hair cells. These hair cells move when the sound reaches them, sending a signal via the auditory nerve to the brain, which interprets the sound.

An important feature of the cochlea is that different hair cells respond to different frequencies. When noise enters the cochlea, it causes the cochlear hairs that respond to high frequencies to react first. As the noise progresses further into the cochlea, it loses energy and thus has less effect on the hair cells that respond to lower frequencies. Put simply, high-frequency noise (for example, compressed air) causes more cochlear hair cell damage than low-frequency noise (for example, noise from generators).

Causes of hearing loss

Hearing loss can be due to many causes. Cochlear hair cells become less sensitive as we get older so we may find it more difficult to understand the speech of people with high-pitched voices. This age-related hearing loss is called presbycusis.

The most relevant cause here is exposure to noise. It can occur for social reasons, such as exposure to loud music, but, more significantly, it can occur due to excessive noise at work.

When we talk about 'excessive noise', it is not only the intensity of the noise that is important but the amount of time that a person is exposed to it. This combination of intensity and duration is often referred to as the 'noise dose'. Excessive noise can be considered a large noise dose.

The amount of hearing loss is also influenced by the overall frequency (pitch) of the noise. In practice, this is likely to result from a mixture of frequencies; for example, being exposed to noise from different machines.

Types of hearing damage

The loss of hearing as a result of noise is known as **noise induced hearing loss (NIHL)**.

Two types of damage result in NIHL:

1 conductive hearing loss, which affects the middle ear, particularly the ear drum; and
2 sensorineural hearing loss, which affects the inner ear, particularly the cochlear hair cells.

Damage to cochlear hair cells can cause both temporary and permanent hearing loss. In **temporary hearing loss**, the cochlear hairs tend to bend over but, if little noise exposure follows, the hairs can return to their normal position (so hearing returns to normal).

In **permanent hearing loss**, the noise dose may be large enough for the hairs to break off or become permanently flattened. They never grow again, making the hearing damage permanent. The resulting deafness does not occur over the whole frequency range; the frequency that is particularly affected is 4kHz.

Tinnitus is another type of hearing damage, often referred to as 'ringing in the ears'. It can sometimes be heard as a continuous buzzing sound. It is due to malfunction of the cochlear hairs. Some studies suggest that if someone has developed tinnitus at work, they are more likely to go on to develop NIHL.

Physical and psychological health

Social and psychological consequences

People may experience difficulty hearing, particularly in an environment where many people are talking or there is background noise, such as a TV or music. There can be a lack of sympathy for people who are hard of hearing, so they can become socially isolated. People may no longer enjoy listening to music because high-frequency notes can become distorted.

Other consequences relating to mental health include stress, general anxiety, the loss of ability to concentrate and mental fatigue, which could all lead to impaired judgement and possibly put workers at risk of accident and disease.

5.1.2 The meaning of commonly used terms

Sound is a series of pressure waves travelling through a medium such as air. The **sound pressure level** (SPL) is the ratio of measured pressure to a reference standard pressure (the reference used is the threshold of human hearing). In simple terms, it is a measure of the perception of loudness of a sound, commonly described as the 'noise level'. The unit of measurement of SPL is the **decibel (dB)**.

Sound intensity is the amount of power (that is, energy per second) being transmitted by the sound per square metre of surface it is falling on (such as the ear) or travelling through. It is measured in **watts per square metre (W/m^2)**.

The decibel scale used for SPL provides values of sound output from different sources or situations. It represents the range of sounds to which the ear can respond, with 0dB being allocated to the softest sound it is possible to hear. Some examples of noise output in SPL from different situations and equipment are shown in Table 1.

Situations or equipment	Typical noise level (dB)
Threshold of hearing	0
Quiet bedroom at night	20
Normal conversation at five metres	60
Underground train	80
Medium-sized printing press	85
Circular saw	90
Steel riveter at five metres	100
Pneumatic drill	125
Threshold of pain	140

Table 1: Examples of noise output

> **TIP**
>
> You do not need to learn values such as those in Table 1, but you do need to be able to outline what is meant by the term decibel.

Frequency is a measure of the number of pressure waves passing a given point per second. The higher the frequency, the higher the pitch of the sound. Frequency is measured in hertz (Hz) or kilohertz (kHz); 1kHz = 1000Hz.

People with excellent hearing (usually young people who have not been exposed to high noise doses) can hear from 20Hz up to 20kHz. The ear cannot hear sounds outside this frequency range. Ears are not very sensitive to hearing low frequency sounds; the ear's sensitivity increases with increasing frequency and reaches a maximum between 2kHz and 4kHz before the sensitivity begins to decrease with further increase in frequency.

The **A-weighting**, with a unit of dB(A), mimics the hearing sensitivity of the human ear at various frequencies, so that measured sound in dB is filtered to a different extent at different frequencies. Put simply, dB is the loudness of the sound at source; dB(A) is the loudness of the sound as the ear detects it. For example, a noise source of 81dB at 250Hz (such as from a generator) would result in a value of 72dB(A), while a noise source of 81dB at 2kHz (such as from compressed air) would result in a value of 82dB(A).

Note that the dB scale is NOT linear; it is in fact logarithmic. The range of noise levels we can hear is very large. Converting to logarithms has the effect of squeezing these down to a manageable scale. For doubling of the SPL value, you add 3dB. For example, if you double an SPL of 84dB, the resulting SPL is NOT 168dB but 87dB.

> **TIP**
>
> You *do not* need to know how these dB(A) values are calculated, but you do need to able to outline the meaning of A-weighting. This shows we are less sensitive to 250Hz (that is, the noise sounds quieter to us) whereas at 2kHz, it sounds a bit louder. 'Loudness' is our perception of sound pressure level (SPL) but is not the same as SPL because we have varying sensitivity at different frequencies.

> **TIP**
>
> The term dB(A) is useful in describing exposure to continuous noise. The term dB(C) describes very short duration noise.
>
> dB(C) is used to measure a peak value for a high-intensity noise of short duration; that is, there may be a high noise level for a short period of time. An example would be an explosion, which might be used in quarrying operations. The C-weighting scale is much flatter (that is, it varies much less with frequency) than the A-weighting scale and the weighting is only applied at very low and very high frequencies.

5.1.3 When exposure should be assessed

The NAW Regulations set out a system of noise exposure limits (above which workers must not be exposed) and noise action values (at which certain control actions must be taken). Noise assessment is carried out to evaluate the health and safety risks to workers and indicate what measures should be used to control such risks. In general, noise exposure should be assessed when:

- noisy equipment is in use in the workplace;
- previous assessments have indicated problems;
- information from the manufacturers or suppliers of work equipment includes a warning regarding noise levels (they may also specify values for normal operation of the equipment); and

- workers have complained about noise levels or, more specifically, where audiometric testing has shown that workers have experienced decreased hearing (see section on health surveillance).

If control measures are in place to reduce noise exposure, you should periodically check these to make sure that they are still effective.

Things you should do when carrying out a noise assessment include the following:

- Review the results of any previous noise assessments.
- Find out how many workers are likely to experience exposure to noise.
- Make a judgement about whether any exposure to noise might damage a worker's hearing. For example, if you need to raise your voice to be heard when speaking to someone who is two metres away, there is a risk of damage to hearing.
- Establish the likely frequency and duration of noise.
- Identify any vulnerable workers; for example, workers who may already be suffering some form of hearing loss.
- Establish how many sources produce noise and how they combine to result in total noise exposure. In work situations, workers may be exposed to more than one noise source. Because doubling noise intensity results in a 3dB increase of that noise, two machines of identical noise output could expose workers to a 3dB increase in total sound level. For example, being exposed to two machines each producing a noise intensity of 80dB would result in a total noise output of 83dB. Table 2 shows what happens if the two machines do not produce the same noise output.

Difference in SPL (dB)	0	1	2	3	4	5	6	7	8	9	10 or more
Add to higher level	3	2.5	2	2	2	1.5	1	1	1	0.5	0

Table 2: Adding different sound levels

For example, two noise sources producing SPL values of 80dB and 83dB would result in a noise output of 85dB (83–80 = 3; so referring to the table, 2dB is added to the louder item = 83+2 = 85). Similarly, where source A produces a noise output of 84dB and source B a noise output of 74dB, source B is not loud enough to affect the total noise level, which would be 84dB (84–74 = 10; so 0 is added to the louder source).

The cumulative effects of noise over time

It is unlikely that workers will be exposed to the same level of noise throughout the working day. So how can you determine the resulting noise level over, for example, an eight-hour working day?

As previously discussed, noise dose is a combination of SPL and time; if you double the noise SPL and halve the time, the noise dose will remain the same. For example, a noise SPL of 90dB for two hours is the same as 87dB for four hours and 84dB for eight hours.

Methods of noise measurement

If there is a significant risk that noise exposure may exceed the limits laid down in NAW, then you need to carry out noise measurements.

5.1 Noise

> **TIP**
>
> You do not need to know how to carry out detailed noise measurements, but you do need a basic knowledge of types of instruments used and the information they can provide (see Table 3).

Type of instrument	Measurements usually provided	Comments
Simple SPL meter	dB(A)	Static meter; moderate accuracy
Integrating noise meter	dB; dB(A); average noise exposure over given time period. The symbol used (for equivalent continuous sound level) is L_{Aeq}	Static meter; usually more accurate than a simple SPL meter
Personal noise dosimeter	L_{Aeq}	Personal meter attached to worker
Frequency analyser	dB values at particular frequency bands	Attached to, or contained within, a static meter

Table 3: Noise measuring instruments

> **TIP**
>
> A portable static meter (sometimes referred to as an environmental meter) can measure noise SPL levels at various locations.

When you have carried out noise measurements in your workplace (or seen the results of such measurements carried out by a competent noise specialist), you want to know whether the noise exposures of workers comply with NAW. This means you are comparing the measurements with legal exposure limits.

NAW identifies certain daily or weekly exposure action values (EAVs). For example, there is a lower EAV, where certain control measures are required for daily (or weekly) personal exposures between 80dB(A) and 85dB(A) and a peak sound pressure of 135dB(C). There is also an upper EAV, where more stringent control measures are required for daily (or weekly) personal noise exposures of 85dB(A) and above and a peak sound pressure of 137dB(C).

There is a general duty to reduce the risk of exposure to noise at source or, if this is not reasonably practicable, to reduce the risk of exposure to the lowest level reasonably practicable.

For exposures at or above the lower EAV, there are the following legal requirements under NAW:

- A noise risk assessment must be carried out (reg 5).
- Information, instruction and training must be provided, describing the hazards of exposure to noise and the control to be used to minimise exposure (reg 10).
- If the risk assessment indicates that exposure to the noise is a risk to workers' health, they must be placed under health surveillance (reg 9).

For exposures at or above the upper EAV, NAW requires the following:

- In addition to the requirements for lower EAV, a programme of technical and organisational measures (not including personal hearing protectors, such as those identified under noise control measures later) must be put into place (reg 6). In other words, you must try to reduce exposure below the upper EAV by other means before relying on personal hearing protectors.
- If the use of technical and organisational measures does not reduce exposure below the upper EAV, then personal hearing protectors must be provided.
- If any workplace area is likely to result in exposure to noise above the upper EAV, the area must be designated as a 'hearing protection zone' and must be identified by the mandatory sign shown in Figure 2.

For exposures at or above the lower EAV, but below the upper EAV, the employer need only provide personal hearing protectors following a request by workers. But for exposures at or above the upper EAV, the employer must provide personal hearing protectors if the programme of technical and organisational measures has failed to reduce exposure below the upper EAV, and workers must then properly use such protectors (reg 7).

There are also 'exposure limit values' that must not be exceeded. These are 87dB(A) for daily (or weekly) personal noise exposures and a peak sound pressure of 140dB(C).

The limits are set at levels that aim to minimise damage to hearing caused by excessive exposure to sound energy.

APPLICATION 1

A worker works in a room containing three large printing presses, X, Y and Z. The worker is responsible for ensuring the smooth running of these three presses. A noise survey has been carried out and the worker was found to be exposed to a noise level of 92dB(A) when working at X, 87dB(A) at Y and 89dB(A) at Z.

What is the next action you should take to carry out a noise risk assessment?

Answer: See end of 5.1: Noise

5.1.4 Basic noise control measures

There are four general approaches for controlling noise exposure.

Reducing noise at source

Here are some examples of how noise can be reduced at source:

- Substitute the current work equipment with quieter equipment. This is likely to be the most expensive measure, but might be best practice if the current machinery is old and/or requires regular extensive maintenance, perhaps due to frequent breakdowns.
- Redesign the process; for example, replacing riveting by cold welding.
- Reduce the speed of the machinery (for example, a printing press). Generally, reducing equipment speed will result in a lower noise output.
- Maintain equipment regularly; for example, lubricate moving parts, replace worn-out parts and tighten screws to reduce noise from vibration.
- Silencers can be fitted to equipment where there is exhaust steam or gas, such as silencers fixed to motor vehicles.

Reducing the noise pathway

The 'noise pathway' could not only affect workers directly through the air, but could also be reflected from surfaces in a workroom. Noise could also travel along a floor surface ('structure-borne noise').

> **TIP**
> Structure-borne noise results from impact on a solid surface, causing that surface to vibrate and resulting in the generation of sound waves.

Ways to reduce the noise pathway include the following:

- **Insulation:** it may be possible to enclose equipment with good sound-insulating material fixed to the walls of the enclosure; a door may be needed for access. But this option may not always be practical; for example, it could lead to equipment inside overheating.
- **Absorbent screens:** where it is not practical to enclose a noise source, it may be possible to place screens between the noise source and the receiver.
- **Damping:** it may be possible to mount machines on insulating material such as rubber mountings to reduce the transmission through the floor.
- **Suitable surface:** where it is not practical to use insulating mountings, work equipment could be placed on a soft surface to reduce excessive noise from vibration.
- **Sound baffles:** hanging absorption panels can reduce noise. In large concert halls such panels are used to absorb reverberated noise from the inside walls of the hall.
- **Isolation:** in some circumstances, a soundproof booth can be provided for workers. For example, in a printing works printing presses can be in use for several hours on a long print run. Workers may only need to make occasional adjustments during this time while spending most of their time in the booth.

Organisational methods

It may not always be possible for you to use technical measures to reduce noise exposure, so instead you might use organisational (administrative) measures. Examples are:

- conducting noisy tasks when few workers are present, such as out of normal hours or at weekends;
- workers taking frequent breaks in a quiet area and/or taking longer breaks to reduce the duration of exposure or how often it occurs; and
- making use of job rotation; it may be possible for workers experiencing exposure to loud equipment to move to other, quieter work, with other workers taking their place at noisy equipment.

Use of personal protective equipment (PPE) for noise

The purpose of using PPE is to protect the person wearing it (the so-called 'receiver' of the noise).

PPE will only be used either when there is no other realistic method of sufficiently protecting exposed workers, or to supplement where other engineering or organisational methods fail to provide adequate protection.

Physical and psychological health

A vital property of any hearing PPE is attenuation (that is, the amount by which it reduces the level of noise getting through to the wearer). A supplier of the PPE is required to provide attenuation data. This identifies the amount of hearing protection provided by the PPE at certain frequencies, usually from 63Hz to 8kHz.

There are two types of PPE to protect against noise: ear plugs and ear defenders. Both have advantages and limitations (see Tables 4 and 5).

Ear plugs

These fit directly into the ear canal. They are often disposable, so are particularly suitable for those visiting noisy areas infrequently.

Non-disposable ear plugs need to be available in different sizes to ensure a good fit, whereas disposable plugs are moulded into the ear (see Table 4).

Ear defenders

These are also known as ear muffs. They consist of rigid cups that fit over the ears and are held in place with a headband. The cups are filled with material that absorbs noise efficiently (see Table 5).

Advantages and limitations of the two types of PPE

ADVANTAGES	LIMITATIONS
Cheap	There may be hygiene problems, for example, inserting plugs with dirty hands
Comfortable to wear for long periods	Difficult for a supervisor to see if the plugs are being worn
Do not interfere with other types of PPE such as respirators	Can interfere with communication
No maintenance required for disposable plugs	Training required: inserting plugs incorrectly will considerably reduce protection

Table 4: Advantages and limitations of ear plugs

ADVANTAGES	LIMITATIONS
More hygienic than plugs	More expensive than plugs
Easy for a supervisor to see if the muffs are being worn	Can be uncomfortable wearing them for a long time, particularly in hot conditions
Can be designed not to interfere with communication	Can interfere with other types of PPE such as respirators
Workers can have dedicated muffs, which may encourage the workers to store and care for them properly	Some maintenance will be required

Table 5: Advantages and limitations of ear defenders

Some PPE will need occasional maintenance (see Table 6).

Type of PPE	Comments/maintenance requirements
Disposable plugs	None needed as this PPE is discarded after use
Non-disposable plugs	• Clean after use • Store in clean container when not in use
Ear defenders (ear muffs)	• Replace seals periodically • Check condition of the cups, for example, for cracks or holes • Avoid twisting or bending the headband • Store in a clean environment

Table 6: PPE maintenance requirements

Where the sign in Figure 2 applies to an area, hearing protection must be worn in that area.

Selecting hearing protection

When you need to select hearing protection, there are a number of things to consider, including:

- noise level (you may have the results of noise measurements);
- main frequencies of the noise (you may have the results of a frequency analysis);
- time of exposure to the noise (you may know this from a production schedule or by simply observing/speaking with workers);
- work patterns, including whether the exposure is constant (such as for a machine operator) or variable (such as for maintenance workers);
- characteristics of workers, for example, whether long hair interferes with the fit of the hearing protection; and
- whether other PPE is worn and if it is compatible with hearing protection.

Hearing protection must be worn

Figure 2: Designated hearing protection zone sign

You should also be aware of why hearing protection might not be providing adequate protection. Some causes might be:

- it may be damaged (for example, ear defenders have damaged cups);
- the attenuation is not sufficient, perhaps due to wrong selection of PPE;
- it is not being worn correctly due to lack of training (for example, on how to insert disposable plugs into the ear);
- it is not compatible with other PPE; and
- workers are removing the PPE before they should, for example, to speak to someone or because it is uncomfortable.

APPLICATION 2

The worker from Application 1 has been found to be exposed to noise levels that could cause significant hearing damage. At present, management provide hearing protection (PPE) on request.

- Is this good enough to control exposure of the worker to the noise?
- If it is not good enough, what would you suggest as possible control measures?

A possible answer is given at the end of 5.1: Noise.

5.1.5 The role of health surveillance (as applied to the protection of hearing)

You may not need to do health surveillance yourself but it is important to know about it.

What is health surveillance?

It is a programme of systematic health checks so that early signs and symptoms of hearing loss can be identified and then for actions to be taken to prevent such signs and symptoms from getting worse. The surveillance process will normally include hearing checks using audiometry.

Audiometry basically measures the sensitivity of a person's hearing over a standard range of frequencies.

The results are then compared to those expected of a normal healthy adult of similar age.

When should health surveillance take place?

It should take place for workers prior to starting work for an organisation where noise exposures are likely to be high. It should also take place when noise exposure reaches levels where hearing loss may occur; these levels are identified in NAW. It may then be carried out on a regular basis to determine whether there has been a decrease with time in the level of that worker's hearing at particular frequencies.

Audiometric testing provides analysis of hearing ability, especially whether there is hearing loss at specific frequencies. For example, where there is gradual increasing hearing loss from about 6kHz to higher frequencies, this may simply indicate hearing loss due to the ageing process. However, if there is a sudden significant hearing loss at 4kHz followed by a recovery at higher frequencies, this is often clear evidence of NIHL. Naturally, some audiograms (a chart showing the hearing loss at various frequencies) could show a combination of presbycusis and NIHL for older workers.

The content of initial health surveillance might include:

- a pre-employment medical examination to find any indications that workers might have pre-existing conditions; and
- pre-employment audiometric testing to determine any existing hearing problems, as workers may not always be aware of such problems if the hearing deficiency is not too great. This deficiency may have been caused by work carried out in previous employment or by social activities.

Further audiometric testing may then be regularly carried out to determine any worsening symptoms.

The results of such health surveillance should be kept in a confidential medical file. Workers should be informed of the surveillance results and the significance of these.

To summarise, four important parts of the role of initial health surveillance are to:

- establish a person's overall baseline hearing ability;
- identify frequencies of their hearing that have already been affected;
- identify the amount of hearing loss already present; and
- enable them or their employer to select appropriate hearing protection. For example, if there is hearing loss in the frequency range 2–4kHz, the PPE selected should have the greatest attenuation in this frequency range.

Note: sometimes another part of the role of health surveillance is that it can identify any deficiencies in noise control measures at the workplace, but this should be viewed with caution. Hearing loss cannot definitively be attributed to noise exposure at work or social noise exposure outside of work, for example, resulting from exposure to loud music.

KEY POINTS

- There are various ways in which exposure to excessive noise levels can cause damage to hearing and result in psychological effects such as stress.
- There are many specific terms used when referring to noise; particularly important are dB, dB(A) and dB(C).
- Assessment of noise exposure will help to inform workers of the level of risk to which they are exposed from noise.
- The preferred method to control exposure to noise is to reduce noise at the source. If this is not practical, then the transmission of noise from the source to the worker should be reduced. The next step is to consider organisational methods of control.
- If these methods are neither practical nor adequate, then PPE needs to be used. This is very much a last resort measure.
- Two types of hearing protection, ear plugs and ear defenders (muffs), are identified and their various advantages and limitations should be considered before selecting the most appropriate type for the worker.
- Health surveillance requires the use of audiometry as a technique for identifying baseline hearing for new workers. The technique can then be used to check their hearing again to find out whether their hearing has deteriorated since starting work.

ANSWERS TO APPLICATIONS

Application 1 answer: Find out how long the worker works at each machine.

Application 2 answer: You first need to consider methods *other than* PPE:

- Totally enclosing the three printing machines could be a good option, although the worker would still need access to the machines at times. Such an enclosure would be expensive.
- A less expensive method would be to construct a 'noise haven' for the worker to be inside when not having to tend to the machines.
- Both of these options would take time to construct, so *in the short term* there would be reliance on PPE. Its use should be mandatory; the employer must provide it and the workers must wear it. The PPE would need to have the appropriate attenuation factor.

References

[1] HSE, *Controlling noise at work. The Control of Noise at Work Regulations 2005. Guidance on Regulations* (L108, 3rd edition, 2021) (www.hse.gov.uk)

[2] HSE, 'Sound Solutions case studies' (www.hse.gov.uk)

5.2: Vibration

Syllabus outline

In this section, you will develop an awareness of the following:

- The effects on the body of exposure to hand-arm vibration and whole body vibration
- When exposure should be assessed; comparison of measurements to exposure limits established by recognised standards
- Basic vibration control measures, including: alternative methods of working (mechanisation where possible); low-vibration emission tools; selection of suitable equipment; maintenance programmes; limiting the time workers are exposed to vibration (use of rotas, planning work to avoid long periods of exposure); suitable PPE
- Role of health surveillance

Introduction

This section will first describe hand-arm vibration and whole body vibration, give examples of where they may occur and identify ill-health effects on the body.

We will examine the factors to be considered when carrying out a health risk assessment for vibration, including the circumstances when exposure should be assessed.

We then consider in detail the control measures (sometimes called precautions) available to help reduce the risks of ill-health when exposed to vibration.

Finally, we look briefly at vibration measurement and consider the role of health surveillance, how it is carried out and how the results may affect the control measures you wish to introduce to minimise ill-health effects from exposure to vibration.

TIP

The main pieces of legislation relating to the control of vibration are:

- the Control of Vibration at Work Regulations 2005; and
- the Control of Vibration at Work Regulations (Northern Ireland) 2005.

The Health and Safety Executive (HSE) has published a range of guidance on controlling vibration risks:

- *Vibration solutions: Practical ways to reduce the risk of hand-arm vibration injury* (HSG170);[1]
- *Hand-arm vibration. The Control of Vibration at Work Regulations 2005. Guidance on Regulations* (L140);[2] and
- *Whole-body vibration. The Control of Vibration at Work Regulations 2005. Guidance on Regulations* (L141).[3]

5.2.1 The effects on the body of exposure to vibration

There are many types of vibration, but the two of interest to us are whole body vibration (WBV) and hand-arm vibration (HAV). HAV is also commonly called hand-transmitted vibration (HTV) but we will use the term HAV for consistency.

Whole body vibration (WBV)

WBV can occur when the body is supported on a surface that is vibrating, which occurs in all forms of transport and when working near vibrating industrial machinery. For example, a worker experiences WBV when they are in contact with a supporting surface (such as the floor or a seat) that vibration is passing through.

An example of a work situation where WBV could occur is the shaking experienced through a (poorly adjusted) seat when driving lorries, tractors or bulldozers across rough terrain.

Figure 1: WBV can result from driving on rough terrain

Effects of WBV
Ill-health effects that can result from exposure to WBV include:

- back pain;
- damage to discs in the back;
- damage to the vertebrae of the back;
- osteoarthritis;
- blurring of vision; and
- nausea.

Hand-arm vibration (HAV)

HAV is experienced through the hands and arms when holding certain types of hand-held power tools, such as rotary grinding tools, chainsaws, angle grinders, hand-held percussion drills and road-breaking tools. It can also occur with hand-guided powered equipment (such as lawnmowers) and even from holding work pieces that are being worked on by machinery (such as holding a work piece against a grinding wheel).

Effects of HAV
The main ill-health effects of HAV are hand-arm vibration syndrome (HAVS) and carpal tunnel syndrome (CTS). The parts of the body affected are fingers, hands, wrists and arms. The effects result from damage to blood vessels and nerves.

The main signs and symptoms of HAVS are:

- numbness and tingling in the fingers (the first sign of HAVS);
- painful joints;
- vibration white finger (VWF): this is the blanching of fingertips. If work continues with the tool causing the VWF, the blanching can spread through the whole finger, and if work still continues, then in extreme cases gangrene could occur, where blood supply has been completely cut off to the finger;
- reduced strength to the hand so the worker may be unable to grip objects properly; and

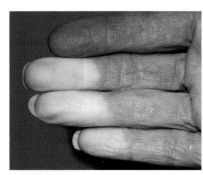

Figure 2: HAVS can cut blood supply to fingers

- reduced sensory perception, meaning the worker may not be able to feel if an object is hot or sharp, which could result in injury, or the worker might be unaware of touching an object at all.

CTS occurs when there is pressure on the median nerve that runs from the forearm to the palm of the hand and can result from an object being repeatedly gripped or from awkward hand movements. This can lead to pain and weakness in the hand.

5.2.2 When exposure to vibration should be assessed

If workers (or others) are frequently exposed to WBV or HAV and obvious steps do not eliminate the exposure, then employers should assess the hazard and risk to health and safety. Following this, employers should introduce prevention and control measures to help reduce risk to the lowest practicable level by all appropriate means.

When exposure to WBV should be assessed

If workers are carrying out activities where there is a risk from exposure to WBV, these activities should be assessed. Circumstances where exposure is likely to be high and should be assessed include when:

- workers are drivers of heavy vehicles such as lorries;
- machine or vehicle manufacturers warn that the machine or vehicle presents risks of WBV;
- it is necessary to drive over uneven ground, for example, in tractors;
- driving takes place throughout the working day;
- vehicles are old;
- workers have complained of suffering back pain, or health surveillance records show that workers have a history of back pain;
- workers are inexperienced in driving the particular type of vehicle they are required to drive; and
- machines or vehicles are not suitable for the tasks for which they are being used; and workers are using unsuitable techniques, such as driving too fast.

When exposure to HAV should be assessed

Circumstances where exposure to HAV is likely to be high and should be assessed include when:

- workers are using hand-held vibrating tools, particularly if this work is of long duration;
- hand-held saws are in use on concrete and metal;
- chainsaws are being used, for example, in forestry work;
- pedestrian-controlled equipment is in use, for example, mowing machines during the maintenance of parks and gardens;
- grinders are being used, for example, in construction work;
- power hammers are in use, for example, in road-breaking operations;
- impact drills are used, for example, drilling into brick;
- workers have complained of numbness or pain in their hands, or health surveillance records show that workers have a history of HAVS; and
- the equipment manufacturer has included warnings, advising that assessment should take place.

There are a number of things you need to do or consider when assessing the risks:

- Identify the sources of vibration (we have already identified many common sources of both WBV and HAV).
- Identify the workers at risk.
- Consider legal limits of vibration exposure and guidance documents.
- Consider the age of the equipment. Older equipment is likely to produce more vibration.
- Consider the time of exposure in terms of the frequency (how often) and duration of each exposure.
- Consider the magnitude, frequency and direction of the vibration. We will discuss the characteristics of vibration later.
- Consider the working temperature. Ill-health symptoms are often worse in cold conditions.
- Look at information contained in manuals produced by the manufacturer of the equipment in use.
- Consider the force needed to use the tool. For example, a drill may need different levels of force to drill through different types of material.
- Consider the adequacy and effectiveness of control measures currently in use.

Measurement of exposure and comparison to exposure limits established by recognised standards

To assess exposure, you may need to estimate actual vibration exposure. This could be done using vibration emission data from the supplier or manufacturer of the vehicle or equipment. But if this is not available, or is unreliable, you may need to measure it. Measuring vibration would need the services of a technically competent person to make sure the measurements are reliable and made using approved methods.

To understand vibration measurement (and the measurement units used) it is worth looking at the nature of vibration. When an object vibrates, it is rapidly accelerating (that is, quickly changing its speed and/or direction). In simple cases you can think of it as the object oscillating about an equilibrium point (the point where it would sit if it stopped vibrating), accelerating first in one direction and then back again.

Real objects can oscillate in different directions all at the same time and at a range of frequencies.

When measuring vibration, there are three basic parameters:

1. Vibration magnitude: a measure of the acceleration and, because it is rapidly changing, a kind of weighted average value is taken over a range of frequencies. The units are metres per second squared, m/s^2, also written as ms^{-2}.
2. Vibration frequency: how many times a second it oscillates, which is measured in hertz (Hz).
3. Vibration direction: conventionally referenced to three orthogonal axes (x, y and z) at right angles to each other.

Acceleration is measured using a device called an accelerometer. Because exposure standards are expressed as daily (eight-hour) equivalent vibration exposure, the measurements taken have to be adjusted to the equivalent exposure over a working day (eight hours), taking account of exposure patterns.

Physical and psychological health

For both WBV and HAV, the daily exposure limit values apply when action to control vibration needs to be taken. That is because you should not just wait until it gets to the limit before you do anything.

Estimated or measured vibration exposure levels are compared against the standards set out in the Vibration Regulations (see earlier Tip box).

There are two standards for both WBV and HAV – these are the daily exposure action value (EAV) and the daily exposure limit value (ELV).

The EAV is the amount of daily exposure to vibration above which an employer is required to take action to reduce the risk of their workers being exposed (or likely being exposed) above this level.

If this occurs, the employer must:

- eliminate the risk at source where reasonably practicable, or reduce exposure to the lowest level reasonably practicable by setting up a programme that includes appropriate technical and organisational measures;
- provide health surveillance for any workers exposed above the EAV; and
- provide information, instruction and training to workers exposed above the EAV about the risks to their health and the measures in place to control those risks.

The ELV is the maximum amount of vibration an employee may be exposed to on any single day. If workers are exposed to an amount of vibration exceeding the ELV, the employer must, in addition to the control measures where the EAV is exceeded:

- immediately take steps to reduce the exposure below the ELV;
- identify the reasons for the ELV being exceeded; and
- modify control measures to prevent the ELV being exceeded again.

For WBV, the ELV is $1.15 m/s^2 A(8)$ and the EAV is $0.5 m/s^2 A(8)$.

For HAV, the ELV is $5 m/s^2 A(8)$ and the EAV is $2.5 m/s^2 A(8)$.

> **TIP**
>
> Some frequencies are more damaging to the body than others. The most damaging frequencies contribute a more significant individual vibration risk to the total vibration risk than other, less damaging frequencies.

5.2.3 Basic vibration control measures

WBV control measures

Control measures to reduce the risk from WBV may include the following:

- Vehicles should be selected that have low vibration levels.
- Vehicles should be well maintained to reduce vibration.
- The seat in the vehicle should be well sprung and adjustable (and adjusted correctly).
- Where possible, vehicle routes should be in good condition and avoid rough ground.
- As it may not always be possible to avoid rough ground, drivers should operate their vehicles at suitable speeds for the ground conditions. They may need to drive slowly.

- Drivers should wear appropriate footwear when operating their vehicles.
- If possible, there should be job rotation for drivers. If this is not possible, then drivers should be provided with suitable breaks. The frequency and duration of their driving should be closely monitored.
- Drivers should receive training on the hazards of WBV, the control measures in place to help reduce WBV and the need to report instances of back pain following driving to their employers.

HAV control measures

Control measures to help reduce the risk of ill-health when using (powered) hand-held vibrating tools include the following:

- In the first instance, try to change the process to avoid using hand-held tools, for example, by mechanisation.
- Operate a purchasing policy that means only tools with low-vibration characteristics are bought for the organisation.
- Consider substitution. If there are tools with a high vibration magnitude in use, replace them with tools that have a lower vibration magnitude.
- If it is not feasible to replace any particular tools with high-vibration characteristics, then identify those tools and label them clearly.
- Ensure there is a planned maintenance policy in force.
- Replace worn parts of tools and sharpen any drill bits.
- Ensure equipment is used correctly; this may require effective supervision.
- Try to introduce job rotation so that each worker experiences less exposure time to vibrating tools.
- Add heated grips to tools.
- Provide warm gloves and instruct workers to wear them to encourage efficient circulation of blood in their hands.
- If any workers display symptoms of HAVS, refer them to medical personnel immediately.
- Provide training to workers on the risks of working with hand-held tools and the precautions they should be taking to minimise the risk of ill-health.

Examples of case studies regarding control measures can be found in the HSE publication *Vibration solutions*.[4]

5.2.4 Role of health surveillance

If a worker is experiencing signs and symptoms of HAVS or carpal tunnel syndrome (CTS), or has been diagnosed with these conditions, then health surveillance should be provided, even if exposure appears to be below the limits set by national/international standards.

Health surveillance should also be provided when workers are likely to be exposed to significant levels (that is, at or above limits) of vibration from hand-held equipment. The surveillance should include asking workers about any symptoms they are experiencing and examining them for symptoms of the possible neurological effects of vibration, such as numbness and elevated sensory thresholds for temperature or pain.

In practice, this can include a simple test such as gently stabbing the hand and fingers of the worker with a blunt needle and asking them to say when they can feel a stabbing sensation.

Physical and psychological health

Where the symptoms of HAVS appear to exist among workers, the employer may be advised that current control measures are insufficient. The employer should then review the vibration risk assessment and improve control measures where necessary.

For WBV, where the risk of developing symptoms, such as back pain, is higher than average, simple health checks are usually done. This can include regular completion of health questionnaires and early reporting of back pain symptoms. During such checks, workers can also be advised about how to avoid things that can make their back pain worse, such as incorrect posture in seated jobs and incorrect lifting techniques.

KEY POINTS

- Two types of vibration in the workplace are whole body vibration (WBV) and hand-arm vibration (HAV).
- Common sources of WBV and HAV include driving across rough terrain, working near vibrating machinery and using hand-held power tools and hand-guided powered equipment.
- The likely ill-health effects include back pain and damage, blurred vision, nausea, hand-arm vibration syndrome and carpal tunnel syndrome.
- A vibration risk assessment would be appropriate when, for example, vehicles are heavy, old or unsuitable for the task; drivers are inexperienced, drive for long periods or over uneven ground; and certain hand-held tools and pedestrian-controlled equipment are being used. Workers' history of back or hand pain or numbness and manufacturer warnings might also prompt an assessment.
- Things to consider in assessing vibration risk include time, frequency and level of exposure, how to measure vibration and comparison with standards that set daily exposure limits.
- Control measures to minimise risks to ill-health from vibration include selection and maintenance of vehicles and tools, training for workers and provision of appropriate clothing.
- Health surveillance with respect to vibration may be required if workers develop related symptoms of ill-health, and this can involve simple health checks and advice.

References

[1]HSE, *Vibration solutions: Practical ways to reduce the risk of hand-arm vibration injury* (HSG170, 1997) (www.hse.gov.uk)

[2]HSE, *Hand-arm vibration. The Control of Vibration at Work Regulations 2005. Guidance on Regulations* (L140, 2nd edition, 2019) (www.hse.gov.uk)

[3]HSE, *Whole-body vibration. The Control of Vibration at Work Regulations 2005. Guidance on Regulations* (L141, 2005) (www.hse.gov.uk)

[4]See note 1

5.3: Radiation

Syllabus outline

In this section, you will develop an awareness of the following:
- The types of, and differences between, non-ionising and ionising radiation (including radon) and their health effects
- Typical occupational sources of non-ionising and ionising radiation
- The basic ways of controlling exposures to non-ionising and ionising radiation
- Basic radiation protection strategies, including the role of the competent person in the workplace
- The role of monitoring and health surveillance

Introduction

Radiation is a complex topic, often handled by specialists, so do not be surprised if you find some of the concepts difficult at first. Firstly, we will explain the nature of radiation and discuss the two forms of radiation (as particles and waves).

We will examine, in some detail, different types of radiation, including a definition of ionising radiation (and radon as an example of this) and the differences between ionising and non-ionising radiation. We will identify common occupational sources of these two types.

We will then cover basic radiation protection strategies and control measures. We will look at the role of the competent persons (the radiation protection adviser and the radiation protection supervisor) in managing ionising radiation.

Finally, we will discuss health monitoring and surveillance.

TIP

The main legislation covering ionising radiation is:
- the Ionising Radiations Regulations 2017 (IRR); and
- the Ionising Radiations Regulations (Northern Ireland) 2017.

The Health and Safety Executive (HSE) has produced the following guidance to supplement the information in these Regulations:

- *Work with ionising radiation. Ionising Radiation Regulations 2017. Approved Code of Practice and guidance* (L121).[1]

The Health and Safety Executive for Northern Ireland (HSENI) has approved L121 for use in Northern Ireland.

> Reference will also be made to the Health and Safety (Safety Signs and Signals) Regulations 1996 and associated HSE guidance:
>
> - *Safety signs and signals. The Health and Safety (Safety Signs and Signals) Regulations. Guidance on Regulations.* (L64).[2]
>
> The main legislation covering non-ionising radiation is:
>
> - the Control of Artificial Optical Radiation at Work Regulations 2010 (CAOR); and
> - the Control of Artificial Optical Radiation at Work Regulations (Northern Ireland) 2010.

5.3.1 The types of, and differences between, non-ionising and ionising radiation and their health effects

The nature of radiation

Radiation is simply energy that is travelling in the form of waves or particles (we look at these two different forms in more detail later). When this radiation strikes matter, it can transfer the energy. A simple example is the way we feel the warmth of strong light from the summer sun when it strikes our skin.

> **TIP**
>
> When we talk about terms used in radiation, we are often considering very large or very small numbers. Therefore, let us also introduce the idea of using powers of 10.
>
> Powers of 10 are used as a shorthand for describing these large or small numbers. Example 1: 1,000,000 (ie 1 million) can be written as 10^6; that means 1 with 6 zeros.
> Example 2: 1/100,000,000 can be written as 10^{-8}; less cumbersome than 0.00000001.
> Example 3: the mass of an electron is about 9.109×10^{-31} kilograms; a very small weight.
>
> We may also come across other terms such micrometre and nanometre. A micrometre is 10^{-6} metres; a nanometre is 10^{-9} metres.

Non-ionising and ionising radiation

To understand the idea of ionisation, it is helpful to first understand the nature of atoms, which are the primary building blocks of all physical matter (including our own bodies).

In simple terms, atoms consist of three types of (sub-atomic) particles: protons, electrons and neutrons. In terms of structure, atoms are a bit like satellites or a moon orbiting a planet. There is a nucleus (like the planet) that contains the protons and neutrons while the electrons orbit around the nucleus (like a satellite or moon).

The two properties of interest of these particles are their relative mass and their electrical charge (see Table 1).

Particle	Relative mass	Charge
Proton	1	Positive
Electron	1/1840	Negative
Neutron	1	Neutral

Table 1: Properties of sub-atomic particles

From the table, you will notice the differences in electrical charge between these particles. All these particles have very little mass in real terms (see the example we gave earlier of electron mass), but protons and neutrons are much heavier than electrons.

Figure 1 shows the structure of the helium atom. The red balls represent protons and the yellow balls represent neutrons, both of which make up the nucleus of the atom. The blue balls are the circulating electrons.

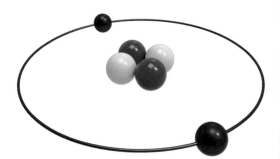

Figure 1: A helium atom

There are over 100 different types of atom that make up matter. We call these 'elements' because they represent the simplest forms that cannot be chemically separated into anything simpler (but they can be chemically combined to form compounds). You will have heard of many of them, such as hydrogen, helium, sodium and oxygen. These elements differ in the number of protons they contain; this is called their atomic number. In their normal states, atoms have the same number of protons and electrons, so are electrically neutral.

Examples, in their normal forms, are the following:

- Hydrogen (chemical symbol H) has 1 proton, 1 electron and no neutrons.
- Helium (chemical symbol He) has 2 protons, 2 neutrons and 2 electrons.
- Sodium (chemical symbol Na) has 11 protons, 11 electrons and 12 neutrons.

Atoms can be broken up if they are exposed to enough energy, ejecting the particles from which they are made. For example, if a sodium atom (symbol, Na) is exposed to energy that is large enough, it can completely separate one of the (negatively charged) electrons of the sodium atom from the rest of the atom; this atom has the same 11 protons and 12 neutrons but now has only 10 electrons, so is now overall positively charged (and is called an 'ion', written as Na^+). This process of removing an electron and forming a positively charged ion is called 'ionisation'. We say that the original sodium atom (Na) has been 'ionised'.

Some radiation (whether it is waves or particles) can cause ionisation in matter (and is therefore said to be ionising) if it has sufficient energy. In terms of human health, the issue is whether it causes ionisation in human cells.

> **DEFINITION**
> *The IRR defines **ionising radiation** as the transfer of energy in the form of particles or electromagnetic waves of a wavelength of 100 nanometres (nm) or less or a frequency of 3×10^{15} hertz (Hz) or more capable of producing ions directly or indirectly.*

Physical and psychological health

> **TIP**
>
> In simple practical terms, ionising radiation has sufficient energy to cause the ionisation of cells in the human body – this can result in undesirable biological effects such as the modification of DNA in the human body.
>
> Note the values in Table 2, where 100nm (10^{-7} metres) and 3×10^{15} Hz are identified as the borderline between ionising and non-ionising radiation.

We now look at two different forms of radiation in more detail: waves and particles.

Wave radiation

Wave radiation has several characteristics. Its wave form has a repeating pattern of peaks and troughs. Going from peak to peak (or trough to trough) is called a cycle (because it brings you back to where you started – the peak or trough). The distance between the maximum of two adjacent peaks (or the minimum of two adjacent troughs) is called the wavelength.

Because the wave travels through space, if you were at a fixed point, you would see the wave keep cycling through the peaks and troughs. The rate at which it does this (the number of cycles passing by per second) is called the frequency. This is illustrated in Figure 2.

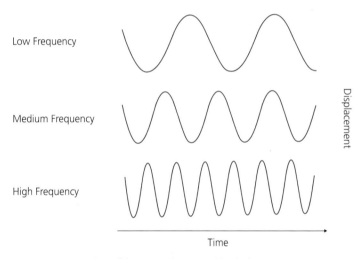

Figure 2: Examples of low-, medium- and high-frequency waves

Figure 3 shows parts of a wave.

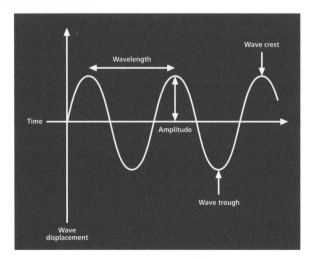

Figure 3: Wavelength and variation of wave displacement with time

For the electromagnetic radiation spectrum (which includes radio waves, infrared, visible, ultraviolet and X-rays) the relationship between frequency (f) and wavelength (λ) is given by:

$$c = f \lambda$$

Where:

c = the speed of light (in a vacuum), which is 3×10^8 metres/second

f = the frequency of the radiation in cycles per second (given the units hertz or Hz for short)

λ = the wavelength of the radiation in metres

c is a constant value; therefore from this equation you can see that as the frequency becomes larger the wavelength becomes proportionately smaller; therefore if you know one parameter you can calculate the other.

Table 2 shows some types of radiation together with typical frequencies and wavelengths.

Type of radiation	Wavelength (metres)	Frequency (Hz)
Gamma rays	10^{-12}	3×10^{20}
X-rays	10^{-9}	3×10^{17}
[Borderline between ionising and non-ionising radiation]	1×10^{-7}	3×10^{15}
Ultraviolet (UV)	Between 1×10^{-7} and 4×10^{-7}	Between 3×10^{15} and 7.5×10^{14}
Visible (violet light)	4×10^{-7}	7.5×10^{14}
Visible (red light)	7×10^{-7}	4×10^{14}
Infrared (IR)	1×10^{-6}	3×10^{14}
Microwave	10^{-2}	3×10^{10}

Table 2: Types of radiation

Table 2 also shows the threshold at which they have enough energy to cause ionisation. Energies with a frequency of greater than 3×10^{15} Hz or a wavelength of less than 1×10^{-7} metres are ionising radiation and those below the borderline in the table are non-ionising radiation. You will notice that ionising radiation has more energy, a higher frequency and shorter wavelength than non-ionising.

Particle radiation

We have considered types of wave radiation (ionising and non-ionising); we now need to briefly discuss some types of particulate ionising radiation:

- Alpha particles: these are helium nuclei containing 2 protons and 2 neutrons, but no electrons; therefore an alpha particle has a double positive charge and can be written as He^{2+}. In atomic terms, this is a heavy particle. Alpha particles are strongly ionising but cannot penetrate very far. For example, they can be stopped by a sheet of paper or the outer layer of skin. They are therefore most dangerous if they get inside the body, for example, if inhaled or ingested. Alpha particles move slowly around the body, causing intense local damage. To summarise: although alpha particles are easily stopped by a sheet of paper in air, the particles cause severe damage if they enter the body, with the damage being concentrated in a small area (see the example of Alexander Litvinenko given later).

- Beta particles: these are fast-moving electrons and penetrate further than alpha particles, needing a sheet of plastic to stop them. If ingested, these would cause less severe but more widespread damage to the body.
- Neutrons: these are very energetic particles and penetrate much further through matter, needing thick concrete to stop them.

Figure 4 shows the penetrating power of the various particles (and some types of wave radiation).

TYPES OF RADIATION AND PENETRATION

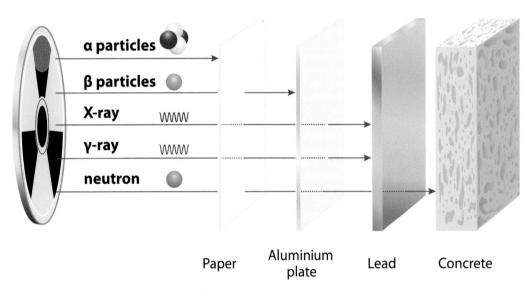

Figure 4: The penetrating power of different particles and waves

Particle radiation is both naturally occurring and can also be made artificially. Some atoms are naturally radioactive. This means they are unstable and can randomly eject particles; we call this process 'radioactive decay'. These decay products are themselves also often radioactive and decay further, again emitting particles.

For example, Alexander Litvinenko (a former Russian spy who became a British citizen) was murdered when his cup of coffee was poisoned with polonium (an alpha particle-emitting compound). Radon gas is another good example of this and is a widespread, naturally occurring source of ionising radiation (particularly alpha particles). The fact that radon is a gas means that it can easily enter our bodies simply by breathing it in. We will look at radon in more detail later.

Health effects from exposure to ionising radiation

- One of the first effects can be erythema (reddening of the skin).
- Further exposure can cause skin burns.
- There is a significant risk of cataracts in the eyes.
- Radiation sickness can occur. This is another of the first signs of radiation poisoning, because the gastrointestinal tract is very sensitive to ionising radiation. The symptoms are nausea and vomiting.

Other possible effects include:

- hair loss;
- cancer: such as cancer of the skin or lungs or of the white blood cells (leucocytes – hence called 'leukaemia'); and
- genetic cell damage, causing birth defects.

Health effects from exposure to non-ionising radiation

The type of ill-health effects that occur will depend on the type of non-ionising radiation and, as discussed earlier, the frequency or wavelength of the non-ionising radiation.

Exposure to UV radiation can cause the following:

- 'arc eye', for example, from exposure to arc welding UV light;
- skin affected by sunburn;
- skin cancer (resulting from long-term exposure); and
- premature ageing of the skin.

Other health effects could be:

- damage to the cornea of the eye (from infrared (IR) radiation);
- cataracts in the eyes (from long-term exposure to IR radiation); and
- damage to internal organs such as the liver and kidneys (from microwave radiation).

5.3.2 Typical occupational sources of non-ionising and ionising radiation

We now turn our attention to possible occupational sources of radiation.

Occupational sources of ionising radiation

- Gamma rays: these are often used to sterilise food so that the food will degrade more slowly when stored.
- X-rays: these are extensively used in X-ray machines found in hospitals and dental surgeries.
  Note: gamma rays and X-rays are similar. The main difference is their origin. Gamma rays are emitted during radioactive decay of the nucleus, whereas X-rays originate from outside the nucleus (energy changes of the electrons orbiting the nucleus). While both can occur naturally, X-rays are most commonly produced artificially using X-ray machines).
- Alpha particles: these are emitted in smoke detectors (widely used in buildings). Another common source of alpha particles is radon gas.
- Neutrons and beta particles: these are emitted in nuclear reactors.

Occupational sources of non-ionising radiation

- Ultraviolet (UV): arc welding. Also, the sun could be considered as an occupational source for lifeguards on beaches. Other examples of UV radiation are used in the printing industry to dry inks and as a sterilising agent in water systems to kill *Legionella* bacteria.

- Visible: artificial lighting. Also display screen equipment for computer operators.
- Infrared (IR): any location where there is a significant heat source, for example, ovens where chefs work in industrial kitchens and furnaces where foundry workers are employed.
- Microwaves: microwave cookers; whereas the risk is quite low for domestic cookers it can be significant for industrial cookers. Also, continuous use of mobile phones may present some risk of microwave exposure. Microwave energy is in general use for point-to-point communication and in electron spin resonance spectroscopy. Finally, if you have ever been caught speeding by the police, it may be that the speed of your vehicle was checked using a radar speed gun emitting microwaves.
- Radiofrequency (RF): aerials on top of vehicles can emit this type of radiation and lead to skin burns. Radio frequency transmitter towers may also be a risk.
- Lasers: the term 'laser' is an abbreviation for 'light amplification by stimulated emission of radiation'. Lasers produce a narrow beam of energy; therefore, a high amount of energy can be directed onto a very small area. The level of hazard is determined by the power output of the laser. The higher the power output of a laser, the greater its hazard. Low-powered lasers are used as supermarket checkout scanners; examples of high-powered laser use are in eye surgery and accurate cutting of metal.

Radon

Although radon is a chemically inert gas, it is radioactive. The main danger is as a source of ionising radiation (particularly alpha particles). Radon is widespread in rocks, soils and water, and so can easily seep up through the ground and accumulate in enclosed workplaces.

As well as occurring in places like mines, radon can accumulate in any indoor workplace, whether above ground, such as an office, or below ground, such as a basement. All these can potentially contain high levels of radon gas.

The fact that radon is a gas means that it can easily enter our bodies by breathing it in. Inhaling radon can cause lung cancer. For non-smokers, it is the leading cause of lung cancer. For smokers, it is the second most common cause of lung cancer. If workers inhale radon, much of it will be exhaled immediately, but radon can produce some radioactive particles and these can be trapped in airways. These go on to produce alpha particles, which can trigger cancerous growth of cells. This is a chronic effect; it can take between 15 and 25 years after inhaling radon for the onset of lung cancer to occur.

> **TIP**
>
> Symptoms of radon poisoning include a persistent cough, wheezing, shortness of breath and chest pain. These symptoms are also found with other substances hazardous to health and will not necessarily confirm radon inhalation.

5.3.3 The basic ways of controlling exposures to non-ionising and ionising radiation

Firstly, we will consider the protection strategies and control measures for protection against ionising radiation.

Controlling exposure to ionising radiation

There are three basic radiation protection principles or strategies: shielding, distance and time.

1. **Shielding:** this uses a material of the density appropriate to the type of radiation that is present. The more penetrative the radiation, the denser or thicker the shielding must be to stop it. For example, a smoke detector emitting alpha particles may only need a thin plastic casing whereas a nuclear reactor will be surrounded by a thick wall of concrete.
2. **Distance:** as we have already mentioned, alpha and beta particles can only travel a maximum of a few metres in air. It therefore makes sense to keep a good distance between the source and the worker. The same principle applies to X-rays and gamma rays, although the distances involved will be much greater. Where a source is emitting radiation in all directions equally, the intensity of the radiation changes in inverse proportion to the square of the distance from the source. For example, consider the situation where a worker is exposed to radiation of intensity X when the worker is at distance D from the source. As a control measure, we now move the worker to distance 2D from the source. The resulting exposure will now be X divided by 2^2, that is X/4, or, put more simply, if we double the distance from the source the worker will receive a quarter of the original radiation intensity; similarly, if we triple the distance from the source they will only be exposed to one-ninth of the original intensity.
3. **Time:** keeping the exposure time to the minimum consistent with the requirements of the task. This is because the received dose will be in direct proportion to exposure time. Put simply, the less time someone is exposed, the less damage the radiation is likely to cause.

5.3.4 Basic radiation protection strategies

Ionising radiation prevention and control measures

Under reg 9 of the IRR employers are required to take reasonable steps to minimise exposure of persons to ionising radiation. To achieve this, employers must:

- where necessary, specify areas to be controlled or supervised; and
- appoint competent persons (radiation protection adviser and supervisor).

Controlled areas

> **ADDITIONAL INFORMATION**
>
> To understand the significance of controlled areas, it may be helpful to understand the types of unit used.
>
> 1. When a person is exposed to ionising radiation, their 'absorbed dose' (AD) is the energy absorbed in the body divided by the body weight; this unit is the Gray (Gy).

> **2** The biological effects of the AD depend on the types of radiation to which the body is exposed. The term 'radiation weighting factor' (RWF) is used. For example, for beta, gamma and X-ray radiation, RWF is 1, whereas for the more dangerous alpha radiation RWF is 20. The unit is the Sievert (Sv), although the unit millisievert (mSv) is generally used. The 'equivalent dose' (EQD) is worked out using the following formula:
>
> $$EQD = AD \times RWF$$
>
> **3** The risk of cancer induction from an equivalent dose depends on the organ receiving the dose. Some organs are more susceptible to ionising radiation than others. The term used to describe this risk is the 'effective dose' (EFD); this unit is also in Sieverts/millisieverts. To calculate the EFD, a 'tissue weighting factor' (TWF) is used, the magnitude of which indicates the level of risk to a particular organ. For example, the TWF for the stomach is 0.12 and is 0.01 for the skin. When taking the body as a whole, the TWF is 1.
>
> $$EFD = EQD \times TWF$$

Employers must designate controlled areas where:

(a) special procedures are needed to minimise exposure to ionising radiation or minimise the risk of radiation accidents; or
(b) any person working in such an area is likely to receive:
- an EFD >6mSv; or
- an EQD >15mSv for the lens of the eye; or
- an EQD >150mSv for the skin (or extremities).

The employer is required to designate workers likely to receive any of the above doses as 'classified workers'.

In addition, classified workers must be at least 18 years of age and be certified by a relevant doctor (that is, a doctor with appropriate qualifications and training) as fit to work with ionising radiation.

Supervised areas

Employers must designate supervised areas where:

(a) it is necessary to keep such areas under review to determine whether they need to be upgraded to controlled areas; or
(b) anyone working there is likely to receive:
- an EFD >1mSv; or
- an EQD >5mSv for the lens of the eye; or
- an EQD >50mSv for the skin (or extremities).

Appointment of competent persons

There are two types of competent person in relation to ionising radiation control.

The radiation protection adviser (RPA)

This person is usually someone appointed from outside the organisation. They will have the necessary competence (knowledge, experience and skills) to be able to advise on such topics as:

- plans for new installations;
- use of new radiation sources;

- design features, safety features and warning devices relating to radiation protection;
- identification of controlled and supervised areas;
- classification of workers;
- personal dosimetry programmes;
- radiation monitoring instrumentation;
- arrangements for prevention of radiation accidents and incidents;
- appropriate training programmes for workers; and
- employment conditions for pregnant and breastfeeding workers.

Where employers need to consult an RPA in order to secure compliance with the IRR, the appointment of the RPA must be in writing.

The radiation protection supervisor (RPS)

Where any work needs to be carried out in a controlled or supervised area, the employer must appoint at least one RPS who will supervise arrangements set out in the 'local rules' – this will normally be an internal appointment.

The local rules need to identify working instructions, compliance with which will restrict radiation exposure in controlled or supervised areas. It is the responsibility of the employer to produce the local rules, which should be appropriate to the nature and degree of radiation risk. The rules must cover not only normal working conditions but also foreseeable radiation accidents.

Local rules will include:

- the effective radiation dose that, when exceeded, requires an investigation to be carried out without delay;
- emergency arrangements for foreseeable radiation accidents;
- the name of any appointed RPS;
- the identification of areas covered, for example, those that are controlled areas and supervised areas; and
- a summary of working instructions.

An RPS will be expected to:

- know and understand the IRR and the local rules relating to work with ionising radiation;
- be allocated sufficient authority to supervise workers;
- understand necessary work precautions;
- have the necessary resources to undertake their functions; and
- know what to do in an emergency.

The protection programme will therefore also include arrangements for monitoring workers and the workplace, the provision of education training and information to workers, health surveillance arrangements, requirements for the assurance of quality and process improvement and the appointment of a radiation protection officer; this is someone appointed within an organisation to oversee the application of any regulatory requirements related to radiation.

Possible practical control measures might include the following:

- a dedicated room or area for carrying out work with sources of ionising radiation. For small sources, work might be carried out in a glove box. When larger areas are needed, there may be a room or area only authorised persons are allowed to enter; this is a form of segregation. This measure might be enforced by entry being available only by means of a key card allocated to authorised persons; and

Radioactive material

Figure 5: Signage for areas where work on ionising radiation sources takes place

Physical and psychological health

- signs placed at the entrances to areas where work on ionising radiation sources is taking place; Figure 5 shows an example.

    Other symbols for radiation can be found in the HSE publication *Safety signs and signals*.[3]

- Sealed sources are to be used where practicable; unsealed radioactive sources such as powders or liquids could present exposure hazards. Sealed sources are sources of ionising radiation that are manufactured in such a way that the radioactive material cannot be dispersed into the atmosphere. For example, radioactive powders may be encapsulated in materials such as metal or plastic. When it is not practicable to use sealed sources, then other controls will be necessary.
- The principle of reduced time exposure should be employed; the frequency and duration of exposure will be carefully monitored.
- Dose monitoring: persons working with ionising radiation sources should be wearing a means of radiation measurement. Such means could be a film badge or a thermoluminescent detector (TLD). These instruments will measure the extent of radiation exposure.
- Vulnerable groups of workers, such as pregnant women and young workers, should be excluded from such work.
- There must be suitable storage for radioactive materials.
- Appropriate personal protective equipment (PPE) must be worn; this is likely to be lead aprons, overshoes and possibly fully protective suits.

    For disposable PPE, there must be correct, carefully controlled disposal procedures. For non-disposable PPE, a specialised, competent laundry organisation must be employed.

- Respiratory protective equipment (RPE) is likely to be necessary where inhalation is a risk.
- Changing room and handwashing facilities must be provided and properly used.
- Any wounds should be covered with waterproof plasters.
- There must be no eating, drinking, smoking or application of cosmetics while working with ionising radiation sources.
- Emergency arrangements must be in place, for example, procedures for clearing up a spillage.
- Those working with ionising radiation sources must receive training in the hazards, risks and control measures of working with ionising radiation sources and in any appropriate emergency procedures.
- Workers must be confirmed by an appropriate doctor as medically fit to carry out this type of work.

Control measures for non-ionising radiation

We now consider control measures for the various types of non-ionising radiation.

Ultraviolet (UV) measures

- Enclosing the source would be the best option if it was practicable.
- When enclosure is not practicable, the use of shielding might be appropriate; for example, in welding operations screens could be placed around the welding area. This measure is to protect people walking by the welding area, who might be exposed to UV radiation.
- Reduced time exposure: controlling the frequency and duration of exposure, perhaps by job rotation or providing sufficient breaks.
- Wearing appropriate PPE: in this case to protect skin and eyes. Therefore, for welding, a welding helmet and gloves would be worn.

Infrared (IR) measures

- Shielding: enclosing a heat source such as a furnace is not practical.
- Reduced time exposure.
- Wearing appropriate eye protection; 'appropriate' meaning protection against the particular wavelengths of the IR source.
- Wearing suitable clothing: again, protecting against appropriate radiation wavelengths; here, light-coloured clothing is more effective than dark-coloured clothing.

Microwave measures

- Some microwave-generating devices, such as microwave ovens, have a metal surround to prevent radiation escape and an interlocked door so that when the door is open the device will not operate.
- Regular inspection should occur to check for any microwave leakage.
- Ensure sufficient distance between source and any persons who might be exposed to microwave radiation.

Radiofrequency (RF)

- The main control measures here consist of keeping a safe distance from the equipment and provision of training.

Lasers

There are many different classes of laser and they are classified according to their power. Most lasers are classified in the class range 1 to 4, where class 1 is the lowest power and class 4 the highest power, the latter presenting the greatest risk to health. You do not need to know about the various classifications, but note that eye protection should be worn when using all but the very low-powered lasers (class 1) and training must be provided on the risks of laser use and the control measures to be employed. For high-powered lasers, protective gloves should also be worn. Also, warning signs must be used; there is an example in Figure 6.

Figure 6: A warning sign for high-powered lasers

5.3.5 The role of monitoring and health surveillance

The basic purpose of monitoring and health surveillance is to help identify potential breakdowns in control measures so that they can be rectified and high standards of health maintained. As regards monitoring, the employer must ensure that all control measures continue to provide the required level of protection with regard to radiation.

Any health assessment procedures may include (but are not limited to) medical examinations, biological monitoring, radiological examinations and a review of health records.

The nature of the medical surveillance for each individual worker should take account of the nature of their work with ionising radiation and the state of health of that individual. Medical surveillance would typically be required for those workers likely to be exposed to significant amounts of radiation; 'significant' here means in relation to any legal annual radiation dose limits that may apply. These workers are commonly called 'classified persons' and would typically work in 'controlled areas'. Medical surveillance for these workers might take the form of full medical

Physical and psychological health

examinations, including blood tests. In contrast, workers in 'supervised areas' might not need medical examinations at all.

For non-ionising radiation, health surveillance might include an eye test and examination at appropriate intervals and skin inspections for those working with UV and IR sources.

KEY POINTS

- Ionising radiation is defined as radiation with enough energy to ionise cells in the human body.
- Frequency and wavelength for wave radiation are interdependent and have an impact on the possible effects of exposure.
- Health effects from exposure to ionising radiation can include skin burns, cataracts, radiation sickness, hair loss, cancer and genetic cell damage.
- The skin and eyes are at the greatest risk from non-ionising radiation, with ill-health effects including 'arc eye', sunburn, skin cancer, cataracts and damage to the cornea.
- Occupational sources of radiation include X-rays, radon gas, artificial lighting, significant heat sources such as furnaces and industrial ovens, transmitters and lasers.
- The protection strategies for ionising radiation are shielding, distance and minimising exposure time.
- The roles of the radiation protection adviser and radiation protection supervisor require specialist expertise and involve helping the organisation comply with legislation relating to radiation.
- Monitoring and health surveillance help identify and rectify potential breakdowns in radiation control measures, and may involve health assessments and medical examination of 'classified persons' most exposed to high levels of radiation.

References

[1] HSE, *Work with ionising radiation. Ionising Radiation Regulations 2017. Approved Code of Practice and guidance* (L121, 2nd edition, 2018) (www.hse.gov.uk)

[2] HSE, *Safety signs and signals. The Health and Safety (Safety Signs and Signals) Regulations. Guidance on Regulations.* (L64, 3rd edition, 2015) (www.hse.gov.uk)

[3] See note 2

5.4: Mental ill-health

> **Syllabus outline**
>
> In this section, you will develop an awareness of the following:
> - The frequency and extent of mental ill-health at work
> - Common symptoms of workers with mental ill-health: depression, anxiety/panic attacks, post-traumatic stress disorder (PTSD)
> - The causes of, and controls for, work-related mental ill-health (see the HSE's Management Standards):
>   - demands
>   - control
>   - support
>   - relationships
>   - role
>   - change
> - Home-work interface: commuting, childcare issues, relocation, care of frail (vulnerable) relatives
> - Recognition that most people with mental ill-health can continue to work effectively

Introduction

In recent times, it has been recognised that many workers experience or have experienced mental health conditions. We start with an overview of some mental health statistics. Different sources report and record such issues in different ways, so direct comparisons are difficult.

We will review some common symptoms of mental ill-health, including depression, anxiety, panic attacks and post-traumatic stress disorder (PTSD).

We then examine in significant detail the Health and Safety Executive (HSE) Management Standards for work-related stress, while considering causes and controls for work-related stress.

We also explore issues that may occur outside work and we note the effect these can have on workplace safety.

Finally, we offer some recommendations to help people continue to work safely with an appropriate level of support.

> **TIP**
>
> The Health and Safety Executive (HSE) has produced a set of Management Standards for managing work-related stress. The Management Standards cover six key areas of work design that have the potential to cause stress if not effectively managed.[1]

5.4.1 The frequency and extent of mental ill-health at work

Firstly, we need to define what we mean by 'mental health'.

> **DEFINITION**
> **Mental health** has been defined by the World Health Organization as "a state of wellbeing in which the individual realises their own abilities, can cope with the normal stress of life, can work productively and can make a contribution to their community".

There are many different organisations providing data on mental ill-health. For example, the Office for National Statistics (ONS) and the National Health Service (NHS). Worldwide, there is the World Health Organization (WHO).

Perhaps not surprisingly, data from different organisations is quoted in different ways. For example, some organisations quote actual numbers of mental health conditions while others quote percentages of the population. As a result, direct comparisons are sometimes difficult to make.

The accuracy of some statistics may be questionable for the following reasons:

- Some people may not wish to report symptoms to a medical practitioner or their workplace due to being concerned about being judged as 'weak'. It has been estimated that 9% of workers disclosing mental health issues to their managers at work have either lost their job or been disciplined or demoted as a result.
- Diagnosis of cases may not always be accurate.
- Reporting organisations may report data in different ways.
- The efficiency of reporting may be different in different organisations and different countries.

Despite this, we can draw some useful conclusions from the available data.

Worldwide, it has been estimated that in 2017, 792 million people lived with a mental health disorder (approximately 10.7% of the global population).[2]

In the UK, depression has been quoted as the leading cause and it has been estimated that one in four people experience mental health problems each year, while in any given week, one in six experience anxiety or depression.[3]

5.4.2 Common symptoms of mental ill-health

Depression: symptoms include feeling restless, irritable and generally unhappy. People often lack confidence and find it difficult to relate to other people. They will often not have the motivation to do anything. There may be a loss of pleasure for anything, poor concentration (for example, in health and safety tasks) and disrupted sleep patterns. Specific events can affect feelings of depression.

Anxiety and panic attacks: symptoms include feeling nervous, restless and tense. People will often feel weak and tired (this may lead to fatigue, which could be a cause of accidents). They will often worry about situations that may appear relatively harmless to people not experiencing this condition. Anxiety can result in panic attacks. A panic attack is a sudden episode of intense fear that can trigger severe physical reactions when there is no real danger or apparent cause. Triggers for panic attacks can include long periods of stress, activities that lead to intense physical reactions (for example, exercise) and a sudden change of environment.

Possible causes of mental ill-health (MIH)

Causes of general mental health issues can include:

- social isolation and loneliness;
- discrimination, for example, based on ethnicity, gender or age;
- levels of debt and an uncertain financial position;
- death in the family;
- divorce or separation;
- long-term physical conditions, for example, diabetes;
- loss of job;
- stress; and
- trauma experienced in early life, such as bullying.

Causes of mental health issues (occupational) for workers might include:

- inadequate health and safety policies, for example, when MIH is not included in the policy, or when the policy has not allocated overall responsibility for MIH issues to anyone, so workers do not know to whom to report MIH problems;
- inadequate communication;
- unsatisfactory management performance, particularly ineffective leadership;
- inappropriate workloads;
- inflexible working hours;
- unclear tasks and objectives;
- role conflict;
- environmental issues;
- inability to influence the way of working;
- high stress levels; and
- workplace bullying.

Post-traumatic stress disorder (PTSD)

> **DEFINITION**
> *Post-traumatic stress disorder (PTSD) is a condition of persistent and emotional stress occurring as a result of severe injury or psychological shock, for example, an attack of severe physical violence on a worker at work.*

Symptoms of PTSD
- Workers may experience sleep disturbance due to vivid recall of a horrific event with resulting nightmares.
- Workers may exhibit dulled responses to others and the outside world in general. Everyday problems can be considered insignificant compared with the traumatic events experienced by such workers.

- Workers may experience flashbacks of the traumatic events experienced, making them anxious. For example, a worker who witnessed the death of a warehouse worker who was struck by a forklift truck may react violently when hearing the horn of a forklift truck coming towards them when they are with another person. They may strongly push the other person away even though the forklift truck is not near them.

Work-related causes of PTSD

Some causes could be receiving a serious injury at work, such as being struck by a heavy falling object, witnessing a traffic accident on the work premises or simply being told about the death of a fellow worker.

It is important to note that stress outside the workplace can also be 'brought into the workplace'. Whatever the cause, mental health conditions need to be managed. We can now see that many factors can cause an increase in stress at work that, in turn, can result in mental health problems.

Stress

For human beings, some pressure can be beneficial; satisfaction can be derived from completing tasks and overcoming challenges, resulting in job satisfaction.

But there is a limit to the pressures that humans can manage. Excessive pressures may be harmful and result in the undermining of work performance and health.

There is no fixed definition of work-related stress, but a definition that has found acceptance is:

> **DEFINITION**
> *Stress is the adverse reaction people have to excessive pressures or other types of work demands put on them when they feel unable to cope with such pressures or demands.*

The initial effect of stress on a worker could be a feeling of not being in control of pressures or demands. At first, the worker might put this down to normal tiredness and, importantly, has not yet identified they may be experiencing stress.

5.4.3 The causes of, and controls for, work-related mental ill-health

One of the causes of mental ill-health is stress. As noted in the introduction, the HSE Management Standards identify six aspects of workplace design that have the potential to cause stress. The terminology in this document refers to stress but could also apply to mental health in general.

Managers and supervisors may not have the expertise to identify specific mental health conditions, especially as some conditions, such as depression and anxiety, may both be present together with, or caused by, stress. Therefore, in the measures that follow, reference is made to stress, as outlined in the HSE Management Standards.

The six potential stressors are identified as demands, control, relationships, support, role and change. Causes of these are now considered along with controls aimed at reducing stress.

Demands

Causes
- Workload: work demands can sometimes be too high and workers are unable to cope with their workloads. Alternatively, the workload could be too low, resulting in boredom and possibly the feeling of being undervalued.
- Work patterns: these may not be reasonable for workers. For example, a working shift of 14.00–22.00 might be followed next day by one of 06.00–14.00, which does not give the worker adequate rest time between shifts.
- The worker's skills and abilities are not matched to the demands of the job.
- Workers have insufficient resources to carry out their work.
- No arrangements are in place for workers to discuss their workloads with management.
- The working environment may not be suitable: the temperature may be too hot or too cold; there may be poor ventilation; or the relative humidity may be unsuitable. If humidity is too low, dry throats could result, and if too high, there may be discomfort due to excessive sweating. There may be too little workspace for workers; there may be excessive levels of noise; or the lighting may be unsuitable. Lighting may be too bright or too dim, or the wrong type has been installed.

Controls
- Provide workers with achievable targets: try to ensure that targets set for workers are realistic, not too easy and not extremely difficult.
- There should be continuous monitoring of workloads to ensure that they remain appropriate: provide workers with reasonable work patterns.
- Ensure the skills and abilities of the workers match the demands of the job and that workers are capable of doing their jobs. Managers should ensure that workers are competent for their job roles and, if necessary, provide additional training as soon as practicable.
- Provide workers with sufficient resources to enable them to carry out their jobs.
- Conduct regular meetings with individual workers to discuss their workload and any anticipated challenges.
- Try to ensure workers can carry out their work in a reasonable environment.

Control

Causes
- Workers have no say in how the work is to be carried out.
- Workers have no control over breaks: this could mean that during a work shift there might be a long period of work without any break.
- Workers have no control over the pace of the work.
- Workers are not consulted about their work patterns.

Controls

- Allow workers to have a say in how their work is organised, for example, about their work patterns.
- Provide suitable work patterns: for example, a reasonable lapse of time between shift time changes should be introduced, with appropriate consultation with workers and their representatives.
- Allow workers to have a say in how their breaks are organised.
- Discuss the skills that workers have and if they believe they are able to use these to good effect.
- Allow and encourage workers to participate in decision-making, especially when it affects them, such as decisions about the way they work.

Relationships

Causes

- There are no systems in place to enable and encourage managers to deal with unacceptable behaviour; therefore bullying and discrimination could take place.
- Workers do not share information that is relevant to their work.
- There may be a poor relationship between workers and their supervisors.
- There may be a poor relationship between workers and clients, perhaps due to clients requiring unreasonable deadlines to complete the work.

Controls

- Develop a written policy for dealing with unacceptable behaviour and grievance and disciplinary procedures for reporting incidents; communicate these to all workers.
- Agree and implement a confidential system for workers to report unacceptable behaviour.
- Discuss how individuals work together and how they can build positive relationships, for example, by sharing information relevant to their work.
- Encourage good communication and provide appropriate training to aid skill development.
- Select or build teams that have the right blend of expertise and experience for new projects.
- Agree and implement procedures to prevent, or quickly resolve, conflict at work and communicate these to all workers.

Support

Causes

- There are no actual policies in place to support workers.
- Workers are unaware of how to access support; they do not know who to go to for support.
- Workers do not know how to access resources to carry out their job, or there may not be sufficient resources provided.
- There is insufficient information, instruction and training to enable workers to carry out their tasks to the required standards.

Controls

- The organisation should have policies and procedures to adequately support workers.
- Systems are put in place to enable and encourage workers to support their colleagues.
- Enable workers to know what support is available and how and when to access it.
- Let workers know how to access the required resources to do their job.
- Provide all necessary information, instruction and training to workers to enable them to carry out their jobs.
- Provide flexibility in work schedules (where practicable) to enable workers to cope with domestic commitments.

Role

Causes

- The job requirements are not clear and workers do not understand what is required of them.
- There may be role conflict: a worker may be responsible for security as well as health and safety tasks (see Case study for an example).
- There may be a change of role, perhaps without further training being provided; this could be due to the organisation expanding or diversifying its business, requiring workers to adopt different roles.
- The role may no longer be required, resulting in the worker fearing redundancy.

> **CASE STUDY**
> **Role conflict leading to stress**
>
> In a particular organisation, a specific worker was responsible for both security and health and safety. This worker locked a fire door to stop theft of the organisation's products. A visiting fire enforcement officer, not surprisingly, imposed immediate requirements on the organisation to leave the fire door unlocked. The worker was then disciplined and was subsequently off work with stress.

Controls

- Clarify job roles so the workers understand what is expected of them. If workers have more than one job role, supervisors should provide clear priorities to workers as to which role takes precedence.
- Provide or revise job descriptions to ensure the core functions and priorities are clear.
- Display team/department targets and objectives to help clarify unit and individual roles.
- Agree specific standards of performance for jobs and individual tasks and review these periodically.
- When a change of role for workers is required, provide suitable and sufficient training for such workers.

Change

Causes
- The change has been put into place without workers being given any reason for it.
- There has been no opportunity for the workers to be consulted about the change.
- There may have been organisational changes, meaning that workers may report to different supervisors or managers.
- The change may make the worker's role redundant and therefore the worker fears losing their job altogether.

Controls
- Make all workers aware of why the change is happening.
- Explain what the organisation wants to achieve and why it is essential that change takes place.
- Define and explain the key steps of the change.
- Fully consult workers about the change and take necessary steps to gain their support for the change. Consultation with workers should take place at an early stage and throughout the change process.
- Make workers aware of the timescales of the change and the impact on their jobs.
- Provide a system to enable workers to comment and ask questions before, during and after the change, particularly workers who want to raise concerns.

Other (more general) control measures include those in the Tip box.

> **TIP**
> - Create a stress policy: this should include the overall aims of the policy (that is, the reduction of mental ill-health at work) and identify managers whose task it is to carry out the policy.
> - Provide suitable leadership from top management: this might include giving appropriate training to other managers and supervisors on recognising signs of stress and what actions to take as a result.
> - If it becomes clear that workers have become stressed, arrangements should be made to offer counselling to these workers.
> - Specific stress awareness training should be provided to workers so that they can report early symptoms to management.
> - Sickness absence should be monitored, especially if it is stress related. Return-to-work interviews should be put in place.

5.4.4 The home-work interface

There are quite a few issues that can start at home but also affect the output and/or temperament of workers while they are at work.

Childcare issues can pose significant problems, especially for single-parent families. Typical problems include being let down by a childminder, maybe because of sickness, causing the parent not to be able to travel to the workplace. Another situation may be that a working parent is asked to collect a sick child from school during their working day. In both cases, the parent has to deal with the stress of caring for their child as well as maintaining sufficient attendance and concentration at work.

5.4 Mental ill-health

ADDITIONAL INFORMATION

Since the Covid-19 pandemic, many office workers are now working at home all the time or are 'hybrid working' (working a few days in the office and the rest at home). While some workers prefer this type of working, for others it can add to their stress. Some people do not like working at home but are forced to do so because of the changing workplace. In these cases, workers may find it difficult to switch between work and home life as both activities are happening in the same place. Other workers miss the interaction of having colleagues around them and the social side of work. In other cases, the worker may not have enough space in their home to set up a permanent workspace.

Homeworking for working parents may also be a challenge. Parents will be juggling the need to get their work done with the needs of their children.

Commuting can also be stressful, especially when there are long journey times and difficulties such as delays, for example, due to adverse weather or industrial action on public transport. Such incidents result in stress about arriving late at work, perhaps missing important meetings, or arriving home late after leaving work, causing family tensions.

Care of frail (vulnerable) relatives: this can also present concerns, especially if carers are unable to attend or there are commuting problems that may lead to nobody being available to provide care at critical moments.

Relocation: moving house is another acknowledged cause of stress. Fortunately, this is a situation reasonably understood by many employers.

Financial worries: the cost of living and possibly debts may be of concern. Apart from thinking about this type of worry while at work, a worker may take a second or even a third job to survive financially. This could produce significant fatigue at their place of employment.

Other issues affecting someone at work may include:

- Stressful events, such as a death in the family: workers may need to take a period of compassionate leave and, on returning to work, may need to undertake low-stress work.
- Experiencing a divorce or separation: this may cause mental ill-health and affect a worker's relationships with others at work.

All these stressful situations may affect a worker's concentration at work and potentially lead to accidents.

5.4.5 Recognition that most people with mental ill-health can continue to work effectively

As discussed earlier, there are many things that can affect mental health both at home and at work. In addition, a worker may have been diagnosed with a mental health condition not covered here, such as bipolar disorder.

For this worker to be effective at work, a number of steps could be taken by the employer:

- The worker could discuss the mental health issue on a one-to-one basis with a person such as a manager at work.
- Discussions must be kept confidential, and the details only shared with those who need the information, such as occupational health or human resources (HR).
- The discussion could, where relevant, include what events have resulted in the worker having the issue.
- There could be agreement of which tasks the worker could carry out and at what pace of work.
- Any need for rehabilitation could be discussed.
- The need for any additional control measures could also be covered.

If the worker has already been on a period of sick leave, other measures could be considered:

- The employer could keep in regular touch with the worker.
- The employer should not put pressure on the worker to return to work too early.
- When the worker does return, there should be a return-to-work interview.
- Arrangements could be made for a phased return to work.
- If necessary, the worker could start by being put on light duties.
- The progress of the worker should be continuously monitored.
- Regular counselling could be made available if needed.

In this way, a worker would perceive that the organisation cared about their welfare and this in itself, plus the measures suggested, would help the worker to continue to work effectively.

KEY POINTS

- Mental health is defined as a state of wellbeing when someone realises their abilities and can cope with normal life stresses, work productively and contribute to the community.
- Statistics indicate that 11% of the world population have had mental health problems and one in six people in the UK experience anxiety or depression.
- General causes of mental ill-health include financial uncertainty, discrimination, long-term physical conditions, job loss, trauma and stress.
- Causes of mental ill-health at work include inadequate communication and management performance, bullying, inflexible hours, inappropriate workloads, unclear objectives and environmental issues.

- Stress is an adverse reaction to excessive pressures or demands, such as workloads that are too high or too low, tasks not suited to the worker, unrealistic targets, not having a say in how work is done and not having any support.
- The HSE Management Standards set out possible causes in detail and suggest control measures such as appropriate policies and procedures, training to raise awareness of mental health issues and better communication and consultation (especially about changes and priorities).
- Post-traumatic stress disorder (PTSD) can result from severe injury or psychological shock from witnessing a serious accident in the workplace, with symptoms including disturbed sleep, dulled responses and flashbacks.
- Stress at home, including due to caring responsibilities and financial worries, can affect someone at work, including their concentration.
- A worker experiencing mental ill-health can continue to be effective at work if given the right level of support by their employer.

References

[1] HSE, *Tackling work-related stress using the Management Standards approach: A step-by-step workbook* (2019) (www.hse.gov.uk)

[2] S Dattani, H Ritchie and M Roser, 'Mental health' (2021) (ourworldindata.org/mental-health)

[3] See note 2

5.5: Violence at work

Syllabus outline

In this section, you will develop an awareness of the following:
- Types of violence at work including: physical, psychological, verbal, bullying
- Jobs and activities which increase the risk of violence, including: police, fire, medical, social workers, those in customer services, lone workers, those working with people under the influence of drugs and alcohol, those who handle money or valuables
- Control measures to reduce risks from violence at work

Introduction

For employers, violence at work can lead to poor morale and a poor image for the organisation, making it difficult to recruit and keep workers. It can also mean extra financial burdens caused by worker absenteeism, higher insurance premiums and compensation payments as well as the costs of incident investigation and, in certain cases, counselling for workers.

For workers, violence can cause pain and distress as well as reduced earnings due to time off work or having to take early retirement due to resulting disabilities. Physical attacks are obviously dangerous, but serious or persistent verbal abuse or threats can also damage a worker's health through anxiety or stress. The constant fear of violence at work can also be extremely stressful.

We will start by identifying the different types of violence at work. We will then move on to review three categories of worker at potential risk of violence: those spending most of their time in one workplace, those visiting many workplaces and those visiting the home of a client.

We will consider circumstances where violence could occur.

Finally, we will describe control measures available to minimise the risk of violence at work.

TIP

The main piece of legislation relating to violence at work in the UK is the Management of Health and Safety at Work Regulations 1999 (MHSWR).

The Health and Safety Executive (HSE) has also issued guidance on general aspects of violence.[1]

We start by considering the meaning of work-related violence.

DEFINITION

Work-related violence *is any incident in which a person is threatened, abused, assaulted or injured in circumstances relating to their work.*

5.5.1 Types of violence

There are several different types of work-related violence.

- Physical: this can occur in almost any workplace setting. For example, police officers and crowd control stewards are at risk of physical attack during demonstrations or at sporting events.
- Verbal: this is perhaps the most common form of violence. For example, public transport workers may be verbally abused when services are cancelled or running late. Garage forecourt workers could be verbally abused if fuel supplies are low, so cars have to queue and drivers become frustrated.
- Bullying: this can occur in many circumstances. For example, new or young workers who have recently joined an organisation could be bullied by older or supervisory workers and threatened with sanctions if they are not more productive. Workers can also be bullied for not conforming to other workers' expectations, ranging from the way the bullied worker dresses to their political views. Often, bullying will happen through a clash of personalities, for example, when a domineering personality takes the opportunity to ridicule or embarrass a less confident worker.
- Psychological: this is a more subtle form of abuse. For example, a worker might be allocated poor-quality tools or other work equipment and told that other workers manage successfully with such tools or equipment (whether this is true or not). Criticism of someone's work in front of other workers (even when there is nothing wrong with their work) can also have a profound psychological impact.

5.5.2 Jobs and activities that increase the risk of violence

There are two broad categories of workers to consider where violence may be a risk: 'stationary' workers, such as those who work in offices, warehouses and manufacturing plants and 'mobile' workers. Mobile workers visit many different environments in the course of their work. For example, the police might go to various workplaces and social workers to clients' homes.

Each category of worker may face different types of risk and need different types of control measure to help protect them against violence.

Table 1 provides some examples of workers who may be at risk of violence at work.

Workers usually at fixed workplaces	Mobile workers (many locations)	Workers who visit people's homes
Shopkeepers	Police officers	Social workers
Office workers	Firefighters	Medical staff
Garage forecourt staff	Medical staff attending emergencies	Police and medical staff attending emergencies
Cashiers, for example, bank staff	Transport staff, such as bus or train drivers	Landlords
Bar staff	Delivery workers	Delivery workers
Waiting staff in a café or restaurant	Traffic wardens	Tradespeople such as plumbers and electricians

Physical and psychological health

Workers usually at fixed workplaces	Mobile workers (many locations)	Workers who visit people's homes
Teachers who work at one location	Teachers who work at several locations	
Door security staff	Politicians	
Workers in warehouses or manufacturing plants	Enforcement officers	

Table 1: Examples of workers at risk of violence

UK law requires managers of organisations employing workers such as those identified in Table 1 to include the risk of violence to such workers in the general risk assessment for the organisation and to introduce reasonable measures to minimise the risk of violence to such workers.

HSE guidance identifies that violence may be verbal (including abuse and threats).[2]

Another type of person who may be at risk is a worker who works alone (see Case study example).

> **CASE STUDY**
>
> A maintenance worker was required to enter the organisation's premises two hours before all other staff to carry out certain tasks.
>
> On one occasion, two thieves entered the premises during this two-hour period, attacked the maintenance worker and stole some tools.
>
> After this, the company installed stronger locks on the outside doors, instructed any maintenance worker to lock themselves in during the period of lone working and also provided them with a personal alarm. A fixed alarm was installed that, if activated, sent a signal to the local police station. A warning notice advising of this fixed alarm was posted in a prominent position outside the building.

Situations that might lead to workers experiencing violence could include:

- Being refused entry to a premises: this may be due to a person trying to enter a premises outside opening hours or because the person behaved in an unacceptable manner previously and was not allowed future entry. A security worker may then be threatened.
- Receiving unsatisfactory or unhelpful answers to questions: for example, if there were delays to rail services and a traveller was unable to get information on the availability of future train services, the rail worker dealing with them might then receive abuse.
- Robbery: a worker might be threatened with physical violence if they did not give a robber cash from a till.
- Influence of drugs or alcohol: a person might display unacceptable behaviour if under the influence of drugs and/or alcohol. Someone in a deep state of intoxication cannot always be reasoned with, so might physically assault a worker. For example, if a doctor refused to prescribe further addictive pain medication to a patient, the patient may resort to physical assault to try to get what they want.
- Enforcement action: an enforcing officer may have ordered a process to stop. The employer may see this production delay as costing them a considerable amount of money. They may then threaten or assault the enforcing officer.

- Unsatisfactory service: a person may not be happy with the service they have received. For example, they may have had a long wait to be served in a restaurant or the food may not be of good quality. A member of the waiting staff may then receive abuse.

Figure 1: Customers may abuse waiting staff if unsatisfied with service

- Debt collection: a worker attempting to collect money could be subject to physical violence if the debtor is unable to pay.
- Attempted theft: a person attempting to steal goods from a store may be stopped by a security worker. This worker may then be abused or threatened.
- Lone worker situation: a lone worker, such as a security person patrolling a premises at night, might be seen as an easy target for robbery with violence. This subject is covered in more detail in 8.4: Lone working.

The outcome of an incident investigation will have resulted from gathering facts, analysing them to determine the root causes and then deciding on appropriate control measures to help prevent a recurrence. This process is discussed in 4.2: Investigating incidents.

There are a range of control measures that employers could use to minimise violence to workers.

5.5.3 Control measures to minimise violence

- For large organisations, employ adequate numbers of security staff who are well trained.
- For small organisations, install an alarm system and take advice from the police on the type of alarm to install and its location(s).
- Install CCTV around the perimeter of the work site, also covering the car park if the organisation has one.
- Restrict access to buildings with the minimum number of entrances.
- Create a procedure for incident reporting. Reports would include the nature of the incident, the time and location of the incident, names of any persons injured, identification of any loss or damage to property and the names (if established) of any violent persons involved. Recommendations to help prevent a recurrence should result from the incident investigation outcome.
- Train workers in de-escalation techniques to encourage calm discussion with any potentially violent person.

Physical and psychological health

- Ensure any car park on the premises of the organisation is well lit.
- Provide any lone workers with personal alarms.
- Minimise instances of keeping cash on the work premises and put up a sign warning that cash is not kept on the premises overnight.
- Ensure there is a procedure for any necessary home visits.

> **APPLICATION**
>
> (a) Where in your organisation do you think you and other workers are most at risk of violence?
> (b) What has been done so far to help reduce this risk?
> (c) What more do you think needs to be done to help reduce this risk even further?

Some workers are 'mobile' and frequently travel to different workplaces or areas. Mobile workers will need other control measures.

Examples of control measures for mobile workers

Bus drivers: on some routes where there is a significant risk of violence, buses are fitted with barriers of hardened glass between the driver and passengers, so that passengers with violent intent find it almost impossible to access the driver or their cash.

Figure 2: Hardened glass can protect bus drivers

Delivery drivers with valuable cargo: these drivers will often vary their route and timings to make it difficult for any robbery with violence to be planned.

Politicians: in democracies at least, it is relatively easy for citizens to meet in person with politicians who represent them at a national or regional level. But this accessibility also brings with it the risk of violence. Police protection may be provided in these meetings, especially in the case of high-profile politicians.

Lone workers: as noted earlier, they should carry personal alarms. If working alone away from the main premises, they should arrange to phone a prearranged contact at the organisation at regular intervals. Although this might not help prevent a violent attack, failure to call would indicate a lone worker in trouble and help could be dispatched quickly.

Control measures for home visits

We now examine some controls that would be appropriate for workers visiting clients' homes.

For example, in the case of a medical worker (such as a doctor or nurse) making a necessary routine home visit, controls might include the following:

- Examine notes from past visits to determine any risk of violence.
- If a risk of violence exists, be accompanied by another person.
- Try to arrange the visit in daylight hours, but if this is not feasible, park in a well-lit area.
- Wear appropriate clothing that is not obviously expensive.
- Do not have expensive jewellery on show.
- When inside the client's home, sit or stand somewhere a quick escape is possible if an uncontrollable violent episode occurs.
- Be trained in techniques for calming a difficult situation.
- Park as close as possible to the client's home to allow a quick exit from the area.
- Ensure the medical centre knows where the medical worker will be, along with their expected times of arrival/departure at/from the client's home.
- The medical worker should carry both a mobile phone and a personal alarm with them to the appointment.

KEY POINTS

- Violence can cause adverse problems at work, particularly worker stress. It can harm morale and the organisation's image and mean extra financial burdens.
- There are different types of violence: physical, verbal, bullying and psychological.
- Workers who might be at risk of work-related violence can be categorised as stationary because they usually work from one place, or mobile, because they work in various locations. Workers who visit people's homes can also be at risk.
- Circumstances in which work-related violence might occur include when people that workers deal with are refused entry, receive unhelpful information or unsatisfactory service, are subject to enforcement action or are under the influence of drugs or alcohol.
- Control measures that could be used to minimise the risk of work-related violence include providing training and personal alarms, installing alarms and CCTV at premises and creating procedures for incident reporting and home visits.

References

[1] HSE, 'Work-related violence' (www.hse.gov.uk/violence); HSE, *The health and safety toolbox: How to control risks at work* (HSG268, 2014) (www.hse.gov.uk)

[2] HSG268, see note 1

5.6: Substance abuse at work

> **Syllabus outline**
>
> In this section, you will develop an awareness of the following:
> - Risks to health and safety from substance abuse at work (alcohol, legal/illegal drugs and solvents)
> - Control measures to help reduce risks from substance abuse at work

Introduction

Substance abuse (involving alcohol, drugs or solvents) can lead to adverse, and even catastrophic, effects at work.

It is well known that driving under the influence of alcohol or drugs can lead to accidents, sometimes resulting in serious injury or death. Operating machinery under such influences may also result in serious injury.

Firstly, we will look at the risks to health and safety from substance abuse at work. We will identify types of substance abuse and some of the associated signs and symptoms. We will also explore some reasons and situations relating to the abuse.

Finally, we will examine the control measures that can reduce instances of, and risks from, substance abuse at work.

> **TIP**
>
> We will refer to the following legislation:
> - Health and Safety at Work etc Act 1974 (HASAWA); and
> - Transport and Works Act 1992 (TAW).
>
> The following guidance has also been produced in relation to substance misuse at work:
> - *Managing drug and alcohol misuse at work* – guidance from the Chartered Institute of Personnel and Development (CIPD);[1] and
> - *Managing drug and alcohol misuse at work* – guidance from the Health and Safety Executive (HSE).[2]

> **DEFINITION**
>
> ***Substance abuse** is the use of illegal drugs, prescription drugs or alcohol for purposes other than those for which they are meant to be used, or use in excessive amounts.*

Before we examine the health and safety risks from substance abuse, we need to appreciate how substance misuse by workers can result in undesirable consequences.

For example, the use of drugs or consumption of alcohol can reduce physical co-ordination and reaction times, and can affect rational thinking, judgement and mood. The consequences could be that users injure themselves or their fellow workers at work.

5.6.1 Risks to health and safety from substance abuse at work (alcohol, legal/illegal drugs and solvents)

Firstly, we will consider common effects of these types of substance.

Some signs and effects of substance abuse are:

- Variable timekeeping: this could mean experiencing difficulties arriving on time for work.
- Declining job performance: this might include making many mistakes, causing loss of productivity and efficiency.
- There might be a tendency to become confused.
- Possible theft: this could include stealing from colleagues or the company to fund a substance abuse habit.
- Sudden changes of mood: this could mean being friendly and co-operative one moment then suddenly becoming aggressive and argumentative the next moment.
- Taking a lot of days of sick leave: this might be to recover from previous episodes of substance abuse.
- Loss of morale among other workers might happen because of the loss of a bonus as a result of the low productivity of the substance-abusing worker or perhaps due to the fact that the organisation is not applying sufficient disciplinary measures against this worker. This may also lead to deterioration of relationships with both other workers and managers.
- Loss of workers: this could occur when a substance abuser has lost their job due to their substance misuse habit.
- Workers may suffer deterioration in health as a result of the effect of a substance on their body and this ill-health may contribute to the risks already identified.

There can be other effects from specific types of substance. For example, illegal use of drugs can result in visible puncture marks on arms and the discovery of drugs paraphernalia at the workplace; abuse of alcohol can also result in such workers smelling of alcohol during work hours.

> **ADDITIONAL INFORMATION**
>
> **Solvents**
>
> It is unlikely that a worker would ingest a solvent during substance abuse, but they may wish to inhale the vapour. It can be difficult to determine whether this type of abuse is taking place, as a wide range of solvents may be used.
>
> Some examples are xylene (found in solvent-based paints and permanent marker pens), amyl acetate (which has a sweet, fruity smell), dichloromethane (found in paint stripper), chloroform, 1,1,1-trichloroethane (used to be the solvent in 'liquid paper' correction fluid, although now banned on environmental grounds) and acetone (found in nail varnish remover).
>
> There may of course be bottles of solvents in use at work. Therefore, there can be easy access to solvents. The risk to health and safety will depend on the nature of the solvent (and its OEL/WEL, explained in 7.3: Occupational exposure limits), the frequency and duration of inhalation and the vapour pressure of the solvent at the temperature of inhalation.

Physical and psychological health

There can be extra problems for workers who need to drive as part of their employment. It is self-evident that if you are not in proper control of a vehicle due to excessive alcohol consumption, the risk of death or injury is increased.

Reasons/situations why workers may engage in substance abuse

Understanding reasons for substance abuse may help employers formulate control measures to help reduce the abuse. The CIPD guidance[3] provides a list of possible reasons for abuse.

Some reasons for the abuse by workers include:

- relationship problems at work;
- a culture at work of general acceptance of substance abuse;
- easy access to alcohol, drugs or solvents;
- workers feeling undervalued at work;
- workers not liking their type of work, which could be boring and monotonous;
- workers having little control over the way of carrying out work;
- time of day: workers may work night shifts when there is less supervision; and
- workers suffering stress.

CASE STUDY 1

A worker had worked in a manufacturing organisation as a supervisor for 14 years. Their performance had been generally good but had deteriorated over the previous nine months. The work pressure had increased significantly over this time and the supervisor had been consuming significant quantities of alcohol during this period to help them get through the day.

The supervisor's manager, concerned about the supervisor's work performance, called them in for an interview. The supervisor acknowledged that they had a drink-dependency issue, but it transpired that they were already being counselled for a workplace incident that had occurred nine months previously in which a worker in the supervisor's department had been badly burned. The supervisor was being counselled for the effects they experienced due to this accident.

Arrangements were made for the supervisor to take 10 weeks' compassionate leave and to receive further counselling. During this time the manager kept in regular contact with the supervisor, which demonstrated a caring attitude.

The supervisor recovered sufficiently during this time to the point that they no longer felt the need to drink alcohol. The organisation arranged a return-to-work interview during which a phased return to work was agreed; this was that the supervisor worked part time to start with, then gradually increased their hours, eventually being able to work full time.

This case had two positive outcomes:

1 the recovery of the supervisor; and
2 the supervisor became a member of a counselling panel. Having been through difficult times due to alcohol dependency, they were in a good position to provide advice and understanding to other workers who might find themselves dependent on alcohol.

5.6.2 Control measures to reduce risks from substance abuse at work

It is important for an organisation to create a substance abuse policy. This will probably be a subsection of the health and safety policy of the organisation.

There should be a general statement of the aims of the policy, which might include:

1 Help workers understand the rules relating to drug and alcohol misuse.
2 Provide a greater awareness in the workplace of the effects of drugs and alcohol, which in turn will aid early recognition of any problems.
3 Have necessary procedures and support packages in place in the event of any problems arising.
4 Have key staff (often managers) trained to understand any issues involved, for example, whether the source of a problem is occupational or domestic, and be competent to deal with them.
5 Make workers aware that the organisation can support them if they have a dependency problem and need help.

The people responsible for the policy being carried out will be identified (the key staff specified in point 4). These will perhaps include managers, supervisors and, if the organisation is large enough, on-site occupational health personnel.

The policy will include arrangements for carrying out the policy, some of which are described later. It is vital that the policy is made known to all workers.

Before a policy is put in place, worker representatives should be involved in creating it. This will help ensure the process is open and transparent because the emphasis should be on supporting workers rather than policing them. This communication will assist in raising awareness of substance misuse.

Other control measures might include:

- Produce a statement of specific rules: these might include, for example, no drinking of alcohol during lunch breaks.
- Create a substance-awareness training session, with a requirement that all workers must attend the session.
- Identify safety-critical jobs in the organisation.
- Carry out drug and alcohol screening and testing. This can be a sensitive issue and must be approached carefully (see Tip box). An important principle is that testing must be justified and appropriate for the nature of the business.

> **TIP**
>
> **Drug and alcohol testing**
>
> Here are some practical points on carrying out screening and testing.
>
> **The legal situation (criminal)**
>
> Under HASAWA, because an employer must ensure, so far as is reasonably practicable, the health, safety and welfare of employees, it follows that if an employer knowingly allows a worker to continue working under the influence of drugs or excess alcohol, the employer commits an offence.
>
> Other more specific legislation includes TAW, whereby it is a criminal offence for particular workers to be unfit due to drugs or alcohol while working on transport systems such as railways and tramways.

> **The legal situation (civil)**
>
> The requirements regarding testing will normally form part of the contract between the organisation and its workers.
>
> In the HSE guidance,[4] there is a recommendation that employers should obtain written consent from employees for drug/alcohol testing.
>
> Other things to consider include:
>
> - The least intrusive form of testing should be used. For example, can a sample of exhaled breath be taken rather than a blood sample?
> - Workers must be advised what they are being tested for.
> - Workers must be fully aware that substance testing is taking place.
> - Testing must be limited to the specified substances and the extent of exposure that will have a significant bearing on the purposes for which the testing is conducted.
>
> Information on testing protocol is provided in the CIPD guidance.[5]

A screening and testing programme will need to be developed, which will include the circumstances in which the screening and testing should take place. For example, in safety-critical roles such as a driver of public transport, such circumstances could be when there has been a significant increase in time off for sickness and when workers have returned to work following time off for an identified abuse problem.

- Provide counselling for workers when substance abuse has been identified. Initial steps are suggested in the Tip box on managing workers with a problem.

> **TIP**
>
> **Managing workers with a substance problem**
>
> - Encourage the appropriate manager to identify the performance issue causing concern.
> - Discuss the issue with the worker.
> - Encourage the worker to acknowledge they have a substance problem.
> - Try to ensure the manager is seen to be firm but fair, demonstrating qualities of concern and empathy but also providing non-judgemental advice and direction.

- In large organisations, the manager may seek medical advice through the organisation's occupational health department. In smaller organisations, which are unlikely to have such a department, the worker may be advised to seek advice from their doctor or be referred to a specialised counselling agency.
- The worker may also be referred to a self-help group that provides support to group members who recognise they need help with their substance abuse dependencies.
- Guarantee confidentiality regarding the results of substance screening/testing and of discussions during counselling sessions.
- Create a system for carrying out spot checks for substances.
- Offer time off for rehabilitation when a substance abuse problem has been identified and discussed.
- Use disciplinary measures if, despite help and support being given to workers, they continue to engage in substance abuse.

5.6 Substance abuse at work

It can sometimes be difficult to find out initially if a worker has a substance abuse problem as people can often hide this fact, as illustrated in Case study 2.

APPLICATION

Find out the substance abuse policy in your organisation.

CASE STUDY 2

In a pharmaceutical company, the preparation of a particular liquid medicine included the automatic mixing of ingredients in a large stainless steel mixing vessel in the manufacturing area. The vessel was equipped with a hinged lid that could be opened occasionally to carry out some testing procedures. Access to the lid was from a platform. The ingredients of the medicine included a small quantity of chloroform, the vapour of which was easily detectable with the vessel lid open.

The manufacturing area was under the general supervision of the manufacturing manager. On a particular day, the manufacturing manager was called to a short meeting with some other managers. During this time, a young worker from another department who had a liking for inhaling solvents (a fact unknown at that point to any other workers) approached the mixing vessel, opened the lid and leant forward to sniff vapour from the mixture. He was soon overcome by the vapour and rendered unconscious.

Fortunately, an experienced and competent first-aider happened to enter the area and noticed the situation. By this time, the worker was showing signs of cyanosis (turning blue due to lack of oxygen) but the first-aider was able to save the worker. The first-aider commented that if there had been a two-minute delay in finding them, the worker would have died.

Subsequently, an enforcing officer interviewed various personnel, including the worker. Recommendations were made to the company, but no enforcement action was taken. The enforcing officer commented that, had the worker died, the outcome would have been very different, with a court case being likely.

The actions taken by the company afterwards included:

- short term: fixing a lock to the lids of all mixing vessels;
- medium term: providing substance training to all workers; and
- longer term: creating a substance abuse policy, carrying out substance screening and testing and setting up a counselling service where needed.

KEY POINTS

- Types of substance that may be abused at work include alcohol, drugs and solvents.
- It can be difficult to initially identify a worker who might be involved with substance abuse.
- Substance abuse by someone at work may result in signs and effects such as variable timekeeping, declining performance, sudden mood changes, significant amounts of sick leave and a deterioration in the worker's health.

- The reasons for substance abuse at work can relate to the work culture, easy access to substances, not liking the job or having no control over it, feeling undervalued or stressed and working night shifts.
- Risks to health and safety that might be present with substance abuse include increased possibility of accidents when driving.
- Control measures to help reduce the risks from substance abuse at work include having a specific policy developed with worker representative involvement, running awareness training, clearly stating specific rules and developing a drug and alcohol screening and testing programme.

References

[1] CIPD, *Managing drug and alcohol misuse at work: A guide for employers* (2020) (www.cipd.co.uk)

[2] HSE, *Managing drug and alcohol misuse at work* (www.hse.gov.uk)

[3] See note 1

[4] See note 2

[5] See note 1

ELEMENT 6

MUSCULOSKELETAL HEALTH

6.1: Work-related upper limb disorders

Syllabus outline

In this section, you will develop an awareness of the following:

- Meaning of musculoskeletal disorders and work-related upper limb disorders (WRULDs)
- Possible ill-health conditions from poorly designed tasks and workstations
- Avoiding/minimising risks from poorly designed tasks and workstations by considering:
    - task (including repetitive, strenuous)
    - environment (including lighting, glare)
    - equipment (including user requirements, adjustability, matching the workplace to individual needs of workers)

Introduction

It has been estimated that 1.71 billion people worldwide have musculoskeletal disorders (MSDs), with lower back pain causing the highest burden with a prevalence of 568 million people. While the prevalence of MSDs varies by age and diagnosis, globally people of all ages are affected.[1]

Firstly, we will define ergonomics, MSDs and work-related upper limb disorders (WRULDs). We will then look at the various ergonomic aspects that need to be considered and specific risk assessment and control measures for display screen equipment (DSE).

Musculoskeletal health can be affected directly or indirectly. For example, a direct effect could result from a worker sitting in a badly designed chair offering them no back support, thus causing the worker to suffer back pain. An indirect effect could result from a worker being exposed to glare, causing them to adopt an awkward posture to avoid the glare, leading to the worker suffering back pain.

TIP

The main pieces of legislation we will cover are the:

- Manual Handling Operations Regulations 1992/Manual Handling Operations Regulations (Northern Ireland) 1992; and
- Health and Safety (Display Screen Equipment) Regulations 1992/Health and Safety (Display Screen Equipment) Regulations (Northern Ireland) 1992.

Musculoskeletal health

> There are also many publications from which we can draw information, a few of which are:
>
> - a World Health Organization (WHO) fact sheet on musculoskeletal health;[2]
> - *Ergonomics and human factors at work: A brief guide;*[3]
> - a UK online guide to musculoskeletal disorders;[4]
> - *Manual handling. Manual Handling Operations Regulations 1992. Guidance on Regulations (L23);*[5] and
> - *Work with display screen equipment. Health and Safety (Display Screen Equipment) Regulations 1992. Guidance on Regulations (L26)*[6]

6.1.1 Meaning of musculoskeletal disorders and WRULDs

Ergonomics

We will start by examining what we mean by ergonomics, and situations that might result in good or bad ergonomics.

When workers are carrying out tasks, sometimes they may experience more discomfort than expected, resulting in physical aches and pains to the body, mental fatigue and, in the worst cases, physical disability or mental ill-health.

Work should be organised so that the work equipment in use, the work procedure and the environment in which the work takes place cause minimum discomfort to workers, whether they are sedentary or mobile. This work can then be described as being conducted in line with good ergonomic principles.

> **DEFINITION**
>
> ***Ergonomics*** *is the relationship or interface between the worker, the equipment they are using, the way in which they are using the equipment and the environment in which the work is taking place.*

When good ergonomic principles are not followed, MSDs can result.

> **DEFINITION**
>
> ***Musculoskeletal disorders (MSDs)*** *are injuries or disorders of muscles, nerves, tendons, joints, cartilage and intervertebral discs.*

6.1.2 Possible ill-health conditions from poorly designed tasks and workstations

The ill-health effects that can occur if good ergonomic principles are not followed include:

- aches and pains in the neck, shoulders and upper limbs;
- swollen joints;
- stiffness in the hands and reduced mobility;
- numbness in the hands and fingers;

- weakness and loss of dexterity in the hands;
- tenderness in limbs; and
- low-quality sleep and disturbed sleep patterns.

We will now discuss some of the factors to consider in an ergonomic risk assessment and classify these into three groups: work equipment and work procedure; the working environment; and the workers themselves. You also need to consider adverse ergonomic features.

Work equipment and work procedure

- Appropriate equipment: if inappropriate equipment is used to complete a task (due to the design of the equipment, or its size or shape), this can result in stress on the body.
- Position of controls: these may be difficult to reach or located at a low level, so the worker has to stretch or stoop to access them. This can be a particular problem for machine operators.
- Seating position: workers doing sedentary work might be seated continuously in one position with bad posture, perhaps because the seating is not adjustable, resulting in some workers being uncomfortable.

Application 1 shows the possible impact of working while seated.

APPLICATION 1

Workers in a packing department were loading components into a cardboard box from a seated position. The work involved bending forward and the workers eventually complained of stiff and aching backs.

As a short-term measure, they were advised to spend some time sitting and some time standing while carrying out the work. A longer-term measure was eventually put in place. This was a conveyor belt positioned in front of the workers so there was no need to bend forward. This eliminated the problem.

Consider where a similar adjustment may be helpful in your own workplace.

- Level of force needed: excessive force may be required to complete a task, resulting in strain. In one organisation, a worker was required to force batteries into a tightly fitting cardboard tube. The work was continuous and put strain on their wrist. Eventually, the worker suffered an injury to their wrist and was off work for four months. The process could not be automated, and the only solution found was to create a system of frequent job rotation. This did not remove the risk but did reduce it to a certain extent.
- Repetitive work: workers continually using the same set of muscles to complete tasks can feel fatigue in those muscles.
- Work rate: workers might be required to operate at a high work rate with insufficient breaks.
- Clarity of gauges and dials: there may be important information to be read from gauges or dials, such as machine speed, temperature and pressure. If the information from the gauges is not clear, this can lead to mistakes and accompanying stress. For example, leaning forward to read dials can put stress on the body, but if there are several to read in a short space of time, and the dials are inappropriately spaced and contain more than one parameter, rapid and frequent neck movements may be required; this can result in further stress.

Musculoskeletal health

- Manual handling operations: manual handling may be required, leading to strain on the body.
- Vibrating equipment: workers may be using vibrating equipment such as drills; without adequate controls, this could result in them suffering hand-arm vibration syndrome (HAVS) (see 5.2: Vibration).

The working environment

An inferior working environment may lead to ergonomic problems, for example:

- Space: some mobile workers might have to complete tasks such as manual handling in a restricted space, which could result in potentially harmful twisting and turning while working. Back pain may result.
- Ventilation: if there is too little ventilation and high relative humidity, this could cause considerable discomfort for workers. But ventilation directly onto the necks of some workers could lead to significant neck pain.
- Lighting: this may not be suitable or sufficient. If the lighting level is too low, this could cause eye strain, but if it is too high this could result in glare when trying to read gauges or information on a computer screen, particularly if there is glare on the screen.

The workers

- Physical characteristics: workers may not possess the individual physical characteristics, such as height, strength or level of fitness, to carry out required tasks.
- Mental characteristics: workers may not have the individual mental characteristics, such as the ability or experience, to carry out tasks.
- Psychological characteristics: workers may not have the appropriate psychological characteristics, such as maturity and patience, for a task.
- Dexterity: workers may lack the dexterity required for a task.
- Health problems: impaired eyesight or hearing could affect the ability of a worker to complete a particular task.
- Training: some workers might not have been fully trained to do a task.

APPLICATION 2

Workers were seated at workstations. Their task was to turn to the right, pick up a small screw (with a component on it) from a box of screws, fit a small nut from a box directly in front of them onto the screw and put the assembled part in a box placed to the left of them.

Lighting was provided by overhead lighting in a tall building.

Workers were complaining of pains in their fingers and eye strain.

Outline some measures you would recommend that might reduce the risk of finger pains and eye strain.

A suggested answer is provided at the end of 6.1: Work-related upper limb disorders.

We now turn to a condition known as work-related upper limb disorder (WRULD).

6.1 Work-related upper limb disorders

> **DEFINITION**
> ***WRULD*** *is a medical condition that affects muscles, tendons, ligaments, nerves, joints and other soft tissues in the upper part of the body from the neck down to the hands and fingers. It is caused by work or can be made worse by work.*

You need to be aware of ill-health effects that could result from WRULDs. These are similar to the effects caused by substandard ergonomics in general.

Possible WRULD ill-health effects

The effects of WRULDs can include:

- pain in the neck, shoulders, back and upper limbs;
- swollen joints;
- numbness in the hands and fingers;
- weakness of the hands, possibly along with a loss of dexterity;
- general stiffness in joints, leading to a loss of mobility;
- tendon strains;
- muscle fatigue; and
- difficulty sleeping.

Causes of WRULDs

Causes of WRULDS could include:

- Repetitive actions: using the same muscles continually will inevitably put a strain on those muscles.
- Uncomfortable working positions: an uncomfortable position can result in workers having to overreach to carry out their work, such as having to reach forward to get a component from a conveyor or twisting their bodies to get a component to the side of them. This type of action can result in bad posture, particularly if workers' chairs are not adjustable.
- Use of excessive force: the need to exert excessive force could be due to a difficult manual handling operation, such as placing an awkwardly shaped object on a high shelf, which might lead to back pain. Having to unscrew a tightly fitting container lid might result in wrist pain.
- Long duration of work: this can lead to muscle fatigue, particularly if workers are not provided with sufficient breaks.
- High work rate required: this might include having to keep up with the speed of a conveyor.
- Tools provided for the work: these might be the wrong type or may not have been well maintained.
- Inadequate training for the task: this could mean workers adopt bad postures when carrying out their work.
- Environment (space): workers may not have sufficient space to carry out their tasks, so they turn round awkwardly or twist the trunk of their body.
- Environment (lighting): if the level of lighting is too low, workers may have to bend forward to be able to see what they are doing, leading to backache. If it is too high, there could be excessive glare or reflections from a working surface, resulting in visual fatigue. The lighting might be the wrong colour. For example, an electrician who needs to see the colour of various sheaths of wires when working under sodium lighting may perceive all wires as orange, again resulting in visual fatigue.

- Environment (other): the working temperature may be too hot or too cold and the relative humidity too low or too high. Ventilation might not only be too low or too high, it may also be badly directed, so some workers have none and others get a draught directly onto their necks.
- Capabilities: some workers may not have sufficient strength or dexterity to carry out tasks. Others may have existing conditions; for example, a worker asked to carry out work using a highly vibrating tool may already suffer from Raynaud's syndrome.

> **ADDITIONAL INFORMATION**
>
> Raynaud's syndrome (also known as Raynaud's phenomenon) is common and does not usually cause severe problems. Individuals can often treat the symptoms by keeping warm. However, it can sometimes be a sign of a more serious condition and can develop as a result of using vibrating tools.
>
> Raynaud's affects your blood circulation. When a person with Raynaud's is cold, anxious or stressed, their fingers and toes may change colour. Other symptoms can include:
>
> - pain;
> - numbness;
> - pins and needles; and
> - difficulty moving the affected area.

6.1.3 Avoiding/minimising risks from poorly designed tasks and workstations

There are a number of control measures you could use to minimise WRULDs. Consider the following strategies for different areas of risk.

- Repetitive work: if the task cannot be redesigned, for example by automation, then using more workers could cut down the amount of repetition. Failing this, job rotation might be considered so that workers can carry out different tasks that use different muscles. If this is also not feasible, then workers would need to be given extra breaks.
- Uncomfortable working positions: it may be possible to alter the layout of the workstation to reduce overreaching. If this is not feasible, consider providing adjustable seating.
- Use of excessive force: it may be that the wrong hand tools have been provided, so the correct tools should be provided, or, even better, power tools, if this is feasible. But providing power tools may introduce other risks, such as HAVS. To avoid replacing one type of risk with another, any power tools should be of low vibration magnitude if possible. Where the force needed is due to manual handling, then the provision of lifting aids could be a solution. Failing that, team handling might be possible. Finally, whatever risk reduction measure is used, there needs to be adequate training so that workers know how to use the control measure, and adequate supervision so they use it properly. Training should include an explanation of how workers can achieve good working postures.

6.1 Work-related upper limb disorders

Figure 1: Substandard ergonomic method for moving boxes

Figure 2: Better ergonomic method of moving boxes using manual handling aid

- Excessive hours/long duration of work: if practicable, allow workers to spread the work over more time. If this not practicable, introduce extra breaks.
- High work rate: employ more workers if this is practicable. If not, use job rotation so that the highly paced work rate is shared among more workers.
- Environment (space): clear the work area of any unnecessary equipment or materials and change the work layout to create more space. If space remains inadequate, consider carrying out the work in a different area where there is more space.
- Environment (lighting): the level of lighting should not be too low, causing eye strain, or too high, resulting in glare or reflections. Surfaces should be matt where possible to avoid reflections. Place workstations so that light from the sun does not cause glare. Install a suitable type of lighting, for example, of the right colour.

- Environment (other): the working environment should be a suitable temperature with sufficient ventilation.
- Individuals: select individual workers who do not have health conditions that may increase the risk of WRULDs. Workers with health conditions should be subject to risk assessment before doing any task that may result in them experiencing WRULDs; they should also receive health surveillance at regular intervals. Any tasks given to workers should match their skills and their health needs.
- Administrative: there should be a reporting system available to any workers who believe they have WRULDs.

You need to be able to identify a few occupations where the workers may be at risk from WRULDs.

Examples of at-risk occupations include pneumatic drill operators, production or assembly line workers, painters and decorators, checkout operators working in supermarkets and DSE users.

We now move on to a specific type of work, where the risk of occupational ill-health is significant – the use of DSE. More and more tasks are being allocated to computers. Computers can solve many problems, perform calculations quickly and are a valuable work tool. We will review the hazards and risks associated with DSE and what to look for in a DSE risk assessment.

The main safety hazards associated with DSE work are tripping over cables and possible electrical hazards, such as a worker attempting electrical repairs without being competent to do so. However, with DSE work, exposure to health hazards is far more likely.

Ill-health effects from DSE work

There are four main health hazards associated with display screen equipment (DSE) work:

1. WRULDs: these can include pain in the fingers, swelling of joints, stiffness in the arms and wrist pain. The severity can range from temporary pain to more severe, well-defined conditions such as carpal tunnel syndrome.
2. Musculoskeletal disorders (MSD): these are characterised by pain in the neck, back and/or shoulders.
3. Visual problems: medical evidence has shown that DSE work does not normally cause permanent damage to the eyes or make existing conditions worse. However, temporary visual fatigue can occur, resulting in symptoms such as tired or sore eyes, temporary short-sightedness, headaches and migraines. Visual symptoms may result from issues such as staying in the same position and concentrating for a long time, flickering images on the screen, inadequate lighting, glare or reflection.
4. Stress: this is acknowledged to be one of the leading causes of occupational ill-health. Evidence suggests that many symptoms described by DSE users relate to stress. There should be management of undesirable aspects of DSE operation, such as users having minimal control over their work and working methods, repetitive and monotonous tasks and demands of the work being perceived as excessive.

Figure 3 shows the potentially harmful set-up of a DSE workstation. Note the user's posture, leaning forward with the neck bent forward to enable them to read the screen.

6.1 Work-related upper limb disorders

Figure 4 demonstrates a much-improved posture. The chair is supporting the user's back and the screen is at an appropriate level to be comfortably read.

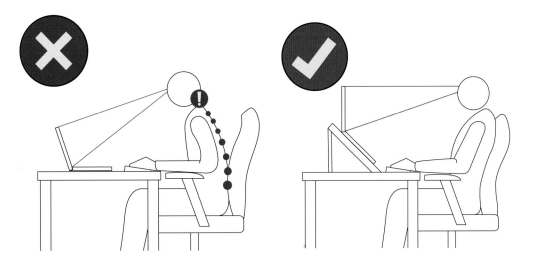

Figure 3: An unsatisfactory workstation set-up

Figure 4: A better workstation set-up showing good posture

Figure 5 demonstrates good general principles for a DSE workstation.

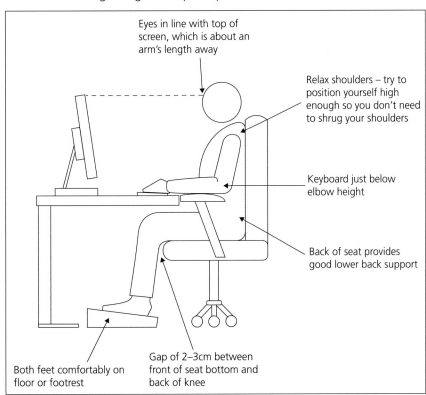

Figure 5: General principles for setting up a workstation

DSE risk assessment

The person carrying out a DSE risk assessment must be competent to do so; that is, they should have appropriate knowledge about DSE workstations and experience of carrying out such assessments and possess appropriate personal qualities, such as good communication skills, as it will be necessary to discuss assessment factors with users. The assessor may be an internal worker or an external consultant.

Frequency of assessments

There is no specific time interval identified in the DSE Regulations[7] for carrying out DSE risk assessments. However, assessments should be carried out in the following circumstances:

- when a new DSE workstation is set up;
- before a new user works at a DSE workstation for the first time;
- when a workstation undergoes substantial change – this could be a change of equipment (such as a different chair), a change of task (such as an increase in hours worked at the workstation) or a change of environment (such as a change of lighting experienced, which could be caused just by moving the workstation in the workplace);
- if a user is working from a remote location or from home;
- if a user believes that their DSE work is causing discomfort or ill-health issues; for example, the user may be experiencing visual fatigue that they believe is resulting from their DSE work, in which case the user may request eye and eyesight tests: under reg 5 of the DSE Regulations, an employer is then required to ensure that these tests are provided; and
- apart from these points, at regular intervals. In such cases, the time interval between assessments will be determined by various things; for example, intervals may be identified in the arrangements section of the organisation's safety policy.

Here are some of the things to look for when carrying out a DSE risk assessment:

The screen:

- Can information on the screen be easily read? Is the screen clean?
- Is the screen at the correct height? Is it adjustable?
- Is the screen free from any flicker?
- Can the screen swivel and tilt?
- Are there brightness and contrast controls?
- Is the screen free from any glare or reflection?

The chair:

- Does it have a stable base?
- Is the height adjustable?
- What is its general condition like? For example, is the covering material torn?
- Does it have good lumbar support?
- Does it have a swivel facility?
- Does the chair have armrests? Some users prefer chairs without armrests; what is important is whether the DSE user feels comfortable. Armrests should not prevent the chair being close enough to the desk.
- Is the seatback adjustable?
- Is a footrest available if required?

The desk:

- Is the desk large enough in terms of having sufficient space on the desk surface?
- Is the desk positioned near enough to electrical sockets? A DSE workstation may need power to various equipment such as the screen, a printer and a desk lamp.
- Do any cables pose a trip hazard?
- Is there sufficient space under the desk? Check the space is not being used to store materials.

- Does the desk have a matt surface? Is it causing the DSE user to be exposed to any glare or reflection?
- Has a document holder been provided and, if so, is it in a suitable position?

The keyboard:

- Is the keyboard separate from the screen?
- Does the keyboard have a matt finish so the user does not experience any glare or reflection?
- Does the user need to have a wrist support?
- Is the keyboard in the correct position?

The environment:

- Is there sufficient space at and around the workstation?
- Has storage been provided away from the workstation for paperwork such as files and for other equipment?
- Are the working temperature and relative humidity reasonable?
- Is there any significant level of background noise?
- Is lighting suitable, sufficient and well positioned? If the level is too high, there may be difficulty in reading the screen. If the light is the wrong type, such as unshaded fluorescent bulbs, these could be reflected onto the screen. If there are no blinds and the user faces a window, there could be glare into their eyes. If the window is directly behind the user, there may be glare onto the screen. If blinds are provided for windows, the assessment should include checking that they are in good condition and are being used properly.

The software:

- Is the software easy to use? You would need to consult the user about this.

The mouse or trackball:

- Do users find this device easy to use? Again, you would need to consult them.
- Look at the position of the hand and arm. The arm needs to be relaxed and should not be stretched out when using the device.

The user:

- How easily can the user view the information on the screen? Does the user need glasses specifically for screen work?
- Consider the physical characteristics of the user. For example, an operator of limited height may require a footrest and a user of significant height might need a higher-than-average desk surface.

Apart from asking the DSE user about the mouse or trackball and the software they use, much of a DSE risk assessment may be carried out by observation. However, some important information will also need to be obtained by asking the user questions. These would include:

- Whether the user had experienced any MSDs resulting from DSE work; of particular interest would be any instances of back pain, numbness in hands or fingers, neck pain or shoulder stiffness.
- Whether the user had experienced any eye strain, headaches or migraines; you should find out if the user wears glasses and, if so, whether these glasses are suitable for DSE work.
- If the user had experienced any of these adverse effects and whether they had reported them to their supervisor or manager.

Musculoskeletal health

- The amount of time that the user undertakes DSE work without taking a break. It is usually considered more beneficial to take a 5–10-minute break every hour rather than a 20-minute break every 2 hours.
- The number of hours on an average day the user carries out DSE work.

Here is a simple example for you to consider.

APPLICATION 3

A DSE user has complained of significant neck pain (but not any arm or wrist pain). Their task is to input information into the computer. The information to be inputted has been placed in written form on the desk. The desk is a reasonable size. The screen is positioned to the right of the user. The chair is in good condition and properly adjusted. The user has a good sitting position in the chair. The environment is totally suitable. The user has good eyesight and does not need glasses to carry out the work. The user does not work for more than 1 hour without taking a break.

Why do you think the operator might be suffering from neck pain resulting from the DSE work?

A suggested answer appears at the end of 6.1: Work-related upper limb disorders.

Training

Training is required to be able to successfully conduct DSE risk assessments. However, rather than just stating that the training consists of being able to identify hazards and the control measures necessary to minimise risk, we can be more specific about the control measures required following a DSE assessment.

DSE-related training should include:

- familiarisation in the use of the software;
- proper use of the keyboard – in particular, not bending the hands at the wrist;
- keeping the mouse or trackball close to the body and not overreaching;
- not cluttering the desk but keeping it clear to create sufficient working space;
- using the chair's adjustment controls;
- ways to report any MSDs, visual or stress problems associated with DSE work and who to report these to; and
- the need for regular eye and eyesight tests.

CASE STUDY

A worker was required to read data while standing in front of a computer screen for a period of 30 minutes. The worker would then sit down at a desk and fill in a form in writing for 15 minutes, and then have a 15-minute break. This cycle was repeated throughout the working day of 7 hours and 45 minutes. There was also a 45-minute break for lunch during the day. This does not seem to be a particularly high-pressure schedule, yet the worker complained of a stiff neck and shoulders.

It was found that the worker had not received an eye and eyesight test for eight years and was having to bend forward to read the computer screen due to extreme short sight. The worker was sent for an eye and eyesight test, received a pair of glasses and was then able to continue the work without any further problems.

6.1 Work-related upper limb disorders

KEY POINTS

- Ergonomics is about the relationship between a worker, the equipment they use and where they work. Musculoskeletal disorders (MSDs) can be a consequence of not following good ergonomic practices.
- MSDs and related ill-health effects include aches and pains, stiffness, fatigue, reduced mobility, weakness, muscle fatigue, loss of dexterity and disturbed sleep.
- Work-related upper limb disorders (WRULDs) can result in similar ill-health effects and may be caused by repetitive actions, uncomfortable working positions, using excessive force or the wrong tools, the work environment and having to work too fast or for too long.
- Control measures to minimise occurrences of WRULDs include redesigning or automating tasks, changing the work layout and environment, providing the correct tools and introducing job rotation to share tasks.
- The main health hazards associated with using display screen equipment (DSE) are MSDs, WRULDs, visual problems and stress. A DSE risk assessment should involve consulting the operator and providing suitable training as well as looking at the screen, chair, desk, keyboard and related devices and work environment.

APPLICATION 2: POSSIBLE ANSWER

- If feasible, automate the process.
- If automation is not feasible, employ job rotation so that the workers are not carrying out the same task throughout their shift.
- If job rotation is not feasible, provide extra breaks.
- Provide adjustable swivel chairs to reduce the effort of twisting the trunk to turn to the left and right from the sitting position, and train workers to adjust their chairs.
- Provide local lighting at each workstation to reduce the risk of eye strain; each DSE operator could then adjust their lighting to their own preference.

APPLICATION 3: ANSWER

The worker had placed the written material to be inputted into the computer to the left of them on the desk. The chair was directly in front of the desk. The screen was placed to the right of the computer at approximately eye level.

During the task, the worker was continually turning their neck left and right between the input document and the screen. Also, because the input data were flat on the desk and the computer screen was at eye level, there was a vertical movement of the neck occurring at the same time as its horizontal movement.

The problem was resolved by providing a document holder for the input data and placing it next to the computer screen. This eliminated the need for the neck movements.

References

[1] World Health Organization, 'Musculoskeletal health' (2022) (www.who.int)

[2] See note 1

[3] HSE, *Ergonomics and human factors at work: A brief guide* (INDG90, 3rd edition, 2013) (www.hse.gov.uk)

[4] HSE, 'Musculoskeletal disorders' (www.hse.gov.uk/msd)

[5] HSE, *Manual handling. Manual Handling Operations Regulations 1992. Guidance on Regulations* (L23, 4th edition, 2016) (www.hse.gov.uk)

[6] HSE, *Work with display screen equipment. Health and Safety (Display Screen Equipment) Regulations 1992 as amended by the Health and Safety (Miscellaneous Amendments) Regulations 2002. Guidance on Regulations* (L26, 2nd edition, 2003) (www.hse.gov.uk)

[7] See note 6

6.2: Manual handling

> **Syllabus outline**
>
> In this section, you will develop an awareness of the following:
> - Common types of manual handling injury
> - Good handling technique for manually lifting loads
> - Avoiding/minimising manual handling risks by considering the task, the individual, the load and the working environment

Introduction

Manual handling can be defined as the movement of a load by human effort alone. This includes lifting, pushing, pulling, carrying and putting down objects.

> **TIP**
>
> The UK's legal requirements for managing manual handling risks are laid out in the Manual Handling Operations Regulations 1992 (MHO); much of the information we will discuss is drawn from MHO and the related HSE publication *Manual handling. Manual Handling Operations Regulations 1992. Guidance on Regulations* (L23).[1]
>
> Simple practical guidance on manual handling (including practical requirements and controls) can also be found in the HSE document *Manual handling at work*.[2]

In the UK, the Health and Safety Executive (HSE) has estimated that more than 25% of all accidents result from manual handling operations.[3]

All types of workers are at risk, whether regular manual jobs are carried out or the work is mostly done while seated.

Back injuries are the most common consequences of bad manual handling, and we start by providing a brief anatomical overview of the back before progressing to common types of manual handling injuries.

We then move on to good manual handling techniques.

We will largely concentrate on avoiding or minimising manual handling risks, including possible control measures.

6.2.1 Common types of manual handling injury

We will start with a simple overview of some relevant anatomical consequences of badly carried-out manual handling actions.

Back injuries are one of the most common types of injury following manual handling.

Here is an outline of the structure of the back.

The spine is a series of bones called vertebrae; seven cervical, 12 thoracic and five lumbar. The lower part of the abdomen is supported by two more bones, the sacrum and coccyx (**each of these is made up of smaller bones fused together**). Each

Musculoskeletal health

vertebra has a thick rounded body at the front, while the back part surrounds and offers some protection to the spinal canal (which contains the spinal cord). There are 31 pairs of nerves that branch out from the spinal canal.

Figure 1 illustrates the general layout of the vertebrae.

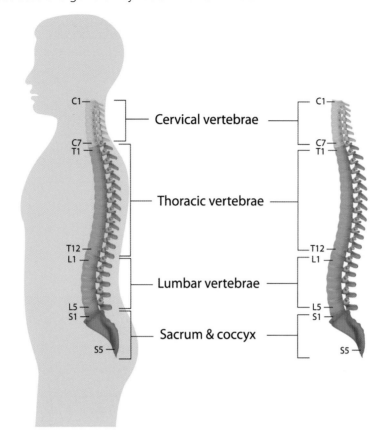

Figure 1: The vertebrae

Each vertebra is separated from its adjacent bone by a fibrous intervertebral disc. When force is applied to the spine, the discs act as shock absorbers to prevent damage.

The disc consists of two parts: a jelly-like nucleus (containing about 80% water) and an outer shell containing several layers of fibres.

As people get older, discs lose their flexibility and are more prone to damage. Improper manual handling can cause prolapsed or herniated discs.

A prolapsed disc occurs when the disc bulges outwards and makes contact with the spinal cord. This in turn can cause compression of a nerve root, leading to significant pain. Note: this is sometimes referred to as a 'slipped disc' even though this is not technically correct because discs do not slip.

Another similar condition is a herniated disc. In this case, the liquid in the nucleus breaks through the fibrous layers of the disc and spurts onto the spinal cord. Again, the nerve root can be affected, causing pain.

Next we need to briefly consider ligaments. These are bands of fibrous tissue that connect two or more bones. They lengthen during tension, then normally return to their original state when the tension is removed. If ligaments are overstretched or twisted, they become torn or strained. Pain and swelling can result from this. Ligament tear often results from a joint twist caused by repetitive movements such as continuous improper lifting of loads. Figure 2 illustrates several types of ligament.

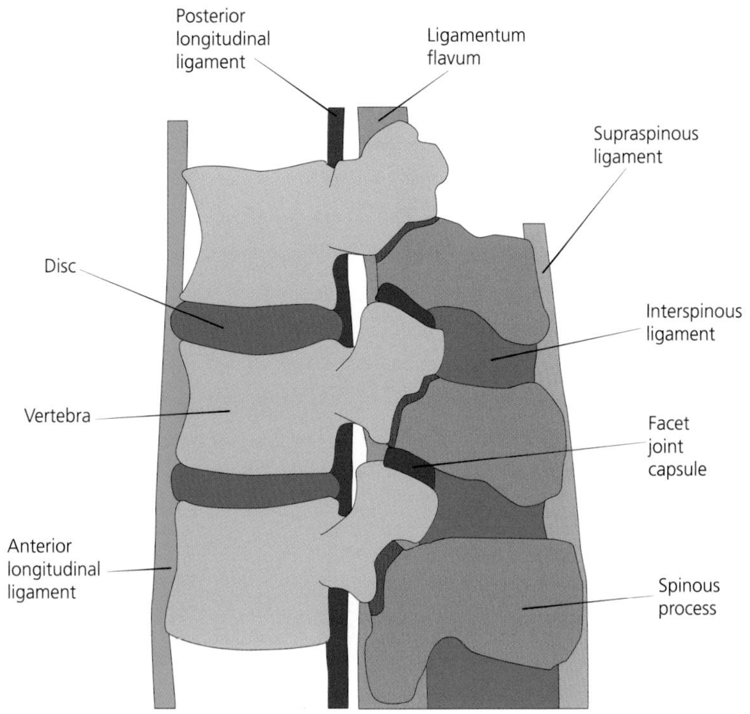

Figure 2: Types of ligament

Meanwhile, tendons are fibrous cords of tissue that link muscles to bones. A tendon strain is a stretch or tear in the tendon that can often result from overuse or repetitive use. The condition of tendinitis (or tendonitis) results from tendons becoming swollen. This can lead to joint pain.

We now return to the common types of manual handling injury. These are:

- prolapsed discs;
- herniated discs;
- torn/strained ligaments;
- tendon strains;
- muscle strains; and
- hernias.

An example of a herniated disc is shown in Figure 3. Note the red bulge showing the herniated disc (the pen is pointing to this in the image), which can now press against spinal nerves and cause pain.

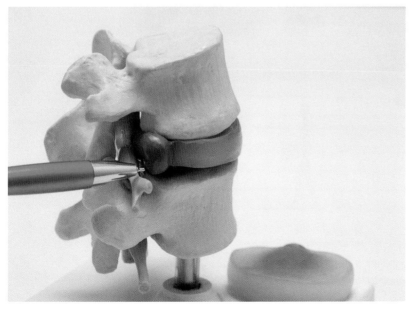

Figure 3: Herniated disc

Musculoskeletal health

Figure 4 shows areas where muscle strain may be felt – see the yellow areas with the orange centres either side of the spine.

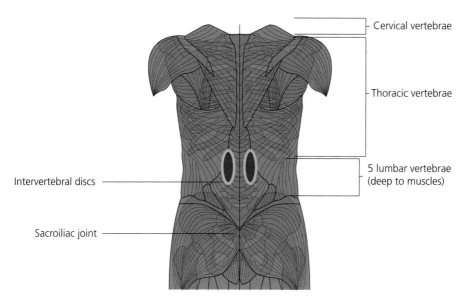

Figure 4: Inflammation/strain in back muscles

It is also possible that improper manual handling could lead to heavy loads being dropped on fingers, legs or feet, so other possible injuries include:

- cuts;
- abrasions;
- crushing; and
- fractures.

6.2.2 Good handling technique for manually lifting loads

Figure 5: Good technique for manual handling[4]

A good manual handling technique should incorporate the following steps.

- Firstly, plan the lift. Think about aspects such as the weight and size of the load and whether it has any sharp edges. Also consider potential lifting aids, rest stops and the distance it must be carried.
- Stand with the feet close to the load and slightly apart, with one foot slightly forward for good balance.
- Grip the load firmly.
- Lift the load smoothly with a slight bending of the back, hips and knees, as opposed to stooping forward or squatting.
- Do not bend the back any further forward while lifting. This tends to happen if the legs straighten before the load is lifted.
- Keep the load close to the body, at waist height, while lifting and travelling.
- Keep the head up and move smoothly.
- Do not twist the back or lean sideways. Keep shoulders level and in line with the hips.

> **TIP**
>
> The physical effort needed to lift the load should be felt in the thighs and not the back.

6.2.3 Avoiding/minimising manual handling risks

Before assessing manual handling risks, you need to consider the following, which identifies employers' duties under the MHO Regulations:

- What activities to assess: avoid assessing trivial manual handling risks, such as lifting a small cup from a table at waist height and placing it on a nearby desk that is also at waist height. It would not be reasonable to allocate resources to managing this particular manual handling operation. If all risks are trivial, they can be ignored, and no further action is needed.
- Avoiding manual handling if reasonably practicable: you may be able to use some form of lifting aid, such as a trolley for lighter loads or a forklift truck for heavier loads. More detailed information is provided in 6.3: Load-handling equipment.

It will not always be possible to make use of a lifting aid, so manual handling will need to be carried out. Therefore, you must assess the manual handling risks and put in place controls to help minimise the risk of injury.

There are many factors you need to consider when assessing manual handling risks to help highlight any conditions that could increase the risk of injury. However, there are four main areas from which risk can arise and these can be remembered using the simple acronym TILE:

1 **T**ask
2 **I**ndividual
3 **L**oad
4 **E**nvironment

We will now examine what needs to be considered for each of these four areas.

Task

- The distance the load is carried from the trunk (torso) of the body
- Twisting of the trunk (torso), stooping or excessive horizontal reaching is required
- Excessive lifting or lowering (that is, bending down or reaching up to grasp a load) is required
- The distance a load must be carried
- Strenuous pushing or pulling
- The load requires precise positioning
- Sudden movements of the load – these could cause loss of control of the load
- The frequency and duration of the task (that is, how often and for how long)
- Breaks available for the workers
- Repetitiveness – this would lead to the same set of muscles being used all the time
- Whether the rate of work is being imposed by the process – the task might involve working at a conveyor travelling at a specific rate
- Whether the load is moved while seated – reduces space to move the load, cannot use stronger leg muscles

Individual

- Specific strength or height required – whether workers are capable of carrying out the task
- Health issues of workers (for example, asthma)
- Vulnerable workers (for example, pregnant workers)
- Extra or special training is required

There are two additional factors that could be considered but should be approached with caution – age and gender. Allocating loads to individuals based on their gender or age without consideration of their personal characteristics (in particular, their fitness and experience) could lead to accusations of discrimination.

Load

- The weight of the load
- The size and shape of the load
- The ease or difficulty of grasping the load
- The load's stability; for example, a liquid load is likely to be unstable
- The load's centre of gravity
- Whether the load has handles and, if so, their position
- The temperature of the load
- Sharp edges
- Whether the load might move (for example, if it is a person or animal)

Environment

- The amount of space available
- Condition of floors or ground surfaces (for example, slippery or damaged)
- Variations in floor level (for example, does the load need to be carried up an incline or steps?)
- Obstacles in the direction of travel
- Extremes of temperature and relative humidity

- Ventilation (for example, there may be none, causing excessive sweating of workers' hands and possibly leading to loads being dropped)
- Weather conditions if loads need to be moved outside (for example, strong winds)
- Lighting – is it suitable and sufficient and is visibility good?

There is one further aspect that is occasionally included with the individual elements but is probably best considered as a stand-alone point. This is the possible effect of wearing personal protective equipment (PPE) or other clothing that might lead to awkward movement while lifting.

Control measures to minimise risk from manual handling

APPLICATION 1

(a) Identify three manual handling operations in your workplace that might cause injury.
(b) Identify the manual handling operation of highest priority; that is, the one most likely to cause significant back (or other) injury.
(c) Using the TILE steps described, carry out a risk assessment of this highest-priority operation.
(d) Describe appropriate control measures to minimise the risk of injury in this highest-priority operation.

Here are some general risk control measures relating to the four risk areas identified under TILE.

Task

Distance of load from trunk: a simple load like a cardboard box may be easy to hold close to the body, but an awkward load may have one end heavier than the other. In this case, keep the heavier end close to the body.

Stooping/twisting/overreaching: you may be able to change the system of work. For example, if there is twisting due to moving an object from one surface to another at right angles to the first, perhaps one surface can be moved to remove the need to turn through 90 degrees.

Excessive vertical movement: if objects that need to be moved are stored on shelves above head height, at waist height and at floor height, perhaps all objects can be moved to waist height. If this is not feasible, put light loads at high level, heavy loads at waist level and medium load at floor level. It is not an ideal solution, but at least the risk has been reduced. This has assumed that no lifting aid is available.

Distance: try to shorten it.

Pushing/pulling: try to push loads rather than pull them. If this is strenuous, a lifting aid such as a pallet truck is probably needed.

Frequency/duration/breaks/repetitive work: job rotation might be possible so that workers carry out less manual handling work. If this is not feasible, perhaps extra breaks could be provided.

Work rate imposed by process: for example, slow the rate of the conveyor being used. If this is unacceptable to production management, consider automation of part of the process.

Individual

Unusual strength and height required: you may have to select specific individuals to undertake these tasks.

Workers with health issues or vulnerabilities: these workers must be carefully assessed to see if they need to avoid manual handling in future or should only work on limited tasks for limited times.

Training: any special training needed must be provided. It is wise to keep a record of such training.

Age and gender: as stated, to avoid accusations of discrimination, it is important to focus on the selection of individuals and not make assumptions based on their age or gender, for example.

Load

Weight: firstly, if material is being delivered in heavy containers, ask the supplier to deliver in smaller containers. For example, at one time cement was delivered in heavy bags but, following requests from users, smaller bags are now available. If there are heavy containers generated by the work, try to split loads. Finally, if no reduction in load is possible, you may have to resort to team lifting (by two or more workers). Suppliers should have marked the load with its weight.

Size and shape: these are not easy to manage; split loads if possible.

Grasping the load: if loads are slippery, it could be of great help to attach handles. It may be possible to ask suppliers to deliver loads in containers with handles.

Stability: a badly packed load might result in contents shifting around, so good packing of the load might reduce instability. Where there is inherent instability, a warning should be given to encourage careful handling.

Centre of gravity: where the centre of gravity is not central, the supplier should mark the heaviest side.

Temperature: it may be possible to delay moving the load until its temperature has stabilised to room temperature. If not, devices may need to be used to transfer the load.

Sharp edges: these should be covered before attempting to move the load.

Environment

Space: when there is restricted space, find out whether the manual handling can be carried out in another area. If not, see if the current space can be cleared; it may be possible to improve the layout of the area.

Floors: damaged floors should be repaired and slippery floors cleaned.

Variations in levels: if steps are involved, a goods lift would be the best answer. If no lift is available, then team lifting (by two or more workers) may be an option. Failing this, some other form of lifting aid may be needed.

Obstacles: clear movable obstacles out the way. If obstacles are fixed, walk the route first to familiarise yourself with the location of the obstacles, then carry out the manual handling task.

Temperature, humidity and ventilation: there may not be much control over these considerations. Therefore, for example, it may be a case of taking breaks in very hot conditions.

Weather: appropriate clothing should be worn, but bear in mind that, in cold conditions, extra clothing may impede movement when carrying out tasks.

Lighting and visibility: care should be taken not to impair visibility by carrying a load so you cannot see over the top of it. Carry less load. The worker may not have much control over lighting but should report any resulting visibility problems to their supervisor.

The following case study illustrates the importance of workers knowing the weight of objects they need to handle and receiving training in manual handling.

CASE STUDY

A worker (a design engineer) was asked by their employer to tidy up a storeroom.

This was not a task that the worker was expected to carry out, nor had they been asked to carry out this task before.

The worker was lifting and moving heavy boxes throughout the morning. There was no indication of the contents of the boxes and their sizes were somewhat different. The boxes were not labelled with their weight. Crucially, no manual handling training had been given to the worker.

By the end of the day, the worker was experiencing significant back pain. The worker carried on like this for a further month, by which time their back pain had become severe. The worker consulted a doctor, who diagnosed strained muscles in the lower back and signed the worker off from work for a month. The employer refused to pay the worker for this month off work. The worker, with the aid of a law firm, took out a claim against the employer, which subsequently agreed an out-of-court settlement. The worker received compensation for the back injury and payment for the month off work.

Table 1 shows a checklist layout that could be used to assess manual handling risks.

Risk type	Risk factor	Low risk	Medium risk	High risk	Comments
Load	Weight				
	Size				
	Shape				
	Centre of gravity				
	Ease of grasp				
	Stability				
	Sharpness				
	Temperature				
Task	Lift height				
	Distance of carry				
	Twist/stoop				
	Distance from trunk				
	Frequency				

Musculoskeletal health

Risk type	Risk factor	Low risk	Medium risk	High risk	Comments
	Duration				
	Repetitiveness				
	Process work rate				
Environment	Space				
	Obstacles				
	Temperature				
	Damaged floor				
	Weather				
	Slippery surface				
	Flat/incline				
	Lighting				
Individual capability	Unusual strength or height needed				
	State of health				
	Fitness				
	Level of training				
	Pregnancy				
Is movement or posture hindered by clothing or personal protective equipment?	Yes/no		Comments		

Table 1: Sample manual handling risk assessment checklist

Let us now review a simple scenario requiring the assessment of manual handling risks and suggest some control measures to help reduce risk.

APPLICATION 2

> An object that is not very heavy needs to be moved onto a high shelf from a table approximately 4 metres away. The surface of the table is at waist height. The object is quite hot and has a smooth surface. The floor is well worn in places so is slippery in sections. A young worker, in their first day of employment, has been asked to move the object.
>
> Using the TILE approach, what would you consider before this task takes place?

The weight of the object and the height of the table present a low risk, whereas, on completing the assessment form in Table 1, the following elements are identified in the 'high risk' column: height of lift, temperature of object, difficulty in grasping object and slippery floor.

The suggested control measures to reduce these risks are shown in Table 2. Note: short-term measures can provide a quick solution temporarily, while a longer-term solution may provide more permanent control.

6.2 Manual handling

Risk factor	Short-term measures	Long-term measures
Height of lift	Use of suitable steps	Change system so that high shelf not used
Temperature of object	Use of suitable gloves	Investigate possible temperature reduction
Smooth surface of object	Affix handles	Purchase object with handles already present
Slippery floor	• Mop floor • Find source of any leak and repair it	• Change floor surface • Install drain

Table 2: Suggested control measures

Some measures may solve two problems. For example, fixing insulated handles to the object may overcome the difficulty of grasping it and the problem with its temperature.

KEY POINTS

- Common types of injury that can occur due to improper manual handling include prolapsed and herniated discs, torn or strained ligaments, tendon and muscle strains, hernias and cuts, abrasions, crushing and fractures from dropping heavy loads.

- A good manual handling technique for manually lifting loads starts with planning the lift and should incorporate standing for good balance, gripping the load firmly, lifting it smoothly and keeping it close to the body at waist height.

- There are many considerations that contribute to a manual handling risk assessment, which can be grouped by task, individual, load and environment (TILE).

- Control measures that could be used to help reduce the risk of injuries from manual handling might include changing systems of work and processes, job rotation, lifting aids, training, selecting appropriate workers for specific tasks, splitting loads, clearing and repairing floors and moving tasks to another area.

- A manual handling risk assessment should identify types of risk, specific considerations and whether these are low, medium or high risk.

- The risk assessment will inform control measures to help reduce the risks, including short-term measures to provide a quick, temporary solution and long-term measures to provide more permanent control.

References

[1] HSE, *Manual handling. Manual Handling Operations Regulations 1992. Guidance on Regulations* (L23, 4th edition, 2016) (www.hse.gov.uk)

[2] HSE, 'Manual handling at work' (www.hse.gov.uk)

[3] See note 1

[4] Source: https://www.hse.gov.uk/msd/manual-handling/good-handling-technique.htm

6.3: Load-handling equipment

> **Syllabus outline**
>
> In this section, you will develop an awareness of the following:
>
> - Hazards and controls for common types of load-handling aids and equipment: sack trucks and trolleys; pallet trucks; people-handling aids; forklift trucks; lifts; hoists for loads and people; conveyors and cranes
> - Requirements for lifting operations:
>   - strong, stable and suitable equipment
>   - positioned and installed correctly
>   - visibly marked with safe working load
>   - lifting operations are planned, supervised and carried out in a safe manner by competent persons
>   - special requirements for lifting equipment used for lifting people
> - Periodic inspection and examination/testing of lifting equipment

Introduction

When loads are too heavy, an awkward shape, too bulky or have other properties that mean you need to avoid manual handling, then you should turn to mechanical assistance to enable you to safely move them.

Mechanical assistance can be classified as either manually operated (non-powered) equipment or powered equipment.

We shall consider hazards and control measures for a wide range of both non-powered and powered work equipment.

We will conclude by explaining that legal requirements for inspecting, testing and examining lifting equipment are necessary to prevent the potentially serious consequences of equipment failure.

> **TIP**
>
> The main pieces of legislation relating to load-handling equipment in the UK are the:
>
> - Lifting Operations and Lifting Equipment Regulations 1998 (LOLER); and
> - Provision and Use of Work Equipment Regulations 1998 (PUWER).
>
> LOLER places duties on organisations and persons who operate, own or have control over lifting equipment. The maintenance and inspection of such equipment is also subject to PUWER.
>
> Both pieces of legislation are supported by Health and Safety Executive (HSE) guidance:
>
> - *Lifting Operations and Lifting Equipment Regulations 1998. Approved Code of Practice and guidance* (L113);[1] and
> - *Provision and Use of Work Equipment Regulations 1998. Approved Code of Practice and guidance* (L22).[2]

6.3.1 Hazards and controls for common types of load-handling aids and equipment

We shall be considering the hazards and controls for a range of equipment, including trolleys, hoists and pallet trucks.

1. Sack trucks and trolleys

The main hazards of this type of equipment are: manual handling hazards when placing loads in or on the equipment; loss of control, including overturning in the case of trolleys; falls caused by moving the equipment over slippery or damaged floors; encountering obstacles due to insufficient space or inadequate housekeeping; overloading the equipment; having to move loads on inclines; and poor visibility due to inadequate lighting, resulting in collisions with other workers or equipment.

Control measures will include:

- ensuring surfaces are suitable for transportation of the equipment, avoiding unsuitable surfaces such as damaged floors and using smooth, level, non-slippery surfaces where practicable;
- avoiding traffic routes that vehicles may use;
- reporting instances of defective lighting;
- ensuring that the equipment is not overloaded; and
- providing any necessary training.

2. Pallet trucks

Manual effort is used to push trucks along while hydraulics enable the truck to be raised.

There are three main hazards associated with pallet trucks: back strain from overloading of such trucks and, if continued over a long period of time, disc injury; bruised feet from loads dropping off the trucks; and possible injury due to collisions with other workers or equipment.

Control measures include: a visual inspection before use to check for any obvious signs of damage; raising the forks to check for any oil leaks and to check that the hydraulic pump is operating correctly; ensuring that the load is securely stacked; ensuring that there are no pedestrians in the way of forward travel; and not using the trucks in vehicle traffic routes.

3. People-handling aids

Hoists are used widely in hospitals and care homes; for example, to lift patients into and out of beds and baths (see Figure 1).

'People-handling aids' include special slide sheets that are used to transfer hospital patients from a trolley to a bed. The major part of such a movement is a horizontal movement rather than a vertical lift, so that the load (in this case, a person) is partially supported by the surface on which they are lying. Emergency evacuation chairs (such as those found in hotels and office buildings) are a further example of people-handling aids.

Musculoskeletal health

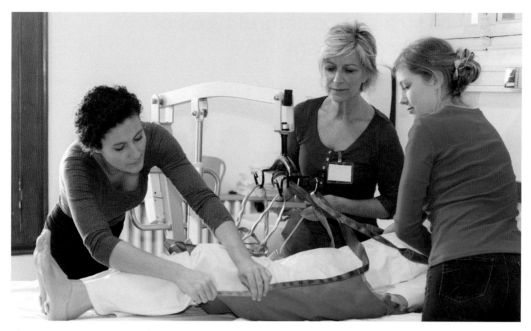

Figure 1: Using a hoist to lift a patient

The main hazards for people-handling aids are: possible unexpected movement from people, especially if being moved from a hospital bed; any faults in the equipment; the floor surface indoors, for example it may be slippery; and adverse weather conditions outdoors, for example a strong wind causing instability of the equipment.

Control measures would include:

- producing a lifting plan before the lift takes place;
- ensuring that the equipment is subjected to thorough testing, examination and maintenance at regular intervals by competent persons;
- ensuring the equipment is stable before use;
- ensuring that workers using the equipment are trained and competent;
- if moving a patient in hospital, explaining to them that they are to be moved and seeking their consent for the move;
- ensuring a sufficient number of workers are present before conducting the lift;
- ensuring that the floor surface indoors is in good condition, level and non-slippery; and
- ensuring that work is carried out in the safe working load of the equipment.

4. Forklift trucks (FLTs)

There are lots of FLTs in use in industry and unfortunately there are many accidents related to them. There are many fatalities, many more serious injuries and even more non-serious injuries each year due to FLT accidents.

There are many different types of FLT, designed for a range of activities. These include reach trucks, counterbalance trucks and pedestrian trucks. However, it is not necessary to go into detail on any one type of truck; a generic treatment of hazards and precautions is mostly sufficient. FLTs can be diesel, gas or electric.

The hazards relating to FLTs include:

- overturning (a key hazard) – 42% of FLT fatalities are due to workers being crushed by a truck tipping over;[3]
- collisions with other vehicles or stationary objects;

- fire and explosion due to charging electric trucks in an unventilated area (the charging process gives off hydrogen gas);
- injuries resulting from falling objects;
- injuries relating to drivers – drivers may not be physically fit, in particular their eyesight may not be adequate for FLT driving; and
- a manual handling hazard for LPG trucks, resulting from lifting gas containers onto the truck.

If FLTs are being used outdoors, further hazards are:

- weather conditions – icy surfaces may result in skidding;
- general visibility if working outside during hours of darkness – obstructions may not be seen; and
- lighting contrast issues – workers may be driving between poorly lit yards and brightly lit warehouses; driving from warehouse to yard may result in a temporary inability to see properly while yard to warehouse may result in temporary glare.

We now consider two of the major hazards in more detail.

Overturning due to instability of FLTs

This can occur due to:

- the load on the FLT being unstable or not properly secured;
- the FLT being overloaded;
- hitting obstructions such as kerbs;
- driving down slopes with the load forward (or reversing up slopes);
- speeding, especially around corners;
- mechanical failure (as a result of inadequate maintenance);
- sudden braking;
- unsuitable ground conditions such as an uneven or icy surface;
- not stacking or destacking loads in the correct manner; or
- turning around on slopes or crossing them at an angle.

Collisions

There may be many causes of collisions. It is convenient to classify these in four categories: workers, equipment, environment and workplace layout. As mentioned earlier, collisions can be caused by overturning trucks. Here are some examples of possible causes classified by category.

Worker causes might include:

- inadequate physical fitness; for example, struggling to get in and out of the FLT cab;
- impaired eyesight – drivers might not have had an eye test; they may not have seen a pedestrian and struck them with the FLT;
- bad driving; for example, driving too fast;
- worker failed to carry out any basic inspection procedures before using the FLT; and
- human behaviour – drivers may be reckless in nature, or they may have carried out the same tasks day after day and become accustomed to cutting corners without being involved in any incidents, then finally been involved in one.

Musculoskeletal health

Equipment causes might include:

- the FLT had not received regular, thorough inspection from a competent source; resulting in, for example, faulty brakes; the driver saw the pedestrian but was unable to stop in time.

Environmental causes might include:

- the floor for inside work was damaged or slippery;
- the ground for outside work was uneven and there were potholes not easily seen;
- the lighting was not suitable and sufficient (in terms of brightness and type);
- adverse weather outside resulted in limited visibility; and
- the environment was noisy and a pedestrian who was struck did not hear an electric FLT approaching. As the environment was noisy, the pedestrian may have been wearing hearing protection and electric FLTs are quiet.

Workplace layout causes might include:

- no separation of traffic routes for drivers and pedestrians;
- no speed restriction signs;
- no junctions with priorities clearly indicated;
- no road markings;
- no turning areas to reduce the need for reversing;
- the traffic routes were not of adequate width;
- no crossing points for pedestrians;
- no guardrails or barriers;
- no suitable hazard signs;
- inadequate lighting (in terms of positioning of lights);
- too many obstructions in the area; and
- no mirrors placed at blind corners.

Having considered the many ways in which FLT driving can go wrong, it is time to discuss measures to control the risks associated with FLT work.

To do this, we shall examine the three stages of this type of work:

1 What must you do before you start to drive?
2 What precautions should you take while driving?
3 What do you do once you have finished driving?

TIP

Before driving – you need to be sure your truck is in the condition it should be, so you need to carry out your pre-use routine inspection.

- Check for any new bodywork damage – if there is some, it is possible that the truck has been involved in a recent incident and requires a thorough check before use.
- Assuming no new bodywork damage, check the lifting mechanism to ensure it is working as it should be.
- Check to make sure there are no oil leaks.
- Check the proper functioning of brakes, horn and lights.
- If the truck is electric, check the battery; if it is a diesel truck, check the fuel level; if it is a gas truck, check the LPG container for leakage.
- Ensure the windows are clean, allowing good all-round visibility.

6.3 Load-handling equipment

> - Ensure mirrors are not damaged.
> - Check the steering works well.
> - Ensure the seat restraints are in good condition.
>
> If your pre-use inspection gives satisfactory results, you are now ready to drive.

> **TIP**
>
> **While driving**
>
> - Ensure the load is securely on the forks and is well balanced.
> - Drive at a safe speed.
> - Check whether the work surface is slippery; if so, reduce speed accordingly.
> - Ensure the cab rotating light is on.
> - Keep to recognised routes.
> - When carrying a load, drive forward up slopes and reverse down slopes.
> - Use the horn when necessary, particularly when driving near doorways or blind corners.
> - Use all controls correctly.

> **TIP**
>
> **After driving**
>
> - Park in the designated area.
> - Do not leave any load on the FLT.
> - Do not block any other vehicles from moving.
> - Ensure you have parked clear of any walkways and particularly any emergency exits.
> - Ensure the FLT is parked on a firm, level surface.
> - Ensure the brakes are on.
> - Ensure the forks are parked on the ground.
> - Ensure the mast is tilted slightly forwards.
> - Remove the key from the ignition.
> - Return the FLT key to its dedicated, secure location – it is essential that no unauthorised person can get access to the key.

5. Lifts

Lifts are commonly used for transporting both goods and people.

The main hazards associated with lifts are:

- Falls from height: this may occur if the landing gates are not securely locked when the lift is in motion and someone on the landing can open the gates (see Case study).
- People can get crushed between the lift doors (when closing).
- Injury can result from lift malfunction; for example, faulty brakes or an uncontrolled change of speed.
- Lift failure due to overloading: a person in the lift could be injured from an uncontrolled fall of the lift.
- Tripping hazard: the lift does not stop flush with the floor, resulting in a step between the lift floor and landing floor.

- Workers struck by falling objects: there could be materials falling from a goods lift with unprotected sides.
- Stress: small lifts may be regarded as confined spaces; a person stuck in a lift that breaks down between floors could panic.
- Infection when using passenger lifts: many workers (and possibly visitors) may use the lift, meaning that infections could be spread between passengers sharing the lift or through many people touching the surfaces inside the lift.

Control measures for lifts may include the following:

- Lifts should be subject to thorough examination, testing and maintenance at regular intervals by competent persons.
- While this examination, testing and maintenance is taking place, all practicable measures must be taken to prevent anyone (apart from the competent persons) gaining access to the lift; for example, by placing barriers on all landing stages by lift entrances and displaying warning signs.
- Lifts should clearly display their safe working loads (for example, the maximum weight they can carry).
- All passenger lifts should include an emergency button installed in the control panel that can be activated if the lift breaks down between floors.
- There should be emergency procedures in place in the event of a worker being stranded in a broken-down lift.
- Lifts should be thoroughly cleaned at regular intervals to reduce the risk of biological infection.

CASE STUDY
A worker was moving a loaded lift trolley on the second floor of a building when they fell into a lift shaft and landed five metres below, sustaining arm and pelvis injuries. Subsequent investigation revealed that the lift doors had a fault. Use of a door release key meant that the lift doors could be opened with the floor of the lift in any place, therefore the safety sensor had been overridden. The organisation responsible for the lift received a large fine in the resulting court case.

6. Hoists

Hoists are often used to transport heavy loads, such as moving bricks from ground level up onto a scaffold. Figure 2 shows a hoist in action.

Hazards associated with hoists include incidents relating to incorrect use, failure of the hoist due to loading above its safe working load (SWL), loss of balance due to a slippery floor and falling materials causing injury and failure of the hoist due to insufficient maintenance.

Control measures should include having a lifting plan in place before using the hoist. The correct slings should be selected and a pre-use check carried out on those slings. The hoist should be properly secured to its supporting structure. If a person is to be lifted, there should be enough workers available to carry out the lift and there must be good communication between them while conducting the lift.

It is essential that any lifting operation is carried out within the SWL of the equipment. Any lift should not be carried out over a long distance. The floor should be free from obstruction and not slippery. Types of hoist do vary, so it is important that those carrying out the lifting are trained in the particular hoist being used. There should be a regular maintenance schedule for the hoist, the maintenance should be carried out and a record of this should be retained.

Figure 2: Operating a hoist

7. Conveyors

There are several types of conveyor. Some examples are:

- Screw conveyors: materials are pushed forward on a rotating screw. These can be particularly useful for raising materials to a higher level. The screw needs full guarding.
- Belt conveyors: materials are transported on a moving belt. Guards are needed at in-running nip points (see 9.1: General requirements).
- Roller conveyors: these may be powered or non-powered. Powered conveyors need guards at in-running nip points where practicable. Non-powered conveyors do not have such nip points, but injuries can occur if workers walk over them. The provision of appropriate walkways may help prevent this.

The main hazards relating to conveyors are 'nip points'. A nip point is a dangerous pinch point between the roller and the moving conveyor belt on the in-running side of the roller. This can result in hands or clothing being drawn in. There is also the possibility of entanglement with the rollers themselves.

Other hazards are injuries to feet from loads falling from conveyors. If the conveyor is operating at height, there could be injury to many parts of the body. Manual handling hazards can occur when loading and unloading conveyors. Also, conveyors can be rather noisy.

General control measures for conveyors:

- Most importantly, nip points must be guarded where practicable to do so.
- It should be noted that nip guards sometimes have to be removed to clear spillages or for cleaning and maintenance purposes. In such cases the power to the conveyor must be isolated to prevent accidental start-up.
- There should be a guarding system for the conveyor itself. In the case of a screw conveyor, the screw should be enclosed. In the case of a roller conveyor, the conveyor may be quite long, so, rather than installing a fixed guard on either side of the conveyor belt, a trip wire may be used as a guarding system.
- If a trip wire system is used, it may be necessary to install edge guards to prevent materials falling off the conveyor.

- There should be guarding of transmission machinery (see 9.1: General requirements for details).
- To minimise manual handling injuries, the conveyor should be located at an appropriate height. This may be around 0.9 metres from the ground for light loads, around 0.7 metres for heavier loads, and only just above ground level for much heavier loads such as large drums.
- To avoid ergonomic hazards, the reach distance to the conveyor must be comfortable for workers.
- The conveyor should be adjusted to produce a suitable speed appropriate for workers.
- Controls for starting and stopping the conveyor should be clear.
- Emergency stops should be very clearly seen and in easy reach of workers.

8. Cranes

Cranes vary enormously in size and type, from the small lorry-mounted crane, which is little more than a hydraulic hoist, to mobile cranes, through to tower cranes used in construction.

Mobile cranes

The term 'mobile crane' includes any crane capable of travelling under its own power. Mobile cranes provide a versatile, reliable means of lifting on site.

Figure 3 shows an example of a mobile crane.

Figure 3: A mobile crane

The main hazards associated with the use of mobile cranes include:

- soft, sloping or uneven ground conditions that could cause a crane to become unstable and overturn;
- underground voids, drains or culverts that may collapse when the crane is in position;
- overhead obstructions, including power lines;
- loads exceeding the SWL of the crane;
- collision with other nearby cranes; and
- damaged lifting accessories or incorrect slinging.

Tower cranes

These cranes are fixed in position and are the type that is commonly seen on large construction sites; Figure 4 shows an example of two tower cranes.

Figure 4: Tower cranes

The main hazards with these cranes are:

- defective mounting and anchorage, so the crane may topple when under load;
- loads in excess of the SWL of the crane;
- site obstructions, such as scaffolds;
- overhead electric power lines;
- high winds;
- dropped loads; and
- moving loads that may strike people.

Overhead cranes

These can often be found in engineering workshops; an example of an overhead crane is shown in Figure 5.

Figure 5: An overhead crane

Some hazards are:

- overloading the crane so the load exceeds the SWL of the crane;
- a total equipment failure of the crane; and
- the load is not securely attached so workers below could get injured by objects falling from the crane.

Control measures to be considered when using mobile cranes include:

- The crane must be suitable for the task, with the SWL clearly marked.
- There must be a lifting plan in place.
- Lifting accessories should be suitable for the task and must be examined for any damage.
- The crane must be sited on firm, flat ground.
- The crane must be far away from any overhead power lines.
- The crane must be stable.
- Barriers should be erected to keep any unauthorised workers away from the crane.
- All workers associated with the lifting operation must be competent, such as crane drivers and slingers.
- The load to be lifted must be securely attached.
- Account must be taken of the weather; lifting operations should be postponed in strong winds.
- There must be a clear, all-round unrestricted view for the driver.
- There should be a warning immediately before a lift is to take place, for example, a siren.
- The load must be raised slowly and steadily to the correct height.
- The load should be transferred slowly to the landing position, the tension on the slings released and then removed.
- Lifting operations require adequate supervision.

When overhead cranes are in use, control measures will be similar, but workers not associated with lifting operations may be present. Therefore, in addition to using competent operators and slingers, ensuring statutory inspections of cranes and slings are carried out, ensuring that the load is adequately secured, ensuring good visibility, not exceeding the SWL of the crane and sounding a warning when a lift is about to start, you also need to:

- not carry out the lift when workers are working directly underneath the crane;
- require that high-visibility clothing and hard hats be worn in the area; and
- ensure hazard warning signs are clearly visible in the area.

6.3.2 Requirements for lifting operations

- Ensure a lifting plan is in place. This will include checking the weight of what is to be lifted, its centre of gravity, the load dimensions and any height restrictions.
- Before starting a lift, put in place risk reduction measures (where needed) such as barriers around equipment or a permit-to-work system.
- For mobile equipment, check that traffic routes are clear of obstructions where practicable.
- The equipment must be strong, stable and suitable.

- The equipment must be installed and positioned correctly.
- Ensure that the safe working load (SWL) is clearly marked on the equipment.
- Ensure surfaces are suitable. For example, for a forklift truck there should not be any potholes on the traffic route. A crane should be sited on firm, level ground.
- All lifting equipment must be inspected and examined (see 6.3.3 Periodic examination/testing of lifting equipment).
- In addition to testing and inspection, a visual examination should be carried out at the start of a shift before lifting takes place; for example, to identify any signs of obvious damage such as missing guards and to check that controls are working properly.
- Ensure good visibility for workers operating equipment; for example, checking that a good standard of lighting is provided.
- Ensure that all equipment operations are planned, supervised and carried out by workers who are competent persons.
- Ensure that special requirements are in place for equipment used for lifting people; for example, so that workers are not crushed or trapped when the equipment is being used in that way, people do not fall when being transported and that any person trapped in the equipment is not exposed to danger and can be freed.

Note: there may be extra requirements for some types of equipment; for example, for cranes, loads must not be transported over people, there must be no risk of contact with overhead power lines and lifting must not take place in conditions of high wind speeds.

Finally, we will consider testing and examination requirements for lifting equipment.

6.3.3 Periodic examination/testing of lifting equipment

Lifting equipment needs to be thoroughly examined and tested, as failure to do so could result in very serious injuries from accidents.

For example, if a load is dropped, then there will at least be damage to the load, but there could easily be serious injuries to workers or even fatalities. By their very nature, lifting equipment and lifting accessories are placed under load; they are also often used in less than ideal environments where they may be subject to corrosion from adverse weather conditions.

Lifting equipment should be thoroughly examined by a competent person before being put into service for the first time, when reassembled at a new site, when subjected to adverse conditions such as bad weather and, in any case, periodically. For lifting equipment for lifting people or accessories, a thorough examination must be carried out at least every six months. For other lifting equipment, the thorough examination must be carried out at least every 12 months.

Whenever testing and thorough examination has been completed, a written record should be retained.

APPLICATION

Find out the procedures in place for inspecting and testing the lifting equipment in your workplace.

KEY POINTS

- There are hazards and control measures associated with both powered and non-powered lifting equipment, including equipment used for lifting people.
- Different types of forklift truck are in widespread use and many serious accidents have been recorded when using them, often as a result of overturning.
- There are hazards relating to mobile, overhead and tower cranes, such as falling objects. Relevant control measures include ensuring cranes are not overloaded.
- Regular inspection, testing and examination are essential for lifting equipment at intervals specified in LOLER.

References

[1] HSE, *Safe use of lifting equipment. Lifting Operations and Lifting Equipment Regulations 1998. Approved Code of Practice and guidance* (L113, 2nd edition, 2014) (www.hse.gov.uk)

[2] HSE, *Safe use of work equipment. Provision and Use and Work Equipment Regulations 1998. Approved Code of Practice and guidance* (L22, 4th edition, 2014) (www.hse.gov.uk)

[3] Forklift Fails, 'Forklift accident statistics' (2017) (www.forkliftfails.com)

ELEMENT 7

CHEMICAL AND BIOLOGICAL AGENTS

7.1: Hazardous substances

Syllabus outline

In this section, you will develop an awareness of the following:

- Forms of chemical agent: dusts, fibres, fumes, gases, mists, vapours and liquids
- Forms of biological agent: fungi, bacteria and viruses
- Difference between acute and chronic health effects
- Health hazard classifications: acute toxicity; skin corrosion/irritation; serious eye damage/eye irritation; respiratory or skin sensitisation; germ cell mutagenicity; carcinogenicity; reproductive toxicity; specific target organ toxicity (single and repeated exposure); aspiration hazard

Introduction

We are now going to consider occupational health and hygiene with respect to chemical and biological agents. Firstly, it will be useful to outline what is meant by these different terms.

DEFINITIONS

Occupational health is a branch of medicine concerned with health at work and includes matters such as health screening, the study of sickness absence, health education and counselling (for example, for workers suffering from stress).

Occupational hygiene means the identification, evaluation and control of factors that might cause ill-health or discomfort at work.

We will examine the forms of both chemical and biological substances and how different forms influence their levels of hazard, then move on to the meaning of acute and chronic effects and the differences between them, before classifying the different types of health hazard.

7.1.1 Forms of chemical agent

Chemical agents come in many different physical forms, which can influence where, and how easily, they can enter the body and cause damage. For example, forms

Chemical and biological agents

such as dusts, fumes, fibres, gases, mists and vapours can enter the respiratory system, while liquids can affect the skin. The eyes are very efficient at transmitting biological agents, so if you touch a surface contaminated with a biological agent (such as a virus) then rub your eyes, it is likely your body will subsequently be affected by that virus.

> **DEFINITIONS**
> *You will often come across the distinction between 'inhalable' and 'respirable' fractions of solid particles, like dusts. In simple terms:*
> - **Inhalable** *refers to the fraction of the airborne particles that can be breathed in through the nose and mouth.*
> - **Respirable** *refers to the fraction that is breathed in and makes it all the way deep down into the lungs, where the gas exchange (getting oxygen into the blood and removing carbon dioxide) takes place. Only the finer inhalable particles will make it that far.*

Dusts

These are solid particles that are denser than air. The particles will normally settle on surfaces but can remain suspended in the air for a while before settling.

If you inhale air that contains dust, the body must use defence systems (covered in 7.2: Assessment of health risks) to reduce the risk of ill-health. Dust can be hazardous in two ways:

1 It may cause breathing difficulties simply due to its physical nature (because it is a dust). If there is no adverse chemical reaction on the organs of the body, this type of dust is often called 'nuisance dust' and the extent of any problem is related to its particle size.
2 It may cause ill-health because of its chemical nature; for example, cement dust, which we will examine in more detail later.

Fumes

These are very small metallic particles that can reach the gas exchange region of the lung. Fumes often arise from welding operations. For example, zinc oxide fumes can be created from vapourised zinc during welding; the metal oxide cools and condenses back to very finely divided airborne particles. The level of resulting harm is dependent on the metals used in the welding process.

Fibres

Fibres can be regarded as natural or synthetic threadlike substances. Whereas dust size is only considered in terms of diameter, fibre size is considered in terms of both diameter and length. Examples of natural fibres are cotton and wool; examples of synthetic fibres are nylon and rayon.

The main hazard of fibres comes from inhaling them into the lungs. The deeper they are deposited into the lungs, the greater the risk of disease. As with dusts, fibres that are respirable pose the greatest risk. Respirable fibres have a length greater than 5 microns (micrometres) and a diameter of less than 3 microns. An example of a fibre that can cause ill-health when inhaled is asbestos.

Gases

Gases have no fixed shape or fixed volume; they can expand or be compressed to fill different volumes and shapes. Gas is one of the standard states (or phases) of matter; normally a liquid becomes a gas when it boils. These are substances with normal boiling points below ambient temperature. Gases can easily enter the bloodstream through the lungs. For example, oxygen enters the body when you breathe and is carried to the organs that need it. But if a harmful gas enters, your blood can also quickly carry it throughout your body, potentially causing damage to organs.

Mists

A mist consists of very small suspended airborne droplets of liquid. The most likely method of mist damaging the body is by inhalation, although mist can be absorbed through the skin and, without facial protection, can be swallowed. An example of where mist is encountered is paint spraying.

Vapours

A liquid open to the air will evaporate from its surface to produce vapour. Evaporation occurs even when the temperature of the liquid is below boiling point. The higher the temperature, the greater the rate of evaporation. In a closed system, at constant pressure and temperature, there comes a point when the evaporation of liquid to vapour and the condensation of vapour back to liquid reaches an equilibrium. The vapour will be a gaseous state of the liquid with which it is co-existing.

Vapours can be inhaled and reach the lungs. As with gases, inhaling harmful vapours can cause damage to body organs.

Liquids

A liquid is a state (or phase) of matter. It does not have a fixed shape; it forms the shape of the container it is in. But it does have a fixed volume at a given temperature. Unlike gases, liquids are virtually impossible to compress into a smaller volume. Under normal conditions, when a solid is heated it will melt to become a liquid at its melting point (also called freezing point) and then a gas at its boiling point. So a liquid is a physical state of a substance in between its freezing point and its boiling point. The main ways liquids could harm you would be through contact with skin or swallowing (ingestion).

We will now look at biological agents.

7.1.2 Forms of biological agent

We will focus on three types of biological agent: bacteria, viruses and fungi. One key difference between chemical and biological agents is that biological agents are living entities (organisms) whereas chemical agents are not. This means that, even though you might be exposed to small numbers of the organisms, these can rapidly reproduce and increase their population to dangerous levels under the right conditions.

Bacteria

These are very small organisms, each being a living cell. They cannot be seen with the naked eye but require a powerful microscope. Bacteria can be classified into three groups:

1 Beneficial: the body needs these bacteria. For example, some provide the body with vitamins K and B12, both of which are essential for human health.

2 Harmful: these cause the body harm, such as *Legionella* and *Leptospira* (both of which we will discuss in 7.5: Specific agents) and *Salmonella*, which is associated with food poisoning.

3 Neither harmful nor beneficial under normal circumstances: these live in or on the body. Many of these types are found on the skin; however, if these bacteria get into the body, they can cause harm.

Under the right conditions, bacteria can reproduce by dividing into two about every 20 minutes.

Table 1 gives you an idea of just how quickly large numbers of bacteria can develop from just one bacterium.

Time after first division (minutes)	Resulting number of bacteria
0	1
20	2
40	4
60	8
120	64
180	512
240	4096
300	32,768

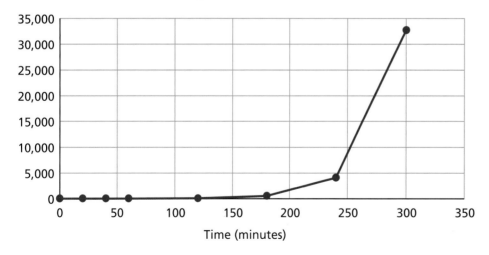

Table 1: Pace of bacteria growth

If you have ever been unfortunate enough to suffer food poisoning after an evening meal, due to ingested bacteria such as *Salmonella*, you may recall feeling very unwell during the night. This is because you had ingested many more bacteria than just one and, considering Table 1, millions of such bacteria were in your digestive system a few hours after the meal.

Bacteria thrive best at temperatures between 20°C and 45°C. Below 20°C, bacteria grow and divide very slowly. Above about 50°C, many bacteria are killed.

Antibiotics can be effective at curing diseases caused by bacteria, although some types of bacteria have become resistant to antibiotics.

Viruses

Viruses are much smaller than bacteria. If you wished to observe them, you would need an electron microscope, which is more powerful than the type of microscope used to observe bacteria.

Whereas bacteria can reproduce outside the body, viruses can only reproduce inside living cells. When a virus finds a suitable cell, it first sticks to the cell membrane and then enters the cell, which is called a 'host cell'. The virus takes over the host cell and causes it to make more copies of the virus. The virus can then be carried around the body (for example, by blood) and can reach certain target cells, resulting in disease.

Examples of diseases caused by viruses are hepatitis, flu and Covid-19.

Antibiotics are not effective at curing diseases caused by viruses. Viruses must be overcome by a defence system in the body itself.

Fungi

These organisms live on dead organic matter. They reproduce by producing spores. Inhaling such spores can result in allergic reactions. For the most part, here we are only concerned with the microscopic version (microfungi), not the large mushrooms you might eat.

Examples of diseases caused by fungi are:

- aspergillosis, including 'farmer's lung', which results from inhaling spores from mouldy hay, perhaps due to storing wet hay in a barn;
- athlete's foot, where a fungus is found between the toes; and
- ringworm.

Antibiotics can often be used against diseases caused by fungi.

7.1.3 Difference between acute and chronic effects

It is important to distinguish between acute and chronic effects when there is exposure to hazardous agents. The differences between these effects are outlined in Table 2.

Acute effects	Chronic effects
Appear either immediately or shortly after exposure has taken place, for example exposure to irritants or to flame	Take longer to develop, appearing later, for example cancer
Occur after a single exposure to the agent	Occur after repeated or long-term exposure

Chemical and biological agents

Acute effects	Chronic effects
The adverse effect may be reversible: if exposure to the agent is removed, there will usually be a fairly quick recovery.	The adverse effect may be irreversible, so even if exposure to the agent is removed, there may be a continued worsening of symptoms.
An example could be one exposure of breathing in a high concentration of ammonia vapour; there will be instant irritation in the nose, probably accompanied by coughing. Removal of the ammonia would generally allow normal breathing after a few minutes.	An example could be repeated exposure to coal dust (for example, from working as a coal miner). This can result in pneumoconiosis, a lung disease associated with inflammation, shortness of breath and coughing that usually results in permanent disability.

Table 2: Differences between acute and chronic effects

Exposure to many substances can lead to both acute and chronic effects, depending on the extent of exposure.

An example is trichloroethylene (TCE). This chemical has been widely used as a degreasing agent and is often found in workplaces such as commercial garages and maintenance workshops.

TCE can produce an adverse reaction in several ways, but we will just consider the effects of inhaling it.

The acute effect of TCE inhalation is to cause drowsiness (sometimes called a 'narcotic effect').

The chronic effect, which might occur from inhaling TCE at low concentrations over a long period of time, could be damage to the liver because the liver is one of the greasiest organs in the body and TCE is a degreasing agent.

7.1.4 Health hazard classification

Substances vary enormously in their hazardous properties. To make things simpler, hazardous substances are grouped into health hazard classifications. There are also classifications for physical and environmental hazards.

The most generally accepted system of classification is known as the Globally Harmonized System (or GHS). Each hazard classification has a specific definition and specific tests and criteria against which the properties of substances are compared. This system has been adopted in the UK and is legally binding.

Information on the classification to which a substance has been assigned is legally required to be found on labels and safety data sheets (discussed in 7.2: Assessment of health risks).

This information is useful when assessing the health risks from using hazardous substances in the workplace, as discussed in 7.2, so it is worth spending some time here to become familiar with the common health hazard classifications. A single substance may fall into one or more classifications. The main classifications are as follows.

Acute toxicity

Acute toxicity refers to the serious adverse health effects (such as lethality) that could occur after a single or short-term oral, dermal or inhalation exposure to a substance or mixture of substances.

> **DEFINITIONS**
> *Oral exposure* means taken in by mouth and reaching the gastrointestinal tract.
> *Dermal exposure* means exposure of the skin.

A toxic substance interferes with the normal functions of one or more systems of the body. Examples of substances in this class are potassium cyanide and ammonia.

If you inhaled irritant vapour, for example from toxic substances such as chlorine, dichloromethane and nitrogen dioxide, it would easily reach the alveoli in your respiratory system and interfere with normal breathing in the gas exchange region of your lung. How hazardous substances enter the body (including the lungs) is discussed further in 7.2: Assessment of health risks.

Skin corrosion/irritation

> **DEFINITIONS**
> *Skin corrosion* refers to the production of **irreversible** damage to the skin occurring after exposure to a substance or mixture.
> *Skin irritation* refers to the production of **reversible** damage to the skin occurring after exposure to a substance or mixture.

Irritants are non-corrosive substances that cause reversible inflammation to the skin, eyes and mucous membranes, and can also interfere with normal breathing.

When the skin is exposed to irritants, its protective oils and greases can be dissolved (known as 'defatting'). The result can be cracking, flaking and reddening of the skin. This condition is known as contact (or irritant) dermatitis. Examples of this type of irritant are white spirit, TCE (used for degreasing) and dichloromethane (used as a paint stripper).

Contact dermatitis usually occurs very soon after exposure to irritants.

Corrosives are substances that attack and destroy skin tissue. Examples are concentrated solutions of sulphuric acid (found in car and forklift truck batteries) and sodium hydroxide (found in some oven cleaners). Normal, as-supplied household bleach is also classified as corrosive.

> **TIP**
> When considering the effects of hazardous substances on skin, you may find it useful to remember the three Ds:
> - **D**estruction (corrosives) – will affect **all** exposed persons
> - **D**efatting (irritants) – will affect **all** exposed persons
> - **D**ermatitis (irritants) – will only affect **some** exposed persons

Serious eye damage/eye irritation

> **DEFINITIONS**
>
> **Serious eye damage** *refers to the production of tissue damage in the eye or serious physical decay of vision that **is not fully reversible**, occurring after the eye is exposed to a substance or mixture.*
>
> **Eye irritation** *refers to the production of changes in the eye that **are fully reversible**, occurring after the exposure of the eye to a substance or mixture.*

Many vapours and dusts can lead to eye irritation, which can result in itching and redness in the eyes. Examples include chlorine, dichloromethane and nitrogen dioxide, but any type of dust may cause eye irritation. A strong acid or alkali solution that goes directly into the eye is likely to act as a corrosive and destroy eye tissue. There could be resulting blindness.

Respiratory or skin sensitisation

> **DEFINITIONS**
>
> **Respiratory sensitisation** *refers to hypersensitivity of the airways after inhalation of a substance or mixture.*
>
> **Skin sensitisation** *refers to an allergic response occurring after skin contact with a substance or mixture.*

Sensitisers are substances that cause an overreaction of the immune system. When the skin is exposed to a sensitising agent, sensitisation dermatitis (sometimes called allergic dermatitis) may result. As with irritants, the skin can show reddening, but the appearance can be in the form of a rash rather than flaking or cracking. However, in contrast to irritants, it may be a considerable time before the symptoms of sensitisation dermatitis occur.

Examples of skin sensitisers are formaldehyde (used in the manufacture of fibreboard and as a preservative for pharmaceutical products) and nickel (used for nickel-plating). Razor blades are often nickel-plated and some people can suffer symptoms of dermatitis from shaving with them, particularly in summer when the pores of the skin are more open in the heat. Wearers of nickel-plated jewellery might also show these symptoms, such as a red ring on the skin underneath a bangle. Latex gloves can also cause sensitisation dermatitis.

You have probably noticed that there are several differences between contact and sensitisation dermatitis, and these are summarised in Table 3.

Contact dermatitis	Sensitisation dermatitis
Caused by irritants	Caused by sensitisers
Causes cracking, flaking and reddening of the skin	Causes rashes and sometimes itching of the skin
Acute effect	Chronic effect
Symptoms occur at site of contact	Symptoms can occur in areas remote from the site of contact

Contact dermatitis	Sensitisation dermatitis
Symptoms will stop when a person is no longer exposed to the irritant and will not return unless the person is exposed to a similar amount of the irritant	Symptoms will stop when a person is no longer exposed to the sensitiser, but in the future, if the person is exposed to a very small amount of the sensitiser, symptoms will return

Table 3: Differences in types of dermatitis

Now consider the respiratory system. A respiratory sensitiser is a substance that will cause extreme sensitivity in the airways after it has been inhaled. It can cause an allergic reaction in the respiratory system and result in occupational asthma.

As with skin irritants, it may be a long time after initial exposure before symptoms appear, but, once sensitisation has taken place, even the very smallest exposure to the sensitiser will result in symptoms occurring. A person may have been working in an environment without experiencing any discomfort and then suddenly have considerable difficulty breathing.

Some examples of respiratory sensitisers are:

- hardwood dust;
- glutaraldehyde: some surgical instruments can be sterilised by heating but others cannot, so glutaraldehyde is used instead, and although effective, this substance has been found to cause both skin and respiratory sensitisation and asthma;
- toluene diisocyanate: one of its uses has been as a hardener in two-part resin/hardener adhesive, resulting in respiratory sensitisation for many workers; and
- various types of tree and grass pollen (which can be a problem for anyone who works with them, such as commercial gardeners): these can result in hay fever that can lead to severe breathing difficulties.

We now discuss types of hazards that, as a result of human cell damage (including damage to DNA) can cause birth defects.

Germ cell mutagenicity

Germ cell mutagenicity refers to heritable gene mutations, including heritable structural and numerical chromosome aberrations (usually induced by a failure of chromosome division), occurring after exposure to a substance or mixture.

Germ cells are those that can produce reproductive cells (male sperm and female egg cells in humans), which are then combined in sexual reproduction to create offspring. Put simply, the germ cells cause DNA to go from parents to offspring. If a substance causes damage to the parent's DNA, that damage can be transferred to their offspring, resulting in a damaged foetus and producing birth defects in humans.

> **DEFINITION**
>
> A **germ cell mutagen** is a substance or mixture that can cause mutations in the germ cells of humans that can then be passed on to offspring.

Specific examples of germ cell mutagens are not easy to find, as they tend to be drugs (including prescribed drugs) with complex chemical structures. However, one chemical that is viewed as a germ cell mutagen is 2-ethoxyethanol, which is used as a solvent in some paints.

Chemical and biological agents

Carcinogenicity

Carcinogenicity refers to the induction of cancer or an increase in the incidence of cancer after exposure to a substance or mixture.

Before we discuss this term, it is helpful to understand why carcinogenic agents (which could include substances or types of radiation) occur. In a healthy human body, cells grow and divide at an orderly rate. However, if something happens to disrupt this, so that the cells in the body grow at a faster rate, this results in extra tissue growth. The extra growth is called a tumour. Tumours can be benign, meaning growth is local and does not spread, or they can be malignant and spread to other parts of the body.

> **DEFINITION**
> A **carcinogen** is an agent that has the potential to cause cancer.

There are several agents that can cause cancer and the cancer can occur in various areas of the body. Table 4 gives some examples of carcinogenic substances and the area of the body affected.

Carcinogen	Type of resulting cancer
Hardwood dust	Nasal
Vinyl chloride monomer	Liver
Ultraviolet light	Skin
Asbestos	Lung

Table 4: Cancers caused by carcinogenic substances

Reproductive toxicity

Reproductive toxicity refers to adverse effects on sexual function and fertility in adult males and females as well as development toxicity in their offspring, occurring after exposure to a substance or mixture.

Specific target organ toxicity (single and repeated exposure)

Where a specific organ is affected by a specific substance, that organ is a 'target organ' for that substance. We noted earlier that TCE can attack the liver with a degreasing action, so we can say that the liver is a target organ for TCE.

You may also recall that a single exposure to TCE tends to result in an acute effect (drowsiness), while repeated exposures tend to cause chronic effects (liver damage).

Here are some other examples:

1 Exposure to lead can result in damage to the kidneys and the brain, so these are target organs for lead.
2 Exposure to ammonia affects the lungs, so the lung is a target organ for ammonia.

Aspiration hazard

Aspiration (inhaled breath) means the entry of a liquid or solid chemical, either directly through the oral or nasal cavity or indirectly from vomiting, into the trachea and lower respiratory system.

Aspiration hazard refers to the aspiration (breathing in) of a substance or mixture that could cause severe acute effects, such as chemical pneumonia, pulmonary injury or death.

Breathing in vomit is likely to cause chemical pneumonia as you will be breathing stomach contents (including acid and enzymes) that can damage the lung/respiratory tract tissues.

> **DEFINITION**
> **Chemical pneumonia** *is an inflammation of lung tissue caused by toxins (unlike 'normal' pneumonia, which is caused by a bacteria or virus).*

KEY POINTS

- Chemical agents come in various physical forms that affect how they can enter the body and cause damage.
- The extent of the hazard from dusts is dependent on their particle size as well as their chemical make-up.
- The ratio of length to diameter (or width) in fibres is a significant factor.
- There are three forms of biological agent – bacteria, viruses and fungi – that reproduce and spread in different ways.
- The less oxygen available to tissues and organs of the body, the greater the hazard.
- Exposure to hazardous substances can have acute or chronic effects, and the differences between these include how quickly effects occur and whether they can be reversed.
- Health hazard classifications are useful in assessing the risks of using hazardous substances in the workplace.
- There are two types of dermatitis – contact dermatitis (caused by irritants) and sensitisation dermatitis (caused by sensitisers) – that have different effects.
- Irritants quickly produce symptoms while sensitisers often have delayed effects.
- A target organ may be affected by a particular substance, through single or repeated exposure or sometimes both.

7.2: Assessment of health risks

> **Syllabus outline**
>
> In this section, you will develop an awareness of the following:
> - Routes of entry of hazardous substances into the body
> - The body's defence mechanisms (superficial and cellular)
> - What needs to be taken into account when assessing health risks
> - Sources of information:
>   - product labels
>   - safety data sheets (who must provide them and information that they must contain)
> - Limitations of information used when assessing risks to health
> - Role and limitations of hazardous substance monitoring

Introduction

First, we will provide a simplified overview of certain functions of the body to help you understand some of the effects of exposure to certain substances.

We will then examine how hazardous agents can enter the body ('routes of entry') and ways in which the body can defend itself to minimise the risk of ill-health.

Then we will review the many things that you need to take into account when assessing the risks to health from using hazardous agents.

Next, we will identify sources of information that can help with health risk assessment and the limitations of such sources.

Finally, we will describe methods of monitoring hazardous substances, including ways to measure various forms of substance. We will note how such measurements can assist you with the risk assessment and to what extent.

Structure of blood

Blood consists of cells and plasma. There are three general types of blood cell:

1 Platelets: their role is to detect any broken blood vessels and start the clotting process to repair them.
2 Red blood cells: these transport materials throughout the body. They contain a large amount of haemoglobin, a protein designed to carry oxygen to various tissues and organs. This will become relevant when you read 7.5: Specific agents, which includes the effects of carbon monoxide entering the blood.
3 White blood cells (also known as leucocytes): these can be considered the body's 'infection fighters'. Their purpose is to prevent disease. They form part of the immune system and there are several different types. This will be important when you consider types of dust and silica in 7.5: Specific agents.

Blood forms part of the body's circulatory system. Organs such as the liver and kidneys use oxygen to help them function and produce carbon dioxide (CO_2) in the process.

7.2 Assessment of health risks

The respiratory system

When we breathe in oxygen (O_2) from the air, it passes through the nose and the trachea (windpipe) and into the lungs, where the oxygen passes through bronchioles into narrow capillaries called alveoli (see Figure 1), and then into the bloodstream, resulting in oxygenated blood.

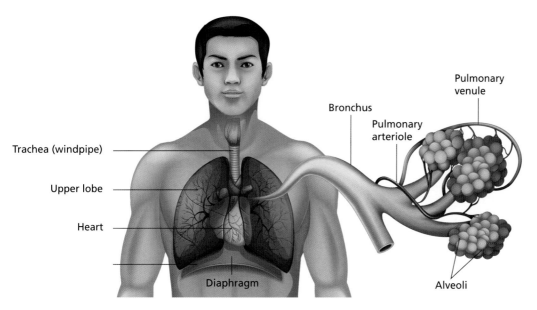

Figure 1: The respiratory system

This blood circulates around the body, providing oxygen to various organs. When the oxygen has been used, carbon dioxide (CO_2) is produced, which passes back into the lungs. The CO_2 is then breathed out. The area where oxygen is transferred from the lungs to the bloodstream and CO_2 is transferred from the bloodstream to the lungs is known as the gas exchange region of the lung.

7.2.1 Routes of entry of hazardous substances into the body

For hazardous substances to harm you, they need to first enter your body. There are four main routes of entry into the body and these routes apply to both chemical and biological agents. These are:

1 **Inhalation:** agents such as dusts, fibres, vapours, gases and airborne viruses can be breathed into the lungs. A spillage of liquid could cause inhalation of its vapour.

2 **Ingestion:** this occurs when agents are taken orally into the digestive system (swallowed). People are unlikely to do this deliberately in a work environment. More likely is 'hand-to-mouth' transmission, when contamination on hands is accidentally transferred to the mouth (for example, on food or from just touching the lips). Clearly, this is more serious with toxic substances (which are harmful in small amounts) and so it is especially important to wash your hands before eating.

3 **Absorption:** this takes place when agents are absorbed through undamaged skin or through the conjunctiva of the eye. Hazardous substances handled without gloves could enter through absorption.

4 **Injection:** this occurs when agents enter the body through a break in the skin, for example caused by a 'needle-stick' injury (from a syringe).

Some substances can enter by more than one route, depending on their physical form. For example, a liquid substance might be swallowed or absorbed through the skin, and its vapour or mist might be inhaled.

7.2.2 The body's defence mechanisms

The body is not defenceless against hazardous agents. Here we will examine the defence mechanisms that are available to the body and the possible limitations of these.

There are three general types of defence mechanism:

- natural barriers;
- non-specific immune responses; and
- specific immune responses.

Natural barriers

The key natural barriers are outlined here.

The skin
The skin consists of three basic layers: the epidermis, dermis and subcutaneous layer. Here we will consider the epidermis.

The epidermis is the outer layer. It consists of dead cells and receives no blood supply. The unbroken epidermis provides protection against invading organisms such as some bacteria. It is less effective against chemicals. Broken skin can provide entry for chemical and biological agents through the injection route.

Mucous membranes
These are covered by secretions that have antimicrobial properties. These secretions can prevent microorganisms attaching to host cells and causing disease.

The gastrointestinal tract
This includes the mouth, tongue, oesophagus, stomach, pancreas, liver and intestines. We will concentrate on the stomach and liver. Stomach acid has a pH of about 1 (ie it is very acidic) and can destroy many harmful microorganisms.

The liver
Liver produces bile, which contains salts and cholesterol. The bile proceeds to the small intestine, where it emulsifies fats so that they can be more easily absorbed. Excess bile is stored in the gall bladder until it is needed.

The liver tries to inactivate toxins so that toxin-free nutrients can enter the bloodstream. Normally, this process is successful, but occasionally the oxidation process may produce a more toxic material.

The liver has two blood supplies:

- the hepatic artery, which carries oxygenated blood to the liver, providing 20% of its blood supply and 50% of its oxygen requirement; and
- the portal vein, which carries deoxygenated blood from the small intestine and the pancreas and contains nutrients, minerals, water and toxins, providing 80% of the liver's blood supply and the remaining 50% of its oxygen requirement.

Leaving the liver is the hepatic vein, carrying deoxygenated blood. The products of the liver functions then proceed to other tissues.

One indicator that the liver may not be in a good state of health is the condition of jaundice, which is visible by yellowing of the skin and the whites of the eyes.

Respiratory system defences

The respiratory system has several defences, which are mostly designed to trap and remove particulate matter (such as dust or fibres). Where particles end up and are stopped depends on their size (in microns or micrometres, shown as µm):

- Greater than 25µm: dust this size will be stopped by the hairs in the nose (nasal hairs).
- Between 10µm and 25µm: particles of this size can pass through the nasal hairs and reach the next line of defence, the mucociliary escalator, which consists of hairs situated in the airway. These hairs are in continuous motion and move mucus that lines the trachea upwards. This results in the particles being trapped on the mucus layer and being coughed out or swallowed and can be accompanied by sneezing.
- Less than 10µm: these particles can pass through the nasal hairs and the mucociliary escalator and end up in the gas exchange region of the lungs. Dust of this size is known as 'respirable dust'. The defence mechanism here is white blood cells known as macrophages, which engulf and attempt to break down these respirable particles.

Figure 1 shows the respiratory system, including the trachea and the alveoli.

> **DEFINITIONS**
> *Here is a brief outline of the main types of white blood cell.*
> - ***Basophils:*** *release certain chemicals that assist in immune response*
> - ***Eosinophils:*** *destroy parasites*
> - ***Neutrophils:*** *swallow and digest bacteria*
> - ***Monocytes:*** *turn into macrophages, which in turn consume unwanted cells and microorganisms*
> - ***B-lymphocytes:*** *secrete proteins called antibodies, which bind to invading microbes that are then destroyed by macrophages; this process is known as phagocytosis*
> - ***T-lymphocytes:*** *destroy invading microbes by binding to them directly*

Non-specific ('innate') immune responses

Macrophages produce substances called cytokines, which initiate an inflammatory response to an injury or infection site. The cytokine response involves fever and increased production of neutrophils. Cytokines affect the growth of all blood cells that help the body's immune and inflammation response.

Specific immune responses

These include the actions of antibodies and lymphocytes. Lymphocytes recognise the unwanted invading microbe and produce a chemical to destroy it (see Definition box).

7.2.3 What needs to be taken into account when assessing health risks

An awareness of routes of entry of hazardous substances, and the body's defence mechanisms, is useful when conducting a health risk assessment. The Control of Substances Hazardous to Health Regulations 2002 (COSHH) and the accompanying Health and Safety Executive (HSE) Approved Code of Practice and guidance (L5) are the authoritative sources of information for this (particularly reg 6).[1] The key points to consider are:

1. The form of the substance: for example, a solid will normally present a higher risk of inhalation if it is fine powder rather than a solid block.
2. The hazardous nature of the substance: for example, whether it is an irritant, sensitiser, carcinogen, etc.
3. The route(s) of entry of the substance: remember that some agents may have more than one route of entry. For example, xylene (used as a solvent in non-aqueous paints) is not only a skin irritant but can also cause nausea and headaches if the vapour is inhaled.
4. The concentration of the substance: dilute sulphuric acid would not normally cause any adverse reaction when in contact with the skin, but the concentrated acid is a corrosive and would quickly cause severe skin damage.
5. The number of workers exposed to the substance: this could be not only those using the agent but those nearby who could be exposed to the agent, for example, through its vapour.
6. Whether any workers exposed are especially vulnerable: for example, those who have asthma or who are pregnant may be more severely affected by certain substances. A check of any health surveillance records might be appropriate here.
7. The frequency (how often) and duration of any exposure to the substance.
8. The occupational exposure limit (OEL) of the substance: the lower the value of the OEL, the greater the risk resulting from exposure (for more details, see 7.3: Occupational exposure limits).
9. How the substance is used, handled or applied: for example, if paint is sprayed onto a surface, there will be a greater risk of inhalation (of the paint mist) than if the paint is brushed on.
10. The suitability and effectiveness of any control measures (to reduce the risk) in place: for example, whether any respiratory protection worn is the most suitable option for protecting the worker.
11. Any environmental factors that may affect the level of risk: for example, if a solvent is being used at a room temperature of 20°C, evaporation of the liquid may be so slow that little of the vapour is inhaled. However, if the solvent is being used at 40°C, there could be a significantly higher concentration of vapour to inhale.
12. Combined effects of substances: in many cases, more than one substance may be in use at the same time and there could be exposure to multiple substances. The resulting exposure may be 'additive'; that is, the effect of the two chemicals is the same as the combination of the two effects of the chemicals acting individually. Of greater concern is a 'synergistic' effect, when the resulting effect is greater than what you would expect from simply adding the individual effects.

7.2 Assessment of health risks

> **TIP**
>
> For additive effects, if the individual effects are X and Y, the combined effect is X+Y, but for synergistic effects the combined effect is much greater than X+Y.
>
> As an example, consider the effects of asbestos and smoking. Both can independently increase the risk of developing lung cancer. However, when combined there is an increased risk over and above simply adding the risks together.

During certain processes, many chemicals may be mixed. This can result in active intermediate compounds being formed prior to the required product being created. The risk assessment will therefore need to take account of potential exposure at all stages, not only to the raw materials (the starting point) but also to any compounds produced during the normal process, the final product and any foreseeable emergencies (such as spillage).

7.2.4 Sources of information

Before starting the risk assessment you need to be able to identify the hazards that may be faced by seeking out sources of information.

Product labels

Any supplier of hazardous substances in packages must label the package to tell anyone handling the package, or using the contents, about the substance's hazards and to give brief advice on what precautions are needed.

UK law requires certain symbols to be displayed to identify the hazards of materials. The UK has adopted the Globally Harmonized System (GHS).

Some examples of these symbols are shown in Figure 2.

Corrosive   Explosive   Oxidising   Carcinogenic   Dangerous for environment

Figure 2: Examples of recognised hazard symbols

Safety data sheets

A safety data sheet (SDS), sometimes called a material safety data sheet (MSDS), is another valuable source of more detailed information provided by suppliers of hazardous substances.

UK law sets out the specification for the SDS content and format. Each SDS has 16 sections, and the titles of these and what they cover are listed here:

1. Identification of the substance (or mixture) and manufacturer's information: such as contact details, including an emergency phone number.
2. Identification of the hazards: such as the GHS classification and hazard symbols to indicate, for example, whether the substance is toxic, corrosive or carcinogenic or has some other hazard.

3   The composition of the substance: including a list of its ingredients and its common name and trade name.

4   First-aid measures: describing necessary measures according to the different routes of exposure; for example, if the substance has entered the eye, the advice may be to flush the eye out with water for several minutes then seek medical attention.

5   Firefighting measures: describing suitable firefighting measures such as the appropriate type of fire extinguisher, and possibly inappropriate types to avoid, such as using a water jet on a solvent fire. This section may also identify any special personal protective equipment (PPE) to be used by firefighters.

6   Accidental release measures: identifying any special measures to manage spillage, for example using granules to absorb the spilled substance.

7   Handling and storage requirements: including precautions for safe storage and handling; for example, the recommendation for many substances is to store them in a cool, well-ventilated area. Any incompatibilities will be identified, such as not storing concentrated acids next to concentrated alkalis.

8   Exposure controls: such as recommendations for both engineering controls and types of PPE to be worn. Occupational exposure limits may also appear in this section.

9   Physical and chemical properties: such as the vapour pressure of the substance at 20°C or its pH value.

10 Stability and reactivity: such as information on any hazardous decomposition products or the possibility of any hazardous reactions; for example, if working with calcium carbide, the addition of water can produce highly flammable acetylene gas.

11 Toxicological information: including likely routes of exposure and possible acute and chronic effects following exposure to the substance.

12 Ecological information: such as hazards to the aquatic environment; for example, some ammonium compounds can be toxic to fish.

13 Disposal information: outlining how to dispose of the substance. For example, the recommended method for many soluble inorganic compounds may be to dilute them with large quantities of water and allow the resulting solutions to be flushed to the drainage system. On the other hand, organic compounds should not be flushed to a drain, and there are specific legal requirements for their disposal.

14 Transport information: including information such as the transport hazard class, the UN shipping name and any special precautions a user needs to be aware of in connection with transport or conveyance inside or outside their premises.

15 Regulatory information: identifying the safety, health and environmental regulations specific to the product in question.

16 Any other information: including, for example, any specific guidance notes.

7.2.5 Limitations of information used when assessing risks to health

The purpose of the SDS is to provide the end user with information on which an assessment can be based. It is important to note, however, that the SDS is not a substitute for a risk assessment. The manufacturer and supplier do not know how the end user will use the substances or under what conditions (for example, how much is used or how frequently workers may be exposed).

7.2.6 Role and limitations of hazardous substance monitoring

When carrying out a health risk assessment, in some situations you need to know the airborne concentration of a substance to which workers are exposed so that you can compare it with acceptable standards. The general term for these airborne concentration standards is occupational exposure limits or OELs, but in the UK they are known as workplace exposure limits (WELs). WELs are listed in the HSE guidance *EH40/2005 Workplace exposure limits* and are legally binding.[2]

These WELs are effectively exposure levels that are considered reasonably safe or, if there is no 'reasonably safe level', at least tolerable. There are some situations when it may be relatively easy to conclude that exposure is always well below such limits. For example, workers might be using only small amounts of a liquid that has a high exposure limit and is not very volatile or toxic, and the exposure is brief. However, in other situations workers are likely to be close to the exposure limit or there is a good deal of uncertainty about it. The concentration of some substances might need to be measured, particularly if there are WELs for such substances identified in HSE *EH40* guidance.

There are two basic types of airborne monitoring for hazardous substances:

1. Static monitoring: placing a measuring device at a particular fixed point, or points; for example, when testing a specific point for leakage of gas or vapour.
2. Personal monitoring: if you need to know the level of exposure of a particular worker, a device can be attached to them with the sampling head close to their breathing zone.

You also need to consider what form of substance to monitor. Here is an overview of monitoring airborne substances such as vapour and dusts.

Monitoring airborne substances

First, it is necessary to understand the units of concentration. For dusts, this concentration is expressed in milligrams of solid per cubic metre of air and written as mg/m^3. For vapours, this same unit could be used, but it is more usual to find the concentration expressed as parts of vapour per million parts (ppm) of air. The 'parts' here are in terms of the volume of the substance being monitored as a proportion of the total volume of air being sampled.

The first monitoring method to consider is the 'stain tube detector'. This is an environmental technique used to measure vapour or gas concentration. This method cannot be used for dust measurement.

Measuring chemical concentration in the air

Suppose you wished to measure the concentration of chemical A in the air at a workplace. To do this, you would use a stain tube and draw the air containing chemical A through it using a pump. The glass tube is graduated and contains a chemical (chemical B). The tube is supplied sealed at both ends (to preserve the chemical inside).

The principle is that when B is exposed to A the colour of B will change. For example, B might start off green and then change colour to red when exposed to A (that is why it is called a 'stain tube').

Chemical and biological agents

There are a number of different types of pump, but a 'bellows pump' is often used.

The method using a bellows pump is as follows:

- The pump is checked to ensure there are no leaks.
- The tube is snapped at both ends and one end is inserted into the pump (the tube usually has a marking on it, such as an arrow, that indicates which way round it should be inserted into the pump).
- The pump is operated using the number of pump strokes required by the instructions that came with the tube (these are also usually marked on the tube itself).
- The borderline between the colour change (in our example, between the red and green colours) is read on the tube scale. This is the concentration of chemical A in ppm.

The method could be repeated in other areas of the workplace, using a fresh tube for each measurement.

The advantages of this method are that it is easy to use, gives results quickly and is relatively inexpensive. The disadvantages are that there can occasionally be interference from other substances, the tubes have a limited shelf life and the accuracy of the results may only be within + or – 25%.

Use of personal exposure measuring equipment

We will now examine a personal monitoring technique; this time, for dust.

The equipment needed for this is a filter, a filter holder, tubing and a pump. The filter and filter holder are together known as the 'sampling head'.

Personal exposure monitoring method:

- The pump is calibrated so that there is a known volume of air passing through it, usually to about 2 litres/minute.
- A filter is weighed and placed in the filter holder.
- The sampling head is connected by tubing to the pump.
- The sampling head is placed as near to the worker's breathing zone as practicable, possibly on the lapel of a jacket or the collar of a shirt/blouse.
- The pump is switched on and allowed to run for a few hours.
- The pump is then switched off and the running time noted.
- The filter is then reweighed, so that the weight of dust can be calculated.

You now have sufficient information to calculate the personal exposure of the worker to the dust in the air.

This method is known as a 'gravimetric' method of analysis as it involves weighing materials.

APPLICATION

If the flow rate of the pump was 2 litres/minute, the pump ran for 4 hours and 2.4mg of dust was collected, calculate the personal exposure of the worker to the dust.

Answer at end of 7.2: Assessment of health risks.

If this equipment is left in an area rather than being attached to a worker, it can provide an environmental reading rather than a personal reading.

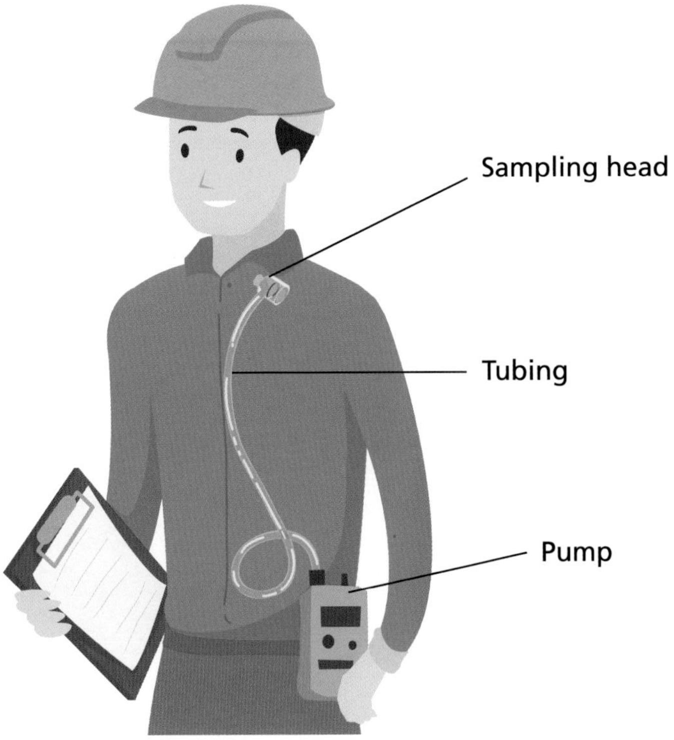

7.2 Assessment of health risks

Figure 3: Personal exposure monitoring equipment

> **TIP**
>
> It is unlikely that you would need to perform this sort of calculation in the workplace. It is included to illustrate how exposures may be estimated quantitatively.
>
> What is important is that you are aware of the general principles of the measurement of vapour or gas (using the stain tube detector method) and of dusts (using the equipment discussed).

Measurements need to be compared with the WELs (or OELs) in the HSE publication *EH40*,[3] which identifies allowable limits of exposure to determine whether the exposures experienced by workers are acceptable. These will be discussed more fully in 7.3: Occupational exposure limits.

Limitations of hazardous substance monitoring

- The airborne types of monitoring only provide results for the inhalation route of entry.
- The dust monitoring technique described will only provide information on exposure to total dust; it will not specify which dust is analysed and there may be a mixture of dusts.
- To determine exposure to a specific dust, more sophisticated and expensive techniques are required, such as atomic absorption spectroscopy. These techniques are often too expensive for most workplaces and workers using them need special training.
- Even with personal monitoring, although the sampling head is placed close to the breathing zone, the dust monitoring method described will not measure what the worker is actually breathing in. For example, the worker could spill dust near to the sampling head, leading to a false high reading. They might mask the head of the filter or forget to wear it for some time, and that would lead to a false low reading.
- Finally, monitoring is not a control measure – it just provides an indication of the level of risk following exposure.

Chemical and biological agents

For some substances, there may not be any limits of exposure quoted, such as new substances still being researched. In this case, organisations would be expected to create their own limits based on their knowledge of these new substances.

KEY POINTS

- The four main routes of entry into the body from exposure to hazardous substances are inhalation, ingestion, absorption and injection.
- Defences available to the body when exposed to hazardous substances, especially dust, include nasal hairs, the ciliary escalator, coughing and sneezing, the action of macrophages and the way lymphocytes act when the body is exposed to biological agents.
- Things to consider when conducting a health risk assessment include substance form, hazardous nature, route of entry, concentration and OEL, who is exposed to it and in what way, the suitability of control measures and additional risks from the environment or exposure to multiple substances.
- The main sources of information that could be used to help in the health risk assessment process are product labels (including hazard symbols) and safety data sheets, but this information is only a small part of the risk assessment.
- Static monitoring and personal monitoring are the main types of monitoring airborne hazardous substances, but these have limitations.

APPLICATION ANSWER

Flow rate = 2 litres/minute

Therefore in 4 hours, volume of air = 2 × 60 × 4 = 480 litres

$1m^3$ = 1,000 litres, therefore volume of air = $0.48m^3$

2.4mg dust was collected, therefore the average concentration to which the worker was exposed = 2.4/0.48 = $5mg/m^3$

References

[1] HSE, *The Control of Substances Hazardous to Health Regulations 2002. Approved Code of Practice and guidance* (L5, 6th edition, 2013) (www.hse.gov.uk)

[2] HSE, EH40/2005 *Workplace exposure limits* (4th edition, 2020) (www.hse.gov.uk)

[3] See note 2

7.3: Occupational exposure limits

Syllabus outline

In this section, you will develop an awareness of the following:
- Purpose of occupational exposure limits
- Long-term and short-term limits
- Why time-weighted averages are used
- Limitations of exposure limits
- Comparison of measurements to recognised standards (*EH40/2005 Workplace exposure limits*)

Introduction

You have already seen that workers can be harmed through being exposed to hazardous substances by four routes of entry into the body. Now we will identify some form of limit for exposure, above which workers should not be exposed. A general term for identifying different types of limits is an occupational exposure limit (OEL), but the term workplace exposure limit (WEL) is also used in the UK and in the Health and Safety Executive (HSE) guidance *EH40/2005 Workplace exposure limits* (*EH40*).[1]

The WEL only applies to the inhalation route of entry and therefore is only concerned with the concentration of hazardous substances in the air.

Here we will focus on the purpose of WELs, why we need different types of limits, the meaning and significance of WELs, the limitations in the use of WELs and the importance of comparing measured concentrations with the standards identified in *EH40*.

TIP

Legislation

Most substances have been allocated WELs. WELs are subject to the Control of Substances Hazardous to Health Regulations 2002 (COSHH).

Note: for substances covered by COSHH, OELs are WELs (as we have indicated), but exposure limits for lead and asbestos are not allocated WELs, as these substances are not covered by COSHH. Lead and asbestos are subject to their own (separate) legislation.

7.3.1 The purpose of occupational exposure limits

DEFINITIONS

COSHH defines **workplace exposure limit** *as the workplace exposure limit for a substance hazardous to health approved by the Health and Safety Executive (formerly the Health and Safety Commission) for that substance in relation to the specified reference period when calculated by a method approved by the Health and Safety Executive, as contained in HSE publication* EH40/2005 Workplace exposure limits *as updated from time to time.*

> *A more practical and useful definition appears in EH40:*
> - ***WELs** are British occupational exposure limits set in order to help protect the health of workers.*
> - ***WELs** are concentrations of hazardous substances in air, averaged over a specified period of time, referred to as a **time-weighted average (TWA)**.*
>
> *Two time periods are used:*
> - *long-term (8 hours); and*
> - *short-term (15 minutes).*

The main purpose of a WEL is to identify a level of an airborne substance that is safe for a person to inhale. The practical effect of a WEL is that it places a limit on the level of airborne contaminant that is permissible in a workplace and to which a worker may be exposed.

This raises two issues: what is meant by 'level' and what is meant by 'safe'.

Deciding on the level of exposure to a hazardous substance will depend on both the concentration of a substance being inhaled and the time that a worker is exposed to it.

Nobody can be completely safe when inhaling a substance, although the risk may be low. A WEL is not a sharp dividing line between safety and danger, but a level below which most people will suffer no ill-effects. WELs also have a safety factor built in. For example, if the level of safety for a substance is stated as x, the WEL may be set at 10% of x.

7.3.2 Long-term and short-term limits

Both the long-term and short-term exposure limits are expressed as airborne concentrations averaged over a specified period of time (usually 8 hours and 15 minutes, respectively). To identify limits in terms of concentration, it is necessary to define the units used.

Airborne substances may be in the form of vapour, fumes, gas or dust. For dust and fumes, the units used for concentration are milligrams of solid per cubic metre of air, written as mg/m^3 (although this appears in *EH40* as mgm^{-3}). For gases and vapours, the usual unit is parts of gas or vapour by volume per million parts of air, written as ppm, although these concentrations can sometimes be quoted as mg/m^3.

> **DEFINITIONS**
>
> *A **long-term exposure limit (LTEL)** is a limit that is intended to control the effects of exposure to a substance by restricting the total intake by inhalation over one or more work shifts, depending on the length of the shift(s). The LTEL is expressed as an 8-hour (a typical working day) TWA concentration. LTELs are intended to help control the chronic health effects arising from prolonged or accumulated exposure to harmful substances.*
>
> *A **short-term exposure limit (STEL)** is a limit intended to control the effects of exposure to a substance by restricting the total intake by inhalation over any 15-minute period throughout the working shift. STELs are intended to protect against acute adverse health effects arising from brief exposure to high concentrations of contaminants.*

7.3.3 Why time-weighted averages are used

The two types of WEL described so far (the LTEL and STEL) refer to averaged concentrations over specific time periods (usually 8 hours and 15 minutes respectively). These limits are time weighted averages (TWAs). TWAs are used because, during a working shift, levels of exposure to an airborne contaminant will not remain constant; they will vary as the worker moves from place to place and performs different tasks.

For example, consider a worker carrying out an 8-hour shift. That worker is also using trichloroethylene (TCE) as a degreasing agent during their working day.

The worker's personal exposure to TCE vapour has been measured during the day and Table 1 shows their exposures for different time intervals.

Activity	Exposure time – in hours (T)	Measured concentration in ppm (C)	C × T
Task 1	3	80	240
Task 2	1.5	200	300
Task 3	2	30	60
Breaks	1.5	0	0

Table 1: Example TCE personal exposure times

There is a formula to calculate the TWA exposure over 8 hours. Multiply each measured concentration (C) by its corresponding exposure time (T). Then, add all the values of C × T together and divide by 8. So, in this case, the 8-hour TWA = (240 + 300 + 60 + 0)/8 = 75ppm. These results will then be compared against the published workplace exposure limits in *EH40* (discussed later) to ensure that the TWA concentrations are safe.

> **TIP**
>
> You might never need to calculate a TWA concentration in your particular workplace. However, the calculation here illustrates that concentrations are averaged over specific time periods to obtain a TWA value.

7.3.4 Limitations of exposure limits

Use of exposure limits, while helpful, is not an exact science.

Limitations include the following:

- Exposure limits are only concerned with airborne contaminants and therefore only take into account the inhalation route. No account is taken of other routes of entry, so the actual level of contaminants in the body may be considerably higher than indicated by airborne monitoring.
- Exposure limits are generic figures that take no account of individual susceptibilities.
- Exposure limits are quoted for individual substances; no account is taken of additive or synergistic effects of multiple substances when combined.
- The ability to set exposure limits is limited by the available scientific and medical knowledge. As knowledge about the effects of substances improves, the limits

will change. It is therefore possible that organisations might use out-of-date figures, resulting in overexposure.

7.3.5 Comparison of measurements to recognised standards

Measurements can be compared with WELs found in *EH40*. To illustrate the idea of comparison of measurements to recognised standards, we will use WELs from *EH40*. These standards can be applied to the specific example of TCE that we looked at earlier.

The WELs for some substances appearing in *EH40* are shown in Table 2.

Substance	LTEL (ppm)	LTEL (mg/m^3)	STEL (ppm)	STEL (mg/m^3)	Comments
Acetone[a]	500	1210	1500	3620	
Flour dust[b]	–	10	–	30	Sen
Hydrogen peroxide[c]	1	1.4	2	2.8	
Silica (RCS)[d]	–	0.1	–	–	
Trichloroethylene (TCE)[e]	100	550	150	820	Carc, Sk

Table 2: Workplace exposure limit examples

a: acetone is often used as a solvent; it is also found in nail varnish remover.

b: workers in a bakery are likely to be exposed to flour dust; 'Sen' means capable of causing occupational asthma.

c: hydrogen peroxide has been used in hairdressing salons.

d: the term RCS stands for respirable crystalline silica; quarry workers are at risk of exposure to RCS. It is examined further in 7.5: Specific agents.

e: this is used widely as a degreasing agent; 'Carc' means capable of causing cancer and 'Sk' indicates that the substance can be absorbed through the skin.

Note: the standards for flour dust and RCS are only quoted in mg/m^3 as these are dusts.

Where personal exposure results exceed the standards, it is evidence that control measures may not be sufficient, are not maintained adequately or possibly are being ignored or improperly used. This indicates the need for some form of action to further protect workers.

To demonstrate how LTEL and STEL limits may be applied, review the results obtained in Table 1 for TCE.

In *EH40*, there are long-term and short-term limits for TCE. The 8-hour LTEL is given as 100ppm while the 15-minute STEL is 150ppm. Looking at the results from Table 1, you can see that the 8-hour personal exposure TWA was 75ppm and therefore does not exceed the LTEL in *EH40*. However, you can also see that, during Task 2, there was a personal exposure well in excess of the STEL. This would mean that the reasons for this must be investigated and remedial action (including, for example, further controls) taken immediately.

For example, Task 2 may have involved manually emptying drums of TCE into an open degreasing tank on a few occasions, the process taking a total of 1.5 hours during an 8-hour working day. The task then resulted in workers experiencing significant vapour inhalation. A future control measure might be to try to automate this task, thus reducing exposure to TCE vapour; it would also reduce manual handling.

As discussed in 7.1: Hazardous substances, the acute effects from TCE include headaches, nausea and drowsiness, and the chronic effects include liver damage. In the example we have given, there would not be too much immediate concern over the chronic effect (although steps should still be taken to attempt to reduce exposure and measurements should be repeated), but there would be immediate concern regarding the acute effects. For example, in view of the acute effect of drowsiness, might this worker be driving a motor vehicle following their TCE exposure?

KEY POINTS

- The purpose of WELs is to protect the health of workers by identifying a level of exposure to an airborne substance that is safe for a person to inhale.

- There are two types of WEL – short-term (STEL) and long-term (LTEL); these protect against acute and chronic adverse health effects. Time-weighted averages for each WEL are based on average concentrations over a specific time period because levels of exposure will vary during a working shift. There is a formula to calculate personal exposure based on measured concentration and exposure time.

- The HSE document *EH40* contains workplace exposure limits (both LTEL and STEL values) for many substances.

- A limitation of WELs is that they can only be used to examine the concentrations of substances in the air and are therefore only applicable to the inhalation route of entry.

- WELs can be used to determine whether the results of personal exposure are satisfactory.

Reference

[1] HSE, *EH40/2005 Workplace exposure limits* (4th edition, 2020) (www.hse.gov.uk)

7.4: Control measures

Syllabus outline

In this section, you will develop an awareness of the following:

- The need to prevent exposure or, where this is not reasonably practicable, adequately control it
- Principles of good practice (see *Control of Substances Hazardous to Health Regulations*, Regulation 7(7) and Schedule 2A, as amended in 2004)
- Common measures used to implement the principles of good practice:
  - eliminate or substitute (hazardous substances or form of substance)
  - change process
  - reduce exposure time
  - enclose hazards; segregate process and people
  - use of local exhaust ventilation: general applications and principles of capture and removal of hazardous substances; parts of a basic system and what can make it less effective; requirements for inspection
  - use and limitations of dilution ventilation
  - respiratory protective equipment: why and when it should be used and how effective it is; types of equipment and the different substances they are best suited for; selection, use and maintenance
  - other protective equipment and clothing (gloves, overalls, eye protection)
  - personal hygiene and protection regimes
  - health/medical surveillance and biological monitoring
- Additional controls that are needed for substances that can cause cancer, asthma or genetic damage that can be passed from one generation to another.

Introduction

7.1: Hazardous substances and 7.2: Assessment of health risks discussed hazardous substances and the factors to be considered when carrying out a health risk assessment. They also noted the usefulness of substance monitoring and how the consideration of occupational exposure limits (OELs), generally known as workplace exposure limits (WELs) in the UK, could help with risk assessment.

You now need to understand the actions that can be taken to help minimise the risk from working with hazardous substances. This involves minimising exposure itself. These actions can be regarded as precautions or control measures (which is the term used here).

We will:

1 review the control measures outlined in Health and Safety Executive (HSE) guidance on the Control of Substances Hazardous to Health Regulations 2002 (COSHH),[1] with particular emphasis on reg 7(7) and Schedule 2A;
2 examine the control measures in more practical terms, including describing some specific control measures in detail, such as substitution, local exhaust ventilation (LEV) and personal protective equipment (PPE); and
3 outline extra control measures appropriate for substances that can cause cancer, genetic damage or mutagenic and reproductive damage.

7.4.1 The need to prevent or control exposure

As discussed elsewhere, a general principle is that first you should at least consider eliminating any risks. Chemicals are no exception. Some common ways of doing that are by:

- eliminating the original chemical hazard; that is, removing the hazardous chemical (for example, by changing the process so it is not needed) or substituting it for something non-hazardous or much less hazardous); and
- preventing exposure to the hazard, for example by totally enclosing the chemical within a process.

Examples are provided later, but in many cases these options may just not be possible, or at least not 100% successful. Indeed, in eliminating one risk, you can inadvertently introduce new risks. So you will often need to introduce one or more control measures to reduce the overall risk to an acceptable level.

The main source of information we will use for the approach to prevent or control exposure to hazardous substances will be the Control of Substances Hazardous to Health Regulations 2002.

7.4.2 Principles of good practice

Regulation 7(7) of COSHH sets out when control measures will be considered 'adequate'. This requires:

1 consideration of the requirements of Schedule 2A (principles of good practice for control of exposure);
2 identification of the requirements associated with workplace exposure limits (WELs); and
3 review of the particular requirements of carcinogens, respiratory sensitisers and 'asthmagens'.

1. Requirements of Schedule 2A

Schedule 2A sets out the principles of good practice for the control of exposure and is subdivided into the following eight general requirements:

"(a) Design and operate processes and activities to minimise emission, release and spread of substances hazardous to health."

Controlling substances hazardous to health (SHH) can be considered to be in three parts: control at source, the path between source and receiver, and control at the receiver (personal protective equipment PPE)). One of the main principles of (a) is

Chemical and biological agents

to aim to control SHH at source wherever possible and to minimise the numbers of sources of emission. This is best achieved at the design stage of a process.

"(b) Take into account all relevant routes of exposure – inhalation, skin absorption and ingestion – when developing control measures."

The common routes of exposure are inhalation, ingestion, skin absorption and injection. For example, if it is not practicable to control at source an SHH such as xylene, which is hazardous by inhalation and skin contact, then both the use of ventilation and the wearing of suitable gloves could be considered.

"(c) Control exposure by measures that are proportionate to the health risk."

This simply means that the greater the risk, the stricter the control measures needed.

"(d) Choose the most effective and reliable control options which minimise the escape and spread of substances hazardous to health."

For example, elimination of an SHH is the most effective option, whereas use of PPE is probably the least effective as it relies on the wearer using it properly.

"(e) Where adequate control cannot be achieved by other means, provide, in combination with other control measures, suitable personal protective equipment."

A clear message here is that, even if PPE is to be used, this does not mean discarding other measures that may contribute to the reduction of risk of exposure to SHH.

"(f) Check and review regularly control measures for their continuing effectiveness."

This may include, for example, inspection, testing and maintenance of engineering control measures, such as those for the continued effectiveness of a local exhaust ventilation (LEV) system.

"(g) Inform and train all employees on the hazards and risks from the substances with which they work and the use of control measures."

Employees need to understand the risks of SHH to which they are exposed and the control measures/precautions used to minimise these risks.

"(h) Ensure the introduction of control measures does not increase the overall risk to health and safety."

For example, enclosing a process involving highly flammable aerosols to minimise exposure to employees may result in an explosive atmosphere.

2. Requirements associated with WELs

Regulation 7(7) states that an approved WEL must not be exceeded.

3. Special requirements for carcinogens (listed in Schedule 1 of COSHH), respiratory sensitisers and asthmagens

For these types of substances, exposure not exceeding the WEL is not sufficient. It is necessary to further reduce exposure to the lowest reasonably practicable level. In practical terms, this means reducing the exposure to a level below which there would be little further benefit without applying a large amount of further resources.

7.4.3 Common measures used to implement the principles of good practice

Elimination

Although eliminating a substance hazardous to health altogether is the preferred method of control, it is probably the method least commonly available. If a hazardous substance is in use, it is probably because it has been identified as the best substance for the process. One possibility is to subcontract the use of that substance to another party, but that simply shifts the risk to someone else. Another possibility is to remove the part of a process that uses the hazardous substance altogether. Although this is not very likely, Case study 1 gives an example of where this was possible.

> **CASE STUDY 1**
>
> A company fabricated metal to make products. It was necessary that the metal was in a clean condition prior to fabrication. Metal sheets were obtained from a nearby foundry. The sheets were delivered on an open-top lorry and unloaded at the company premises, where they were stored under cover, but not in an enclosed building. When the metal sheets were ready for use, they were loaded into a degreasing tank where the degreasing hazardous chemical cleaned the sheets ready for use.
>
> The degreasing process was expensive in terms of the amount of chemical used and the energy needed for this process. A detailed health risk assessment was required due to exposure to the hazardous chemical.
>
> The company then made the following decisions. The delivery company was ordered to deliver the metal sheets in a closed lorry. The company built a temperature and humidity controlled storage facility. When the sheets were delivered, they were transferred immediately to the storage facility until needed. The result was that the degreasing plant and its hazardous chemical were no longer needed.
>
> This was a rare example of the control measure of elimination.

Substitution

This method involves replacing a hazardous chemical with one that is less hazardous. This is a more practical method than elimination.

An example is the selection of industrial paints. Solvent-based paints often contain xylene and butan-1-ol as solvents. The short-term exposure limit (STEL) of xylene is 100ppm and that of butan-1-ol is 50ppm. Therefore, replacing a paint containing both solvents with a paint only containing xylene would be a good substitution. Even better, consider replacing a solvent-based paint with a water-based paint.

Another example is the replacement of leaded petrol with unleaded petrol.

Change of process

One method might be to change the way in which an SHH is applied to a component; for example, instead of spray painting a component, application by brush painting would eliminate exposure to the harmful aerosol. Exposure to harmful vapour due to brushing should present a much-reduced risk.

Another example might be to lower the temperature of use of a solvent. The lower its temperature, the lower its resulting vapour pressure and thus the lower the level of inhaled vapour.

Reduce exposure time

There two general methods that may be considered.

If feasible, introduce job rotation, so that each employee spends less time at a process associated with the greatest exposure of an SHH.

Instead of, or in association with, job rotation, introduce sufficient extra breaks to reduce exposure to an acceptable level.

These methods should be effective in reducing the frequency and duration of exposure.

Enclose hazards

It may be possible to contain an SHH within a given area and/or reduce the need for employees to be present in that area.

A practical example is using a totally enclosed process and handling system: this should prevent any exposure of workers to hazardous substances. Such systems may involve observation windows to enable workers to observe the flow of the process. It may be possible to totally enclose a process and materials-handling system, for example, on conveyor belt systems or on car-body painting lines.

Segregation of processes and people

The segregation of hazardous processes from workers can be done, for example, by remote control. If total enclosure is not practicable, segregation may be possible; for example, screening off an area to separate workers from a process involving hazardous chemicals.

Use of local exhaust ventilation (LEV)

A schematic diagram of an LEV system is shown in Figure 1.

The principle of an LEV system is to extract an airborne contaminant from the working area and transfer it to a safe place, which is normally outside the building.

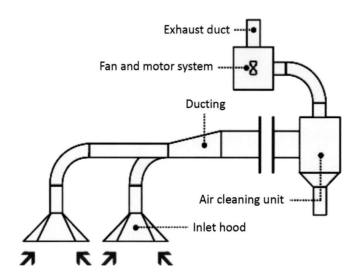

Figure 1: LEV system: exhaust ventilation

7.4 Control measures

The five key components of an LEV system and their functions are:

1 The inlet hood: this is a nozzle or hood and is placed as near as practicable to the source of the contaminant so that it is captured before being breathed in or entering the general work area.
2 Ducting: this transports the contaminant from the hood to the next stage of the system.
3 Air cleaning unit: there are different types for different forms of contaminant. The bag filter unit is widely used when the contaminant is in the form of a dry dust. The dust enters the unit and passes through filter bags, then the cleaned air passes out of the unit at right angles to the inlet flow through more ducting. Other types of contaminant would need different types of air cleaner; for example, hot gases are usually best controlled by a wet scrubber (such as a venturi scrubber) and fumes are best controlled by an electrostatic precipitator.
4 Fan and motor system: this is the driving force of the system and is selected to draw sufficient flow through the system.
5 Exhaust duct (a system outlet): this is where the cleaned air is discharged to the atmosphere.

What can make an LEV system less effective than its design performance?

LEV is a good control measure when it works properly. However, there are things that might cause it to work inefficiently:

- Positioning of the inlet hood: if the hood is not placed close enough to the source of the contaminant, the contaminant will not be sufficiently captured.
- The design of the system may be poor; for example, if there are right-angled bends in the ducting, there will be a loss of flow through the system. Where there needs to be a change of direction in the ducting, it should be in the form of a smooth curve. Another problem may occur if there is a change of process without modifying the system; for example, a new material producing more contaminant or a denser contaminant may require an increased duct velocity for efficient removal of the contaminant. Therefore, the new process would require a changed system.
- There may be a build-up of contaminant in the ducting; this may be a result of insufficient air flow through the system.
- The air cleaner (such as the bag filter) may be full.
- The bag filter itself may be damaged.
- The ducting may be damaged; for example, there may be a hole in it.
- An LEV system requires maintenance; failure to maintain it is likely to lead to inefficiency of the system.
- There can be corrosion of fan blades, especially if the contaminant is an acid gas; this will also reduce air flow rate through the system.
- There may be unauthorised alterations to the system.
- The system outlet may be blocked; for example, birds have been known to build nests in outlets.

Requirements for inspection

Under reg 9 of COSHH, an LEV system must be thoroughly examined and tested at least once every 14 months. In rare cases, the examination/testing period is more frequent, but discussion of such cases is outside of our scope here.

Chemical and biological agents

An inspection would include:

- examining external parts for wear and tear and for damage;
- checking that filter cleaning devices are working correctly;
- inspecting the filter fabric; and
- checking the condition of the fan belt.

The HSE publication *Controlling airborne contaminants at work: A guide to local exhaust ventilation (LEV)*[2] is very comprehensive and provides some information on the basic principles of LEV.

> **TIP**
>
> LEV may not always be a practical control measure. For example, there may be several sources of emission of a contaminant, or perhaps a process is of short duration (for example, no more than a few days) so that it is not reasonably practicable to install an LEV system.

Dilution ventilation

When LEV is not suitable, dilution ventilation (DV) may be an option to consider.

What is dilution ventilation?

It is simply diluting a contaminant with a sufficient quantity of clean (uncontaminated) air; the greater the level of contamination, the greater the quantity of clean air needed.

The simplest method may be to work outside a building or to open windows if working inside. A more sophisticated method would be to install fans in a building; for example, in the wall or ceiling. The density of the contaminant should be considered here; for a contaminant denser than air, the fan would be installed near floor level. For a contaminant less dense than air, a ceiling fan would be more suitable. For protection from any contaminant by this method, the contaminant must be non-toxic and must only be present in small quantities. It must never be used when the contaminant is dust.

Uses and limitations of dilution ventilation

Table 1 shows circumstances where DV is, or is not, suitable as a control measure.

Uses (where DV is suitable)	Limitations (where DV is not suitable)
Non-toxic contaminants, such as steam in a shower room	Hazardous contaminants, such as those assigned a WEL or significant concentrations of dusts
Where the rate of evolution of the contaminant is low and constant	Where the rate of evolution of the contaminant is high or variable
Where the rate of evolution is known	Where the rate of evolution is unknown
Where the vapour pressure of the contaminant is low	Where the vapour pressure of the contaminant is high
Where there are several sources of contamination	
Where there is no other reasonably practicable means of contaminant control	

Table 1: Circumstances where dilution ventilation may or may not be appropriate

Respiratory protective equipment (RPE)

RPE is needed to protect employees against harmful airborne substances.

It will be used when no other reasonably practicable control measure is sufficient to protect employees or when other control measures do not offer sufficient protection.

There are two types of RPE: respirators and breathing apparatus (BA). If a worker is in an environment where there is a plentiful supply of breathable oxygen, a respirator will be the natural choice of RPE for them. However, if the working environment is deficient in oxygen, such as in a confined space or a smoke-filled area (as experienced by firefighters), then BA will be required. BA would also be required for workers entering an area where harmful airborne biological agents are known to be present.

Each type of BA and respirator has its own merits and limitations. It is important to select the right one for the job and know what might cause them to become inefficient. You need to know whether it is capable of reducing the exposure of the worker to the hazardous substance to below the WEL. The term relating to this is the 'assumed protection factor' (APF) of the RPE, although some suppliers also use the term 'nominal protection factor' (NPF).

How do you get all this information in practice?

- The manufacturer or supplier will provide the value of the APF for the RPE.
- The concentration of the contaminant in the worker's environment, that is, the concentration of the contaminant outside the mask (A), will be measured.
- The WEL will normally be available from *EH40* (see 7.3: Occupational exposure limits).

If you know the value of the APF, and have measured the concentration of A, you can calculate the concentration of B. If B exceeds the (known value) of the WEL, the mask is not satisfactory and one with a higher APF must be selected.

Therefore, the purpose of this calculation is to select a mask of the appropriate APF knowing the concentration of A.

> **TIP**
>
> The APF of RPE is worked out as follows:
>   APF = A/B
>
> Where A = concentration of the contaminant outside the RPE
> And B = concentration of the contaminant inside the RPE.
>
> Example:
> If the measured concentration of substance X in the environment (so it could be breathed in by a worker not wearing RPE) is 240ppm and the APF of the RPE is 15, then:
> 15 = 240/B, so 15 = 240/B
> Therefore B = 240/15 = 16ppm.
> In other words, if A = 240ppm, the worker (wearing a mask of APF 15) would be exposed to a concentration of 16ppm, because the concentration inside the mask would be 16ppm.
> If the WEL of substance X is 60ppm, the RPE is sufficient. However, if the WEL of X was 10ppm, then the RPE is not sufficient.

Respirators

A simple face mask is nearly always disposable and is cheap and easy to use, but it has a low APF and is only suitable to protect against nuisance dusts.

A more effective type of respirator is a 'half-mask' type, covering the mouth and the nose.

Half-mask respirators offer a higher degree of protection, since they are made from rubber and therefore fit the face shape and create a better seal. They are also provided with better-quality filters and offer a reasonably good level of protection as a result. As they are only half-mask respirators, no eye protection is provided, which leaves the eyes vulnerable to chemical attack.

When the user inhales, the air passes through a cartridge that removes most of the contaminant; when the breath is exhaled, this passes through an exhaust valve. There are different types of cartridge; the most common offer protection against solvent vapour, dust or acid gas. It is important to fit the correct cartridge for the hazardous substance in the area.

An even more effective option is to use a 'full-face' respirator. Again, these come with a variety of filter options and therefore offer a good degree of protection. The adjustable straps enable the wearer to create a good seal against the face. The full-face respirator also offers eye protection, although this may reduce peripheral vision.

One of the best types is the positive pressure-powered respirator (sometimes referred to as an 'airstream helmet'), which draws air across the wearer's face, thus providing some cooling while at the same time providing protection against leakages.

Figure 2: Example of a respirator

Breathing apparatus (BA)

There are three types of BA:

- Fresh air hose apparatus: fresh air is delivered from a source that is not contaminated. It may require considerable breathing effort by the user.
- Compressed air line type: the air is delivered from a hose connected to a compressed air source. No breathing effort is required by the user, but the inhaled air can contain traces of oil and grit.
- Self-contained BA: air is inhaled from a cylinder. This is a completely sealed system and should provide pure, uncontaminated air.

Selection of RPE

Some of the issues that need to be considered when selecting RPE are:

- the OEL/WEL of the contaminant;
- the required APF of the RPE;
- the amount of exposure, which could be obtained from air monitoring results;

- the frequency and duration of use;
- whether there is a normal supply of oxygen – this might not, for example, be the case if working in a confined space;
- the form of the airborne substances (dust, solvent vapour or other);
- the health of the workers: you need to consider if there are there any vulnerable workers, for example those with pre-existing health conditions (an asthmatic worker might not be able to wear a mask). It is vital to obtain information about a worker's health. The easiest way to obtain this information is to ask the worker when selecting the BA. Information could also be obtained from health surveillance or pre-employment medical screening;
- how the RPE will be maintained and inspected;
- the cost of the RPE – however, this should not be prioritised over the level of protection required;
- the fit testing procedure;
- what training would be required in the use of the RPE;
- the suitability of fit – for example, the worker might have a beard;
- compatibility with other PPE, for example with hearing protection; and
- whether wearing the RPE would introduce any other hazards – for example, reduction in visibility.

RPE can occasionally be defective and not provide the protection intended. You should be aware of this. Some of the reasons include:

- damaged seals in the equipment;
- filters could be the wrong type for the equipment;
- cartridges may have passed their use-by date;
- the battery may be low on the power unit of a powered respirator;
- straps on the equipment may be worn; and
- the RPE may have been inadequately stored, causing damage.

Maintenance of RPE

Parts of RPE (except disposable masks) need to be checked to ensure they are in good working order; in particular:

- Straps should be checked; for example, to ensure they have not lost their elasticity.
- Filters (cartridges) should be replaced when necessary. These sometimes have a frequency of change identified by the manufacturer, but a high level of usage may require a cartridge change before a specified date. Manufacturers sometimes state that cartridges should be changed following operator complaints (for example, of smell); this is of no use if the contaminant is odourless and/or the operator has a poor sense of smell. Keeping accurate records of cartridge changes for particular applications is useful as this may help to identify when a change is appropriate.
- Seals should be checked to determine if they are worn and need replacing.
- For powered respirators, the efficiency of the battery pack should be checked.
- If using a respirator fed by a compressed air source, the inhalable air could be contaminated, for example, by oil or grit, and therefore checks should be made to ensure a suitable in-line filter is installed in the system.
- The guidance associated with reg 9 of COSHH recommends that the quality of air supplied to breathing apparatus should be checked at least once every three months.
- RPE should cleaned and disinfected after every use – this is crucial if any RPE may be used by more than one person.
- RPE should be stored away from high and low temperatures, and also away from dusty environments.

Other personal protective equipment (PPE)

As with RPE, this final control measure in the hierarchy must be used only as a last resort when other controls are not available or where those other controls do not provide sufficient protection. You must first decide what part of the body needs to be protected, which may be the skin, eyes, whole body or respiratory system. For each of these, there are specific factors to be considered.

For all types of PPE, the safety data sheet (SDS) for the hazardous substance(s) being used must be referred to for advice on what is required.

The skin

There are various elements to consider for skin protection:

- The correct size of gloves; a range of sizes for workers should be available.
- The length of time that gloves can be used; for example, before a chemical breaks through the gloves.
- Whether any gloves produce allergic reactions; for example, many workers are allergic to latex gloves.
- The durability of the gloves, including whether they tear easily.
- The level of comfort; for example, whether gloves are too tight.
- The level of dexterity needed; some gloves are highly resistant to chemicals but workers can have difficulty using them due to the stiffness of the gloves preventing sufficient finger movement when wearing them.
- The length of the gloves; in some cases, gauntlets may be preferable to protect arms as well as hands.
- Finally, some training may need to be given; for example, in how to remove the gloves or gauntlets without contaminating the skin, and how and to whom to report any glove defects or damage.

Figure 3: Gloves must fit well   **Figure 4:** Gauntlets protect both hands and arms

The whole body

There are different types of whole-body protection.

Paper disposable overalls could be particularly useful for protection against a toxic dust, as the overalls could then be safely disposed of after completion of the work. Clearly, this type of overall would be unsuitable for protection against an aggressive chemical, such as a corrosive liquid. In this case, a liquid-resistant overall would be appropriate.

Workers may sometimes need to work in hot atmospheres while using aggressive chemicals. In this case a water-cooled overall would be suitable PPE.

The eyes

There are three types of eye protection: spectacles, goggles and face visors. It is unlikely that safety spectacles will provide sufficient protection, therefore goggles should be the minimum protection for eyes against hazardous substances. The goggles should be of the appropriate standard to protect against the substance being used. A face visor should be used to protect the face as well as the eyes.

There are many different standards for eye protection, depending on what the eyes need to be protected against. The standards change from time to time; for example, the standard EN 166 was replaced in 2021 by ISO 16321-1, which covers eye and face protection for occupational use (Part 1: General requirements). For specific requirements there are other standards. For example, for use during welding, the standard is ISO 16321-2.

Figure 5: Goggles must be the right standard

Figure 6: Visors protect the face and eyes

Although the use of PPE is a last-resort control measure, there will be some occasions where PPE is the only option; for example, in emergency situations, for maintenance operations, when the worker is mobile (such as grinding paving slabs) and when there is simply no other practicable option.

APPLICATION

Consider the PPE that is made available in your own workplace.

- What is it protecting workers against?
- How is it stored?
- What training is provided in the use of the PPE?

Personal hygiene and protection regimes

Employers are expected to provide appropriate hygiene facilities, such as:

- Adequate washing facilities: these should be conveniently near to the work area, but not so close that they may become contaminated.
- Appropriate changing facilities: this is particularly important where work clothing is likely to become contaminated. The facilities should be sited so that work clothing does not contaminate personal clothing.
- Facilities for eating and drinking: these should be in uncontaminated areas (smoking should be prohibited except in specified outdoor areas).

Employees should receive information and training on how and why such facilities should be used. Accordingly, employees should comply with instructions provided by their training.

Examples of systems that may assist good standards of hygiene are:

- Reduction of workers and exclusion of non-essential access; reducing the number of workers exposed to hazardous substances should reduce the overall risk of worker ill-health. Keeping processes that use hazardous substances in dedicated

Chemical and biological agents

rooms should assist in keeping non-essential workers away from these processes. If this is not feasible, it may be possible to erect barriers around processes.
- Reducing the frequency and duration of exposure of workers to hazardous substances will help reduce the risk of ill-health.
- Regular cleaning of contaminated surfaces, which could include work surfaces and walls.
- If working with a liquid with a high vapour pressure at room temperature (that is, it evaporates quickly), any rags soaked with such a substance should be stored in a waste bin with the lid kept closed.

Health surveillance and biological monitoring

Regulation 11 of COSHH identifies health surveillance requirements and provides some associated guidance.

The HSE guidance to reg 11 of COSHH states that the objectives of health surveillance are to:[3]

(a) check the health of individual employees by detecting, as early as possible, adverse changes which may be caused by exposure to substances hazardous to health;

(b) collect, keep up to date and use data and information for determining and evaluating hazards to health so that action can be taken to prevent more serious disease from occurring;

(c) check control measures are working by effectively providing feedback on the accuracy of the risk assessment and the effectiveness of control measures to identify where further steps to manage risk are needed.

It would be appropriate to carry out health surveillance when a disease or some form of health effect can be related to the exposure of any SHH and there is a reasonable likelihood of such a disease or adverse effect occurring.

There also needs to be a valid analytical technique available for detecting the disease or adverse effect (any such technique should only present a low risk to an employee). Furthermore, there are a few situations where medical surveillance is required (identified in Schedule 6 of COSHH; for example, work with vinyl chloride monomer (VCM). Exposure to VCM can result in hepatic angiosarcoma (which is a rare type of liver cancer).

Strictly speaking, carrying out health surveillance is not a control measure but part of the risk assessment of SHH. However, actions taken as a consequence of the results of health surveillance are control measures.

Biological monitoring

This involves the measurement and assessment of workplace agents, or their metabolites (a substance made or used by the metabolism), following exposure to these agents or metabolites, and can involve taking samples of urine or blood for analysis. This technique is particularly important if there are several routes of entry of agents (remember that atmospheric monitoring only provides results for the inhalation route of entry). Lung function monitoring is another technique often used following exposure to agents.

Medical examinations may be carried out under certain circumstances and should serve the following purposes:

- They can prevent further deterioration of workers' health.
- They may help evaluate the effectiveness of control measures in the workplace.

- They can act as reinforcement of safe methods of work and health maintenance.
- They can assess the fitness of a worker for a particular type of work; individual susceptibility must be taken into account.

Figure 7: Worker medical checks

Control measures for biological hazards

A somewhat different set of controls for biological hazards is required compared with those for hazardous substances.

These control measures include:

- Disinfection: for example, in a veterinary surgery, the examination surface for animals will be wiped down with bleach between each animal patient. In an office, telephones may be regularly sanitised since many people may use them and a person's mouth is close to the phone. This is particularly important for protection against viruses such as Covid-19.
- Engineering control measures: these might include using a glove box or a fume cupboard.
- Vermin control: for example, control of rats. Exposure to rat urine can cause leptospirosis. Practical examples of control include:
  - food shops stopping rats being on the premises so that food packaging that could be handled by shop workers is not contaminated by rat urine; and
  - controlling the rat population at sewage treatment plants to reduce the risk of workers being exposed to rat urine.
- Water treatment: this might entail adding a biocide or working at a temperature that might kill the biological agent. Stagnant water spray or fine water droplets can contain *Legionella* bacteria, which can cause serious disease or death. A chemical such as chlorine dioxide can be used in cooling towers to control these bacteria. The bacteria multiply fastest between 20°C and 45°C, so in water systems cold water is usually provided at temperatures of less than 10°C and hot water should be provided at temperatures of at least 50°C.
- Taking great care about personal hygiene: in particular, covering wounds with waterproof plasters prior to working with a biological agent. Employers should provide adequate welfare facilities and train workers to use them appropriately.
- Ensuring proper disposal measures: for example, using sharps boxes for used needles. The boxes should then be collected for disposal at an appropriate site. Any PPE used, such as masks or RPE, will either be reusable or disposable. If the PPE is reusable, it must be cleaned and disinfected, then properly stored away from any source of contamination. If the PPE is disposable, it should be placed in a plastic bag and the opening tied. It should then be 'double bagged' by placing the first bag in a second one. Again, the opening should be tied. The

tied bag should be kept in a secure location prior to collection for disposal at an appropriate site.
- Identifying the different types of PPE and RPE that will be required.
- Use of vaccination or immunisation for known effects.

7.4.4 Additional controls that are needed for substances that can cause cancer, asthma or genetic damage that can be passed from one generation to another

> **DEFINITIONS**
> A **carcinogen** is an agent that has the potential to cause cancer.
>
> A **mutagen** is an agent that damages the genetic material of a cell and causes abnormal changes that can be passed from one generation to another.
>
> An **occupational asthmagen** is an agent that can cause asthma in some people who have been previously exposed to the agent. Once sensitised, these people can suffer asthma even when exposed to a very small amount of the substance.

Control measures

Many of the control measures already identified for hazardous agents will apply to the agents defined – carcinogens, mutagens and occupational asthmagens (these will be collectively referred to as 'these agents') – but there are some more stringent requirements.

Again, eliminating the use of these agents is the first option to consider. Substitution should involve not just using a less hazardous compound but a non-carcinogen if practicable.

For the substances discussed earlier, control was deemed sufficient provided you were able to reduce exposure below the 8-hour TWA and not exceed the 15-minute STEL. For these agents, it is necessary to go further and reduce exposure to the lowest reasonably practicable level.

The minimum quantity of these agents must be used. There should be total enclosure of the system these agents are used in as far as reasonably practicable. These agents must be stored in closed containers, which must be labelled with clearly visible warning signs. In addition:

- Areas of use should be restricted for these agents.
- Those areas where use is permitted must be clearly identified.
- The number of workers in areas where these agents are present should be minimised.
- Entry for non-essential people should be prohibited.
- Any waste products should be securely stored and clearly labelled. These waste products may only be removed by a licensed contractor.
- Workers should not eat, drink, smoke or apply cosmetics in areas where these agents are in use.

7.4 Control measures

- Floors, walls and work surfaces should be regularly cleaned.
- Adequate washing facilities should be available.

CASE STUDY 2

Two workers from an electroplating factory were found to be suffering from occupational asthma. They were using a chemical that contained isocyanates. Isocyanates are well known to be respiratory sensitisers.

It was found that the workers had been exposed to isocyanate concentrations exceeding the occupational exposure limit (OEL).

The employer subsequently installed a local exhaust ventilation system that efficiently extracted the isocyanate vapour.

As a result, the concentration of isocyanate in the factory area where the two workers were employed was substantially reduced to a concentration well below the OEL.

KEY POINTS

- The COSHH Regulations require certain actions to be taken with regard to health surveillance and control measures, especially when a risk assessment shows that current measures are possibly inadequate.
- Practical control measures that may reduce exposure to hazardous substances to acceptable levels include elimination of the hazardous substance, substituting it with a less hazardous substance, changing the process, isolation techniques, technical measures, administrative measures and personal protection.
- A hierarchy of control measures presents an order of preference for selecting control measures, with PPE being the last resort.
- There are five components of an LEV system to extract airborne contaminants from working areas and transfer them to a safe place, but an LEV is only a good control measure when it works properly. There are various reasons it might become inefficient, including damage to the components.
- Selecting between different types of PPE and RPE depends on which parts of the body need protecting and against what as well as the nature of the working environment.
- Control measures for biological agents differ in some ways from those for chemical substances.
- Extra controls are necessary for substances that could cause cancer, asthma or genetic damage.

References

[1] HSE, *The Control of Substances Hazardous to Health Regulations 2002. Approved Code of Practice and guidance* (L5, 6th edition, 2013) (www.hse.gov.uk)

[2] HSE, *Controlling airborne contaminants at work: A guide to local exhaust ventilation (LEV)* (HSG258, 3rd edition, 2017) (www.hse.gov.uk)

[3] See note 1

7.5: Specific agents

> **Syllabus outline**
>
> In this section, you will develop an awareness of the following:
>
> - Health risks, controls and likely workplace activities/locations where the following specific agents can be found:
>   - asbestos (excluding removal and disposal)
>   - blood-borne viruses
>   - carbon monoxide
>   - cement
>   - *Legionella*
>   - *Leptospira*
>   - silica
>   - wood dust

7.5.1 Health risks, controls and likely workplace activities/locations where specific agents can be found

In 7.4: Control measures we considered general categories of health risks and control measures to minimise exposure to hazardous agents. It is important to note that the control measures/precautions need to be suitable for the specific substance and conditions of use. For example, if a hazardous airborne dust is being produced rapidly and in high volumes, then general dilution ventilation is not likely to be an appropriate control measure because it simply could not capture and remove the hazardous substance fast enough. Local exhaust ventilation (LEV) would be a better choice if ventilation was being considered as a control option.

This section looks at some specific examples of hazardous substances, including circumstances when such substances might cause ill-health following exposure.

Asbestos

> **TIP**
>
> For asbestos, we will make use of information from the following legislation and guidance
>
> - Control of Asbestos Regulations 2012 (CAR);
> - *Managing and working with asbestos. Control of Asbestos Regulations 2012. Approved Code of Practice and guidance* (L143);[1]
> - *Asbestos essentials task sheets*, which provide details of control measures for carrying out specific non-licensed work with asbestos;[2]
> - *Asbestos: The survey guide* (HSG264).[3]

The term 'asbestos-containing materials' is commonly abbreviated to ACMs, so we will use that abbreviation here.

Asbestos has caused ill-health and fatalities for many years.

It is a fibrous, naturally occurring material. There are several types, the three main ones being:

- chrysotile (commonly called white asbestos);
- amosite (brown asbestos); and
- crocidolite (blue asbestos).

The actual type cannot be established by colour alone (the colours identified here are their natural colours), as it may be mixed with pigments or incorporated into a coloured substrate.

Exposure to crocidolite poses a greater risk than exposure to amosite, which in turn poses a greater risk than chrysotile. But exposure to chrysotile still comes with a great risk of disease.

Health risks

The three main health conditions resulting from exposure to asbestos are asbestosis, lung cancer and mesothelioma. They result from exposure to respirable asbestos fibres. In asbestosis, respirable fibres travel down to the alveoli of the lung and cause scarring of the lung. This does not necessarily cause cancer but, in combination with other adverse effects, cancer can occur. In mesothelioma, asbestos fibres lodge in the lining of the pleura of the lung. This will cause cancer.

There is no known cure for asbestosis or mesothelioma. Common ill-health symptoms are shortness of breath, a persistent dry cough and loss of appetite leading to weight loss.

Circumstances where asbestos may be found

Asbestos was widely used in buildings for many years and can still be found:

- in ceiling tiles;
- on corrugated asbestos roofs;
- in downpipes from gutters;
- in insulation around pipes;
- as sprayed asbestos around structural beams of buildings;
- in electrical panels;
- in internal walls; and
- in textured paints.

Chemical and biological agents

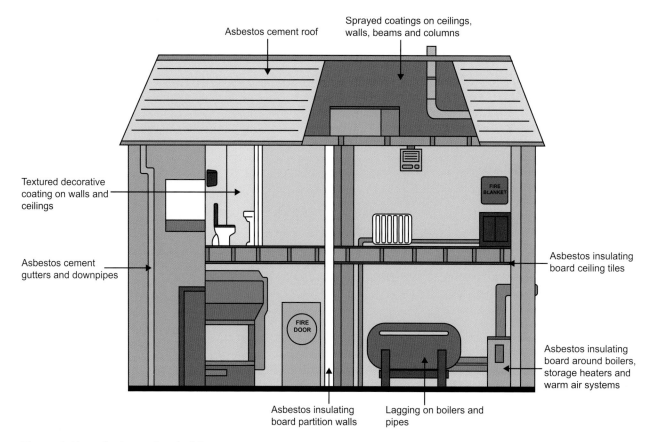

Figure 1: Uses of asbestos in a building

In the UK, amosite and crocidolite were banned in 1985 and chrysotile in 1999 with the effect that most uses of asbestos in the UK were banned after 1999.

However, asbestos is still likely to be present in older buildings (ie those built or renovated before 2000). It may safely be left in place if it is in good condition and is unlikely to be damaged or disturbed. But, because of its widespread use in the past, anyone likely to disturb, renovate, maintain or alter parts of buildings will probably encounter it. For example, if an organisation wishes to knock down an internal partition wall in an old building, this could cause considerable release of airborne asbestos fibres that could be breathed in (leading to various diseases as noted earlier).

Many workers have suffered ill-health as a result of exposure to asbestos in this way, including plumbers, electricians, building contractors, maintenance workers and schoolteachers. Managing asbestos in buildings is mainly about protecting people from exposure to asbestos fibres or dust, rather than removing it (which can be more dangerous than leaving it alone). There are some very specific requirements for removal and disposal of asbestos (such as licences and notification), but this is beyond the scope of the syllabus. Instead, we will limit the discussion to management of asbestos that is (or is suspected of being) present in a building.

Management of asbestos

Under CAR there is a legal requirement for asbestos to be managed in non-domestic premises. The person responsible for this is known as the dutyholder and they must be specifically identified; for example, in the company safety policy.

Who is the dutyholder?

The dutyholder is essentially the person (an individual or organisation) who is *mainly* responsible for maintaining or repairing the non-domestic premises (or common parts of domestic premises). This means that, in the absence of a specific contract or agreement, the dutyholder would normally be the owner if the building is owner-occupied. But in cases where there is a specific agreement (eg tenancy agreement or contract) it could be:

- The occupier
- Landlord
- Sub-lessor
- Managing agent
- Tenant

The exact nature of the duties will depend on the agreement terms; the duties may even be shared between these parties.

> **TIP**
>
> It is not always straightforward to determine whether a material contains asbestos. Therefore, it is convenient to classify materials into four groups:
>
> 1 Materials **known** to contain asbestos.
> 2 Materials **strongly presumed** to contain asbestos; for example, materials commonly used in the past in certain products, such as in cement roofs and insulating boards.
> 3 Materials **presumed** to contain asbestos because there is insufficient evidence to state with certainty that the material is asbestos-free.
> 4 Materials **known** to be asbestos-free, such as metal or stone.

The dutyholder must ensure that:

- reasonable steps are taken to locate ACMs in non-domestic premises and establish their condition, such as the extent of any damage to the ACMs;
- materials are presumed to contain asbestos unless evidence is available that they do not (see point 3 in Tip box);
- the risk of any person being exposed to ACMs/potential ACMs is assessed;
- an asbestos plan (or register) is produced that identifies both the **location** of materials that are known, strongly presumed or presumed to be ACMs, and their **condition**;
- the written record is kept up to date; and
- any ACMs/potential ACMs found are labelled.

It may be that there have been previous survey reports of asbestos in a building. It would be useful to refer to these to help establish where ACMs had been found. The reports may also have defined the type of asbestos present.

Control measures are associated with actioning a management plan, which aims to:

- ensure any ACMs are maintained in good condition;
- decide whether any ACM presenting a significant risk, due to its location or position, is repaired and protected or removed (most removal work requires the use of specialist licensed contractors, but some lower-risk work on ACMs such as asbestos cement sheeting can be done by unlicensed competent contractors); and
- ensure that information and location of any ACM is provided to any worker who is liable to disturb it or otherwise be potentially at risk.

Chemical and biological agents

Controls for working with asbestos (excluding removal and disposal)

As mentioned earlier, asbestos is best left alone if it is in good condition and unlikely to be disturbed. Removal and disposal is beyond the scope of the syllabus. However, there are numerous occasions when asbestos may conceivably be disturbed during even minor building maintenance work, such as drilling through asbestos board to mount something on a wall. While higher-risk work (including removal) must be done by specialist licensed contractors, even unlicensed, lower-risk, smaller jobs must be done by competent contractors (people trained and experienced in working with asbestos).

The HSE has produced guides on a wide range of very specific tasks on ACMs. These are collectively known as *Asbestos essentials task sheets*.[4] The idea is that you find the task sheet that reflects the job you are planning and this sets out the detailed control measures that should be followed. For example, sheet A9 looks at drilling holes in asbestos cement and other highly bonded materials. This is classed as unlicensed work. The task sheet takes you in detail through all the equipment needed as well as the step-by-step procedure. For this task, control measures include:

- protecting nearby surfaces from contamination with polythene sheeting (fixing it with duct tape);
- stopping the surface to be drilled from crumbing away by covering the drilling point with masking tape;
- applying paste or foam on the drilling point (to stop escape of fibres/dust), combined with a class H vacuum while drilling;
- sealing the edges of the hole after drilling;
- use of personal protective equipment (PPE), such as disposable overalls with a hood or respiratory protective equipment (RPE); and
- disposal of waste.

The controls vary with the specific tasks, but the emphasis is on controlling fibres/dust at source, capture of any escaping fibres and scrupulous clean-up and disposal.

Blood-borne viruses (BBVs)

> **TIP**
>
> Provisions relating to BBVs are covered in reg 7(10) and Schedule 3 of the Control of Substances Hazardous to Health Regulations 2002 (COSHH), which provides detailed information about the provisions relating to biological agents, such as their hazard classification (in groups 1 to 4) and the specific control measures required for each group. For example, hepatitis B (which is a BBV) is classified as a group 3 biological agent, which means that exposure to hepatitis B can cause severe human disease and may be a serious hazard to workers. It may spread to the community, but there is effective prophylaxis (that is, preventative measures such as vaccination) or treatment available.
>
> There is also HSE guidance on BBVs.[5]

BBVs are, as their name suggests, carried in the blood. They can generally be spread from one person to another through contact. Routes of transmission of BBVs are mainly:

- inhaling the breath of, or having contact with, another person who is already infected with the virus;
- being exposed to splashes of a BBV-containing material; or
- injection of the virus through broken skin.

Health risks
These will vary according to the particular BBV to which there is exposure. However, common symptoms are headache, nausea, difficulty in breathing and aching muscles.

Circumstances where infection may occur
Such circumstances may include work involving the following types of workers:

- first-aiders;
- paramedics;
- sewage workers;
- workers engaged in research laboratories where BBVs are handled;
- undertakers; and
- refuse disposal workers.

Precautions
- Carry out work in such a way as to prevent cuts or puncture wounds; for example, by minimising the use of sharps such as needles.
- Where sharps are used, ensure there are adequate sharps bags available and that they are properly used. This may require appropriate training in the risks of BBVs and the precautions to be taken.
- Cover all cuts with waterproof dressings.
- Prohibit eating, drinking and smoking while working in areas where BBVs are present.
- Ensure appropriate PPE is worn, such as waterproof protective clothing and suitable gloves. Also, protect eyes with goggles or a face shield.
- Carry out immunisation where possible; for example, against hepatitis B.

Carbon monoxide

Red blood cells (erythrocytes) hold large amounts of oxygen. Oxygen combines with haemoglobin in the red blood cells to form oxyhaemoglobin, which travels around the body providing oxygen to body tissues. If carbon monoxide is inhaled, the haemoglobin will combine with it much more readily than it does with oxygen. The result is the formation of carboxyhaemoglobin. When this circulates around the body, oxygen is no longer available to tissues.

Health risks
The greater the concentration of carbon monoxide, the greater the risk to health. Typical symptoms resulting from exposure to carbon monoxide are:

- fatigue;
- confusion;
- difficulty in breathing;
- loss of consciousness; and
- if exposure continues, death.

Circumstances where there is a risk of exposure to carbon monoxide
Carbon monoxide is a colourless, odourless gas; therefore you would not always be aware of its presence.

One perhaps obvious circumstance is inhalation of vehicle exhaust fumes; therefore workers in garages, such as maintenance staff, may be at significant risk.

Chemical and biological agents

Further examples include generators in confined spaces and faulty gas boilers. These have been the cause of fatalities to many people in the past, as illustrated by Case study 1.

> **CASE STUDY 1**
>
> **Generator causes fatal poisoning**
>
> The area above a shop had been converted into flats.
>
> The electricity supply to the shop and flats had been connected by illegal means and was subsequently cut off by the regional electricity organisation.
>
> To avoid paying for the electricity supply to be reconnected and used, the shop owner and the landlord of the flats decided to pursue a cheaper option by installing a petrol-driven generator in a poorly ventilated storeroom of the shop.
>
> The generator was positioned directly under one of the flats. Soon afterwards, the tenant of this flat died. The post-mortem showed that the tenant had died from carbon monoxide poisoning.
>
> In the subsequent court case, both the owner and the landlord received jail sentences for gross negligence manslaughter.[6]

Precautions

Figure 2: Ensure there are regular inspections of all gas boilers

- Ensure that gas boilers are regularly serviced by a competent person such as a heating engineer. The UK operates a Gas Safe Register scheme, whereby competent workers carry an identification card that specifies the categories of work a worker is competent to undertake. Gas boilers should be inspected once per year.
- Install carbon monoxide alarms at indoor workplaces where carbon monoxide may be present.
- Place an LEV system on the exhausts of motor vehicles when it is necessary for engines to be running during servicing, maintenance or repair, such as in motor vehicle repair garages.

- Ensure there is adequate dilution ventilation in workplaces where carbon monoxide may be present.
- Provide information and training to people who may be in an area where carbon monoxide may be present, covering the risks of exposure to carbon monoxide and precautions to take.

Cement

Cement is widely used in the construction industry. There are many different formulations and they are a complex mixture of many compounds. Two major constituents are calcium oxide (usually at least 60% of the mixture) and silicon dioxide. Exposure to cement is covered by COSHH.

The following examines cement in its powdered form and when mixed with water.

Health risks

Powdered form:

- Inhaling the powder can cause severe respiratory irritation as well as soreness to the nose and throat. (The long-term exposure limit in *EH40*[7] for calcium oxide powder is 2mgm^{-3}.)
- Contact with the skin can cause irritation and contact dermatitis in some cases. A tiny number of formulations contain a hexavalent chromium compound, which can also cause sensitisation dermatitis.
- If any powder enters the eye, severe eye irritation would be experienced.

Wet cement:

- This is regarded as a corrosive due to its strong alkalinity, and skin burns would be likely.
- Wet cement in the eye would cause significant eye damage.

The level of risk for both forms is dependent on the duration of exposure and, for the powdered form, on individual sensitivity with regard to the risk of dermatitis.

Circumstances where harm could occur

The types of worker who would be considered to be at significant risk include stonemasons, those using concrete and plasterers.

Precautions

- Use ready-mixed cement (or concrete) to eliminate the risk of exposure to powdered cement.
- If ready-mixed cement is not available, mix cement in a well-ventilated area if working under cover, or work upwind if mixing outside with no cover.
- Remove jewellery items (for example, watches) when working with cement to avoid cement being trapped underneath them.
- Arrange for a competent person to carry out regular skin inspections of workers who are using cement.
- Ensure workers are wearing suitable body-covering PPE. This will include either waterproof overalls or long trousers and long-sleeved shirts. They should definitely not wear shorts and short-sleeved shirts, even in hot weather.
- Ensure workers are wearing waterproof shoes and alkali-resistant gloves.
- Ensure workers are wearing respiratory protective equipment (RPE) when handling powdered dry cement. The minimum would be a half-mask respirator, but it would

Chemical and biological agents

be advisable to wear a full-face respirator to protect the eyes as well, especially if working in windy conditions.
- Provide information and training to workers regarding the risks and precautions when handling cement, especially to wash their hands regularly when using cement.

Legionella

Legionella is a group 2 biological agent; that is, it can cause human disease and may be a hazard to workers. It is unlikely to spread to the community and there is effective prophylaxis or treatment available.

> **TIP**
>
> Provisions relating to *Legionella* are covered in reg 7(10) and Schedule 3 of COSHH.
>
> Detailed guidance can be found in the HSE approved code of practice *Legionnaires' disease. The control of Legionella bacteria in water systems. Approved Code of Practice and guidance* (L8).[8]

Health risks

Harmful effects occur when people inhale fine droplets of water (a mist) that contain *Legionella* bacteria. The mist can reach the alveoli of the respiratory system and ill-health can result.

There are two diseases that can occur due to exposure to *Legionella* bacteria:

1 Pontiac fever: the symptoms of this condition are headaches, fever, aching muscles and a cough. The symptoms can often be mistaken for mild flu. People affected often recover in a few days without the need for medication.
2 Legionnaires' disease: the symptoms are similar to those of Pontiac fever but can cause greater discomfort. Crucially, there is the possibility that a fatal type of pneumonia can occur (which does not occur with Pontiac fever).

Health risks are considered greatest for smokers, elderly people and people who already suffer from respiratory diseases.

Circumstances where there is a likelihood of exposure

Legionella bacteria are extremely widespread and should be assumed to be present in all water systems. But there are conditions that encourage growth, including:

- where there is stagnant water;
- where there are dead ends in a water system (water can collect and become stagnant);
- where there are sludge deposits in a water system (sludge greatly assists the proliferation of the bacteria);
- where water has not been treated with agents to kill the bacteria:
- where part of a water system has not been in use for some time; for example, if a shower has not been used for a long time, stagnant water can collect in the shower head; and
- where there is water with a temperature of between 20°C and 45°C. Below 20°C, the bacteria proliferate extremely slowly, if at all. Above 45°C, the bacteria start to be killed.

Control measures/precautions

- Design water systems that have no dead ends.
- Keep the temperature of cold water below 20°C and the temperature of hot water above 50°C.
- Test water regularly for the presence of *Legionella* bacteria.
- Treat water with chemicals to kill bacteria; for example, with chlorine.
- Maintain water systems regularly, including removing any build-up of scale, rust or sludge.
- Flush water systems regularly to avoid build-up of stagnant water.

Leptospira bacteria

These bacteria can enter the body through the mucous membranes of the nose and mouth. As with all bacteria, the eyes can also be an efficient route of transmission. The other route of entry is through cuts in the skin.

> **DEFINITION**
>
> *Leptospira* bacteria can cause a bacterial infection called **leptospirosis**.
>
> *The bacteria can pass from animals to humans when a break in the skin comes into contact with the urine of certain animals, particularly rats and cattle. It can also be transmitted through the eyes.*

> **TIP**
>
> Provisions relating to *Leptospira* are covered in reg 7(10) and Schedule 3 of COSHH. It is a group 2 biological agent.
>
> Further details are provided in HSE guidance on leptospirosis, *Harmful micro-organisms: Leptospirosis/Weil's Disease from rats*.[9]

Health risks

There are three diseases that can be caused by *Leptospira*:

1. Leptospirosis: this is caused by exposure to rat urine; the symptoms are headache, fever, muscle aches and sometimes vomiting. In more serious cases, the normal functions of the liver and kidneys can be disrupted and the symptoms of jaundice (such as yellowing of the skin and the whites of eyes) can appear.
2. Hardjo disease: this is caused by exposure to cattle urine; the symptoms are similar to those presented by leptospirosis but are usually less severe.
3. Weil's disease: this is a more severe form of leptospirosis.

Circumstances where there is a risk of exposure to *Leptospira* bacteria

The following occupations/locations are likely to present such risks:

- farm workers; there may be cattle and rats on farms;
- sewer workers; rats are well known to live in sewers;
- veterinary surgeons;
- waste handling depot workers;

Chemical and biological agents

- water-sport centres, especially where such activities take place on still lakes;
- construction site workers, particularly during demolition activities when rats' nests can be disturbed; and
- anywhere likely to support a rat population and therefore pose the risk of contact with rat urine.

Precautions
- Pest control needs to be in place, especially control of rats.
- There needs to be good personal hygiene, for example washing any cuts or grazes immediately.
- Cuts must be covered with waterproof dressings.
- Training should be provided so that workers are aware of the risks, symptoms and precautions regarding leptospirosis.
- Any person suffering symptoms they believe may have been the result of exposure to rat or cattle urine should report these to a doctor as soon as practicable.
- Appropriate PPE should be worn, such as goggles to protect the eyes and waterproof gloves.
- Rats should never be touched with bare hands.
- Surfaces that may have been contaminated with rat urine (such as in veterinary surgeries) should be thoroughly cleaned using a 10% solution of bleach or other suitable disinfectant.

Silica

> **TIP**
>
> Legal requirements relating to silica are found in COSHH, and the WEL for silica is identified in *EH40*.[10]
>
> Further guidance is found in the HSE publication *Control of exposure to silica dust: A guide for employees* (INDG463).[11]

Silica occurs naturally and is found in rocks, sand, clay and quartz. It is also found in concrete. You may recall that silica is a component of cement, which in turn is a component of concrete. When power tools such as drills are used on concrete materials (for example, kerbstones, slabs or blocks) fine dust is produced, and this dust will contain (among other substances) respirable crystalline silica (RCS). The term respirable means that RCS can, when inhaled, reach the alveoli in the gas exchange region of the lung. The white blood cells of the body's cellular defence system attempt to clear the RCS from the lung, but RCS kills these defending white cells.

People exposed to RCS while working may not immediately be aware of the dangers as the effects of exposure are chronic rather than acute.

Health risks
- Silicosis: the disease is progressive; that is, it gets worse with time. Common symptoms are a persistent cough, shortness of breath and difficulty breathing. Some sufferers may also experience fever and fatigue.

- Lung cancer.
- Chronic obstructive pulmonary disease (COPD): the symptoms of this include shortness of breath, a long-lasting cough, wheezing and eventually a loss of appetite and swollen ankles.
- There is also evidence suggesting that the risk of tuberculosis is increased.

Circumstances where silica dust may be present

Risks may result from the following activities:

- demolition, particularly where concrete structures are being demolished;
- drilling and cutting of concrete;
- quarrying;
- work involving stonemasonry;
- craft/commercial activities involving clay and similar materials; and
- sandblasting of buildings.

Figure 3: Demolition can generate silica dust

Precautions

If possible, use silica-free materials. This would be the best method as it eliminates the risk entirely.

If it is not possible to work with silica-free materials, then there are other precautions:

- Use tools with vacuum attachments to capture the dust at source; also equipment such as concrete saws, which spray water as they saw.
- Use water jets to dampen down the area.
- Good personal hygiene is essential, so welfare facilities should be provided and hands washed after working in areas where silica may be present.
- RPE with a suitable assumed protection factor (APF) must be worn; this would normally be a half-mask respirator.
- During cleaning operations, silica dust must never be swept; a vacuum with a high efficiency filter must be used.

- Workers should be trained in the risks of exposure to silica and the precautions to be taken.

It would also be advisable for workers to receive periodic health surveillance, particularly chest X-rays. Silica particles can be seen on X-rays and any workers found to have such particles on their lungs could be removed from this work and precautions reviewed.

> **CASE STUDY 2**
> **Unventilated silica exposure results in fatal lung disease**
>
> A worker was employed as a grinding machine operator and stonecutter. As a result, the worker was exposed to silica dust. There was no ventilation in the work area.
>
> After a while, the worker was referred to hospital with significant breathing difficulties. Medical examination concluded that the disease on the worker's lungs was typical of exposure to respirable silica dust.
>
> The lung condition became worse, and the worker was admitted to an intensive care unit and later died.

Wood dust

> **TIP**
>
> Legal requirements relating to wood dust are found in COSHH, and the WEL for wood dust is identified in *EH40*.[12]
>
> Further advice, including practical detail on LEV is found in the HSE guidance *Wood dust: Controlling the risks*.[13]

When reviewing the hazards of wood dust, it is convenient to classify the dusts as hardwood and softwood dusts. Hardwoods originate from trees such as oak and mahogany. Softwoods come mainly from coniferous trees such as pine, yew and cedar. Dusts from hardwoods are more hazardous than those from softwoods.

Health risks
- Wood dusts can cause irritation in the nose and rhinitis.
- Inhalation of wood dust can cause asthma.
- Wood dusts are known to be skin sensitisers and can cause sensitisation dermatitis.
- Inhalation of hardwood dust can cause nasal cancer.

Circumstances that may generate wood dust
Wood dust may be generated in several situations when working with wood, including:

- sawing;
- routing;
- cutting; and
- sanding.

Control measures to reduce health risks

- Use an LEV system attached to the woodworking machinery; the dust is collected in a bag.

Figure 4: Trap wood dust to reduce risks

- Inspect and maintain the LEV system; for example, frequently changing the dust collection bags.
- Use a downdraft table for sanding operations.
- Wear RPE with an adequate APF, such as a half-mask respirator.
- Wear PPE such as overalls and gloves.
- Use health surveillance records to identify any vulnerable workers; for example, if a worker suffers from asthma.
- Provide suitable welfare facilities.
- Prohibit eating, drinking and smoking when working with wood.
- Use a vacuum cleaner to clean floors (never sweep up dust).
- Provide information and training to workers, including on the health risks of working with wood and the precautions to be taken, in particular the importance of good hygiene, such as washing hands after working with wood.

APPLICATION

Consider your own workplace.

- What do you think is the most hazardous chemical or biological agent in your workplace?
- What has been done so far to reduce the risk from this chemical or biological agent?
- What further measures do you think need to be carried out to further reduce the risk?

Chemical and biological agents

KEY POINTS

- There are hazards, workplace activities and control measures for eight specific agents: asbestos, blood-borne viruses, carbon monoxide, cement, *Legionella* bacteria, *Leptospira* bacteria, silica and wood dust.
- Dutyholders must carry out certain actions to comply with the Control of Asbestos Regulations.
- Particularly important features of asbestos are its type and condition.
- Exposure to exhaust fumes presents a significant risk of exposure to carbon monoxide and using carbon monoxide alarms is important.
- Cement is hazardous both in powdered form and as wet cement.
- Two key situations where *Legionella* bacteria may proliferate are stagnant water being present and situations where water temperature is in the range of 20–45°C.
- Following good hygiene procedures is a key control measure to avoid contracting leptospirosis.
- Exposure to respirable crystalline silica (RCS) can cause diseases such as lung cancer and COPD, which are likely to cause irreversible health effects. Minimising exposure to RCS is crucial to minimise health risks.
- Exposure to wood dust can cause respiratory disease and nasal cancer. Reduction of exposure to the dust is an important control measure.

References

[1] HSE, *Managing and working with asbestos. Control of Asbestos Regulations 2012. Approved Code of Practice and guidance* (L143, 2nd edition, 2013) (www.hse.gov.uk)

[2] HSE, *Asbestos essentials task sheets* (www.hse.gov.uk)

[3] HSE, *Asbestos: The survey guide* (HSG264, 2nd edition, 2012) (www.hse.gov.uk)

[4] See note 2

[5] HSE, 'Blood-borne viruses' (www.hse.gov.uk). There is also specific guidance for healthcare workers available at https://assets.publishing.service.gov.uk/government/uploads/system/uploads/attachment_data/file/382184/clinical_health_care_workers_infection_blood-borne_viruses.pdf

[6] Adapted from 'Greater Manchester Police News' (19 February 2021) (www.gmp.police.uk)

[7] HSE, *EH40/2005 Workplace exposure limits* (4th edition, 2020) (www.hse.gov.uk)

[8] HSE, *Legionnaires' disease. The control of legionella bacteria in water systems. Approved Code of Practice and guidance* (L8, 4th edition, 2013) (www.hse.gov.uk)

[9] HSE, *Harmful micro-organisms: Leptospirosis/Weil's Disease from rats* (www.hse.gov.uk)

[10] See note 7

[11] HSE, *Control of exposure to silica dust: A guide for employees* (INDG463, 2013) (www.hse.gov.uk)

[12] See note 7

[13] HSE, *Wood dust: Controlling the risks* (2nd edition, 2020) (www.hse.gov.uk)

ELEMENT 8

GENERAL WORKPLACE ISSUES

8.1: Health, welfare and work environment

Syllabus outline

In this section, you will develop an awareness of the following:

- Health and welfare:
  - supply of drinking water, washing facilities, sanitary conveniences, accommodation for clothing, rest and eating facilities, seating, ventilation, heating and lighting
- The effects of exposure to extremes of temperature; control measures

DEFINITION

Welfare facilities *are those that are necessary for the wellbeing of workers, such as washing, toilets, rest and changing facilities, and somewhere clean to eat and drink during breaks.*

Welfare is closely linked with both health and safety. For example, working in extremely hot environments can also lead to serious health problems, including heatstroke. Poor lighting can lead to a failure to clearly see a safety hazard. Poor sanitation, such as lack of toilets and poor cleanliness, can lead to serious gastrointestinal problems. The main authoritative reference used here is the Health and Safety Executive (HSE) INDG293 guidance (see Tip box).

TIP

The main UK legislation relating to health and welfare at work is:

- the Workplace (Health, Safety and Welfare) Regulations 1992.

Standards and guidance on welfare for employers can come from a number of sources. Here are some important examples from HSE:

- *Welfare at work: Guidance for employers on welfare provisions* (INDG293);[1]
- *Workplace health, safety and welfare. Workplace (Health, Safety and Welfare) Regulations 1992. Approved Code of Practice and guidance* (L24);[2] and
- *Workplace health, safety and welfare: A short guide for managers* (INDG244).[3]

> HSE has produced guidance on the specific topic of lighting:
>
> - *Lighting at work* (HSG38).[4]
>
> Trade unions have also produced helpful information for workers, such as this guidance from the Trades Union Congress (TUC):
>
> - *Cool it! A TUC guide for trade union activists on dealing with high temperatures in the workplace;*[5] and
> - *Give us a (Loo) break!*[6]

8.1.1 Health and welfare

Supply of drinking water

A basic human need is access to clean, uncontaminated drinking water. In workplaces, clean water needs to be available and employers should consider factors in the working environment such as the temperature of the surroundings and the types of work taking place.

Drinking water will normally be supplied by mains water and supplies should be clearly marked to show that they are suitable as drinking water.

Water should only be provided in refillable enclosed containers when it cannot be obtained directly from a mains supply. The containers should be refilled at least daily (unless they are chilled water dispensers, when the containers are returned to the supplier for refilling). Bottled water or water dispensing systems may still be provided when there is no source of mains drinking water.

In places where drinking water is provided, there should be enough suitable cups, unless the drinking water is in the form of a water fountain from which people can easily drink.

Toilets and washing facilities

> **DEFINITION**
> ***Toilet** is a broad term that includes all the sanitary facilities used in different cultures, such as flush toilets, pour-flush toilets, latrines, composting toilets and urinals.*

Clean and easily accessible toilets, along with washing facilities, are essential for health, dignity and privacy. Toilets and washing facilities may need to be specially adapted so that they are accessible to those with disabilities.

General guidance is available to help determine the minimum numbers of washing facilities required (see Table 1). Showers should be provided where the nature of the work means that workers become particularly dirty or contaminated, especially where there are health implications (for example, for sewage workers).

Number of people at work	Number of washbasins
1–5	1
6–25	2

8.1 Health, welfare and work environment

Number of people at work	Number of washbasins
26–50	3
51–75	4
76–100	5

Table 1: Number of washbasins for mixed use[7]

Washing facilities should normally:

- be near to toilets;
- be near to changing rooms;
- have a clean supply of hot and cold (or warm) water – where possible, it should be running water;
- have soap (or other means of washing or cleaning) and towels (or other means of drying);
- have a basin large enough to wash hands and forearms;
- be well ventilated and lit and kept clean and tidy; and
- have showers, if required for particularly dirty work. Showers should be fitted with thermostatic mixers to prevent scalding.

Separate washing facilities ought to be provided for male and female workers. The exception is where the facilities are provided in a separate room that is intended for one person only and that can be locked from the inside.

All toilets and the rooms containing them should be kept clean and the facilities should have adequate ventilation and lighting. Guidance about the number of toilets needed is shown in Table 2.

Number of people at work	Number of toilets
1–5	1
6–25	2
26–50	3
51–75	4
76–100	5

Table 2: Number of toilets for mixed use[8]

Other considerations when providing toilet facilities should include having:

- enough toilets and washbasins for those expected to use them, so that people do not have to queue for too long;
- clean facilities where the walls and floors should (preferably) be tiled or covered in suitable waterproof material to make them easy to clean;
- a supply of toilet paper and, for female employees, a means of disposing of sanitary dressings;
- rooms that are well lit and ventilated; and
- access for people with physical disabilities, allowing for additional wheelchair access and a way of summoning assistance.

General workplace issues

Accommodation for clothing

Adequate accommodation should be provided for:

- personal clothing not worn at work; and
- work clothing, such as uniforms, overalls, thermal clothing, laboratory coats and food hygiene clothing, that is worn at work but not taken home.

The accommodation ought to:

- be secure against theft when personal clothing and possessions are stored;
- have a separate hook or peg for each worker but, for added security, it may require a lockable locker;
- where necessary, provide separate accommodation for clothing worn at work and other clothing to avoid risks to health or damage to the clothing;
- include facilities for drying clothing where appropriate; and
- be in a suitable location convenient to the workplace.

Facilities for changing clothing

If workers have to change into special work clothing when they arrive at work, changing facilities are normally provided. If a worker has to undress, separate facilities (or separate use of the same facilities) should be provided for men and women. Facilities should be convenient and near to washing and eating areas. They should allow privacy and should have seating. Facilities should also be large enough to prevent overcrowding.

Rest facilities

Rest facilities need to be provided and should be easy to get to from where workers are. This applies particularly to workers who have to stand to carry out their work activities.

Canteens may be used as rest facilities as long as there is no obligation to purchase food or drink.

New and expectant mothers

Suitable rest facilities should be provided for new and expectant mothers. They should be near to sanitary facilities and contain furniture to allow these workers to lie down if necessary.

Eating facilities

Where food eaten in the workplace could become contaminated, suitable facilities for eating meals need to be provided. Work areas can be counted as rest areas and as eating facilities, if they are clean enough and there is a suitable surface on which to place food.

Eating facilities should also enable workers to make a hot drink and to heat food (for example a microwave oven) if hot meals cannot be otherwise obtained.

Smoking

In the UK, smoking is not allowed in indoor workplaces. Rest facilities should have suitable arrangements for protecting non-smokers from tobacco smoke (for example, separate smoking and non-smoking areas). If smoking is allowed outdoors, suitable shelters should be provided to protect workers from inclement weather.

General workplace seating

A workplace should include an adequate number of tables and seats with backs, as well as seating that is suitable in both number and design for disabled workers. Seating is especially important when workers have to stand for long periods.

Room dimensions and space

To ensure the health, safety and welfare of workers, every workroom should have sufficient floor area, height and unoccupied space. Guidance suggests a volume of at least 11 cubic metres ($11m^3$) per person. For the purposes of the calculation, ceiling heights above 3 metres should be ignored.

Workrooms should have enough free space to allow people easy access to and from workstations, to move in the room with ease and not to restrict their movements while performing their work. This includes ceilings being high enough to allow safe access to workstations unless the work lasts only a short time. Obstructions such as low beams should be clearly marked.

> **APPLICATION**
>
> Example: If six people normally work in a workroom that measures 6m × 5m with a height of 3m, then each person has $15m^3$ of space [(6 × 5 × 3) ÷ 6 = 15], which satisfies the basic space requirements.

The space requirements do not apply to certain unusual working environments, such as the cabs of machines (tower cranes or similar) or to parts of the workplace that are used infrequently.

Workstations

Workstations should be suitable for the people using them and for the work they do.

Workstations should be designed so that workers do not have to overreach and to minimise the strain on their bodies. There should be no slip or trip hazards when accessing or leaving the workstation and it should be possible to leave without delay in an emergency.

Workstations might need adapting for disabled workers.

Ventilation

All parts of workplaces need to be adequately ventilated with fresh, clean air from an uncontaminated source outside the workplace. Ventilation should also, where appropriate, remove warm, humid air and provide air movement that gives a sense of freshness without causing a draught.

Typical rates of fresh-air supply can be measured as air changes per hour or air flow rates in litres per hour per occupant. In practical terms though, the presence of bad smells is sufficient to indicate that the air is not being refreshed sufficiently.

Factors that can affect the fresh air required can include:

- the floor area per person;
- the processes and equipment involved; and
- whether the work is strenuous.

If the workplace contains process or heating equipment, or other sources of dust, fumes or vapours, more fresh air will be needed to provide adequate ventilation. Windows or other openings may provide enough ventilation but, where necessary, mechanical ventilation systems should be provided and regularly maintained. Where mechanical ventilation is used, it ought to have a failure warning device to let users know if the air supply has a fault.

Lighting

Appropriate lighting is needed to prevent eyestrain, to reduce glare and to be able to detect and negotiate hazards (note that 6.1: Work-related upper limb disorders also considers lighting in the context of preventing muscoskeletal injuries and upper limb disorders). The type and level of lighting will depend on the tasks to be performed and the hazards that are present. If necessary, local lighting should be provided at individual workstations. Good levels of lighting will be required at places of particular risk, such as on stairs and at crossing points on traffic routes.

The effects of substandard lighting can be:

- too little: eyestrain, fatigue, headaches, stress and accidents; or
- too much: 'glare' headaches and stress.

Either can bring about lower-quality work, increased absenteeism, lower productivity and ill-health.

Common lighting problems at work can be:

- dark or unlit areas, especially near hazards such as unguarded machines, stairs or steps;
- inadequate natural light because of dirty or badly placed windows;
- glare from badly positioned or insufficiently shaded lights, unshaded windows or reflecting surfaces;
- energy-saving programmes leading to reduced lighting levels;
- workers suffering from eyestrain or fatigue from bad posture due to deficient lighting;
- dirty or badly maintained lighting, leading to light loss and flicker;
- unsuitable décor, leading to low lighting levels (such as excessive contrasts or too much glare); and
- security risks at night caused by inadequate lighting.

Lighting – general requirements

Workplace lighting systems can be divided into two categories – general lighting and local lighting.

General lighting aims to provide an approximately uniform light over the whole work surface. Local lighting is over relatively small areas where visual tasks are carried out, alongside general lighting. Figure 1 illustrates the typical differences between the systems.

Some simple measures can improve workplace lighting. For example:

- Make full use of daylight in the work environment.
- Find the best place for the light source to avoid glare.

8.1 Health, welfare and work environment

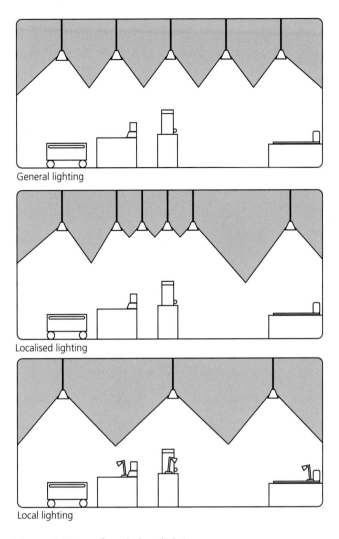

Figure 1: Types of workplace lighting

- Use the most appropriate lighting devices and fixtures, such as suspended ceiling lamps or desk-mounted lamps.
- Avoid working where shadows are present and use as much natural light as possible.
- Make sure that all windows, skylights, etc are clean as this will allow in more natural light.
- Use blinds to cut down direct sunlight and glare.

Minimum lighting levels

Table 3 shows some guidance for minimum illumination levels for safety. These are measured in lux. For example, a very bright day in the Caribbean might be more than 100,000 lux, an overcast day 1000 lux and dusk 100 lux.

Activity	Locations/types of work	Average illuminance (lux) 1x	Minimum illuminance measured (lux) 1x
Movement of people, machines and vehicles (assumes no perception of detail required)	Lorry park, corridors, circulation routes	20	5

General workplace issues

Activity	Locations/types of work	Average illuminance (lux) 1x	Minimum illuminance measured (lux) 1x
Movement of people, machines and vehicles in hazardous areas; rough work not requiring any perception of detail	Construction site clearance, excavation and soil work, loading bays, bottling and canning plant	50	20
Work requiring limited perception of detail	Kitchens, factories assembling large components, potteries	100	50
Work requiring perception of detail	Offices, sheet metal work, bookbinding	200	100
Work requiring perception of fine detail	Drawing offices, factories assembling electronic components, textile production	500	200

Table 3: Indicative lighting levels (for avoiding visual fatigue only)[9]

It is important to be careful in using these figures as they are minimum levels for safety and many people prefer light levels that are significantly above these.

8.1.2 The effects of exposure to extremes of temperature

Temperature in the workplace

Temperature in all internal workplaces during working hours should be reasonable to ensure that workers are comfortable. However, how you experience 'thermal comfort' is not just about the surrounding temperature but is due to a combination of factors. There are six key parameters in relation to thermal comfort:

- Air temperature: this is usually measured by a simple thermometer.
- Radiant heat (thermal radiation): this is the transfer of electromagnetic radiation. The sun emits its energy by radiant energy and can heat the air of an office even through windows. Examples of work processes involving radiated heat include furnaces, pipework containing hot gases or fluids and infrared electric heaters.
- Humidity: this is the amount of moisture in the air, which affects the rate of sweat evaporation and is linked to the body's ability to cool itself.
- Speed of air movement: this is known as 'wind chill' in outdoor environments, but air movement in indoor settings can also significantly affect thermal comfort, either positively or negatively.

- Clothing worn: this includes protective clothing, as it provides insulation and prevents heat loss from the body in the cold. Lightweight clothing can aid heat loss and perspiration in hot environments.
- Work rate: this will affect the heat generated by the body (metabolic heat).

It is the combination of all these factors that determines whether or not workers have a suitable thermal environment.

Minimum indoor temperature

Different people find different temperatures comfortable and it is difficult to find a thermal environment that satisfies everyone. Where the work activity is mainly sedentary (such as offices), the temperature should normally reach at least 16°C one hour after work starts. If work involves physical effort, the workplace temperature should be at least 13°C, unless other considerations, such as those relating to food, require lower temperatures (such as in a cold store).

Extremes of temperature – heat

Where processes generate excessive heat, employers can increase the distance between the equipment and the exposed workers; for example, by using long ladles in furnaces where workers are dealing with molten metal. Other measures can include the use of thermal lagging on hot pipework, water-cooling of hot surfaces and the use of reflective shielding.

In excessive outdoor temperatures, steps to ensure welfare can include halting work during the hottest part of the day and providing cool drinks to keep workers hydrated. It is also possible to provide protective clothing, such as wide-brimmed hats.

Employers can provide rest areas in the shade. It can also be helpful to use the 'wet bulb globe temperature' (WBGT) index to assess the level of 'heat stress'. This is used to estimate the combined effect of temperature, solar radiation and wind speed on the human body. It is recommended that work stops when the WBGT exceeds 32.1°C.

Other measures include preparing a risk assessment for each outdoor site, remembering to consult with workers and training them to identify signs of heat stress. Another approach is to co-operate with other employers on the same site to put in place a site-wide heat stress reduction plan.

Annual medical checks for outdoor workers are also recommended.

Extremes of temperature – cold

Cold temperatures can affect safety because judgement and concentration can be reduced. Where work needs to be carried out at unusually low temperatures, employers should ensure there are cycles of work followed by rest and provide warm shelters for workers. This is especially relevant when work is likely to last for some time and the temperature and wind speed vary.

When working in the cold, work should be planned to allow extra time for tasks and the need for time to take drink and food breaks.

If possible, work rates should be designed to avoid heavy sweating but, if this does occur, workers need dry replacement clothing, with warm changing facilities.

Where it is not possible to eliminate the need for work in cold environments, employers should provide cold weather clothing together with adequate facilities for changing and arrangements for cleaning and drying clothing/footwear between shifts. Headgear needs to be comfortable to wear and windproof (if appropriate), with adequate protection for ears and neck. This needs to be compatible with hard hats, if worn.

> **CASE STUDY**
>
> **Consequences of extreme temperatures**
>
> A security guard tasked with guarding a wind farm in a remote part of Scotland died from hypothermia after being trapped in snow for four and a half hours.
>
> Both the parent company of a construction contractor that was building the wind farm and the security firm that employed the guard admitted safety breaches.
>
> The court was told that the 74-year-old security guard was working a 12-hour shift at the wind farm with a colleague. Heavy snow was forecast to fall across large swathes of Scotland and it was reported that the site lost all power after a snowstorm and there was no heating and lighting.
>
> It was thought that the security guard had left his post in an effort to reach a second cabin just over half a mile away, in the hope that it still had power so he could survive the night.
>
> He was discovered by rescue services but, despite being airlifted to hospital, he later died from hypothermia.
>
> The construction company admitted failing to ensure a safe system of work. It also admitted that it had failed to provide a reliable source of back-up power, such as a generator, for heating. It was alleged that the generator at the main compound regularly broke down and there was no back-up. Neither was there a back-up for a generator at the guard house.
>
> The company admitted that it had not provided the security guards with a means to contact emergency services. Nor did it ensure that there was a plan in place for them to evacuate the site.[10]

Training

Workers exposed to excessive heat or cold, as well as their supervisors, should be trained:

- to recognise symptoms that may lead to heat stress or hypothermia, in themselves or others;
- in the use of rescue and first-aid measures; and
- to recognise that there is an increased risk of accidents due to high and low temperatures.

Workers should also be advised of:

- the importance of physical fitness for work in hot or cold environments;
- the importance of drinking enough water and maintaining salt levels. Salt levels can reduce due to sweating; and
- the effects of drugs and alcohol, which can reduce tolerance to heat exposure.

KEY POINTS

- Many of the elements discussed may seem straightforward but they are the foundation for the everyday comfort and welfare of workers. Having sufficient toilets, hot and cold running water and drinking water are basics of health and safety, irrespective of the industry or the location.

- While UK standards and laws are the main reference points, there is also helpful guidance from trade union sources.

- Hand washing or sanitising and good ventilation are important in the prevention of diseases. Food in the workplace is less likely to be contaminated if there are proper and hygienic rest areas and washing facilities.

- Good welfare facilities that include clean and easily accessible toilets, suitable washing facilities and fresh drinking water are, among other things, essential for health, dignity and privacy at work.

References

[1] HSE, *Welfare at work: Guidance for employers on welfare provisions* (INDG293, 2011) (www.hse.gov.uk)

[2] HSE, *Workplace health, safety and welfare. Workplace (Health, Safety and Welfare) Regulations 1992. Approved Code of Practice and guidance* (L24, 2nd edition, 2013) (www.hse.gov.uk)

[3] HSE, *Workplace health, safety and welfare: A short guide for managers* (INDG244, 2nd edition, 2007) (www.hse.gov.uk)

[4] HSE, *Lighting at work* (HSG38, 2nd edition, 1997) (www.hse.gov.uk)

[5] TUC, *Cool it! A TUC guide for trade union activists on dealing with high temperatures in the workplace* (2017) (www.tuc.org.uk)

[6] TUC, *Give us a (Loo) break!* (2010) (www.tuc.org.uk)

[7] See note 2

[8] See note 2

[9] See note 4

[10] Adapted from: 'Wind farm firms fined almost £900,000 over security guard's death', *IOSH Magazine* (25 November 2021) (www.ioshmagazine.com)

8.2: Working at height

Syllabus outline

In this section, you will develop an awareness of the following:
- What affects risk from working at height, including vertical distance, fragile roofs, deterioration of materials, unprotected edges, unstable/poorly maintained access equipment, weather and falling materials
- Hierarchy for selecting equipment for working safely at height:
  - avoid working at height by, for example, using extendable tools to work from ground level; assembly of components/equipment at ground level
  - prevent a fall from occurring by using an existing workplace that is known to be safe, such as a solid roof with fixed guardrails; use of suitable equipment such as mobile elevating work platforms (MEWPs), scaffolds; work restraint systems
  - minimise the distance and/or consequences of a fall, by collective measures such as safety nets and airbags installed close to the level of work, and personal protective measures such as fall-arrest systems
- Main precautions necessary to prevent falls and falling materials, including proper planning and supervision of work, avoiding working in adverse weather conditions
- Emergency rescue
- Provision of training, instruction and other measures
- General precautions when using common forms of work equipment to prevent falls, including: ladders, stepladders, scaffolds (independent tied and mobile tower), MEWPs, trestles, staging platforms and leading edge protection systems
- Prevention of falling materials through safe stacking and storage

DEFINITION

Work at height is defined in the Work at Height Regulations 2005 to be:

(a) work in any place, including a place at or below ground level;

(b) obtaining access to or egress from such place while at work, except by a staircase in a permanent workplace,

where, if measures required were not taken, a person could fall a distance liable to cause personal injury.

Workplace fatalities and major injuries are often caused by falling from height. Many injuries involve falls from comparatively small heights and control measures should be based on an assessment of the risk.

One aspect of risk will be the vertical distance, but there may be others, such as what people may land on (or in) if they fall.

Control measures are based on a hierarchy of avoid, prevent and minimise, with collective measures taking precedence over individual (personal) ones.

We will provide an overview of methods to control the risks that arise from working at height, including mobile elevating work platforms (MEWPs), scaffolds and ladders. We will also review work restraint, work positioning and fall-arrest systems.

> **TIP**
>
> The main relevant legislation in this area is the Work at Height Regulations 2005. However, as the focus here is on practical ways to work safely at height (and so comply with the law's objectives), a good overall Health and Safety Executive (HSE) guide is:
>
> - *Working at height: A brief guide* (INDG401).[1]
>
> In addition, there is HSE guidance on specific types of work at height, including:
>
> - *Health and safety in roof work* (HSG33);[2]
> - *Fragile roofs: Safe working practices* (GEIS5);[3] and
> - *Warehousing and storage: Keep it safe* (INDG412).[4]
>
> There is also useful guidance on the safe use of common forms of access equipment, such as:
>
> - *Using ladders and stepladders safely: A brief guide* (LA455) – jointly produced by the HSE and the Ladder Association;[5] and
> - *Preventing falls in scaffolding operations* (SG4:15) – published by the National Access and Scaffolding Confederation (and endorsed by the HSE).[6]
>
> We will draw on the advice in these guides.

8.2.1 What affects risk from working at height

Some examples of work at height are:

- gaining access to the loading area on a road or rail vehicle or container;
- working on top of a road or rail tanker, vessel or container (to check inspection hatches);
- using ladders or stepladders to gain access to vessels, tanks, silos and storage bins; and
- using working platforms such as scaffolds, tower scaffolds, cherry pickers, scissor lifts and podium steps.

In the construction industry, scaffolders, bricklayers, roofers and steeplejacks will normally spend considerable amounts of time at height. Most other construction trades (painters, joiners, electricians, glazers, etc) will be at risk for some of the time.

A person works at height if they:

- work above ground level;
- could fall from an edge of a raised area (vehicle, container or cargo bay);
- could fall from ground level into an opening in a floor or hole in the ground; or
- could fall from one level to another.

Work at height needs to be properly planned and supervised to ensure that all relevant risks are considered and suitable controls are put in place. There are some specific issues that will affect the level of risk when working at height.

Vertical distance

The higher the worker is from the ground, the higher the risk is of a serious injury should they fall. This risk will need to be mitigated as much as possible so the hierarchy of control and specific control measures should be considered when planning work at height.

Fragile roofs

When planning for working at height, the roof structure must be considered. Some things to think about here are:

- What is the roof constructed of? Many roofs will be solid structures (for example, built of tiles over a frame), whereas other roofs could use weaker materials such as corrugated iron or plastic.
- Some solid roof structures may also have either glass or plastic windows (commonly known as rooflights) built into them. These are a common cause of injuries, with workers either not seeing them and then stepping on them or workers not being given information about the rooflights or overestimating the strength of them and then stepping on them.

Unprotected edges

This could relate to either a roof or other parts of the workplace (loading bays, for example). Edges should be protected to stop materials falling from height and to stop workers from taking a step backwards into empty space.

Deterioration of materials

All materials will deteriorate over time. This should be considered when planning the work, especially when roof work is required. It is important that the structure of buildings is regularly maintained and repaired when necessary.

Unstable and/or poorly maintained access equipment

It is important to use access equipment on flat ground whenever possible. Some work equipment will have 'outriggers' that will help to make the equipment as stable as possible.

Poorly maintained equipment could lead to failure at critical times. It is essential, therefore, that the equipment has a planned maintenance programme to help prevent these critical failures.

Weather

Weather conditions can increase the risk from working at height. For example, hot weather can mean that workers become dehydrated and fatigued more quickly, which could affect their ability to work safely. Regular supervision/checks on workers should happen in these circumstances.

Bad weather (eg rain and high winds) will make working at height a high-risk activity. Workers could slip on surfaces or may lose dexterity in their hands due to the cold or be blown off structures. Working at height should be avoided wherever possible in bad weather.

Falling materials

Falling materials will affect workers (and others who may be in the area at the time) below the work area. For example, if a worker on the top level of a scaffold drops a hammer, it could strike a worker on a lower level. Likewise, if materials are dropped over the edge of a scaffold, they could injure someone on the ground beneath.

8.2.2 Hierarchy for selecting equipment for working safely at height

The level of risk associated with a task affects the degree of control needed to reduce the risk as much as possible. When working at height, the Work at Height Regulations 2005 specifically require consideration of whether there is even a need for the work to be done at height; that is, can it be avoided and the work instead done safely in some other way, by using extendable tools, for example?

If that is not reasonably practicable, then the duty is to prevent workers falling a distance likely to cause injury.

Where this is not reasonably practicable (or does not eliminate the risk entirely), you also need to minimise the distance that someone might fall and/or the consequences of such a fall.

A combination of these measures might have to be used to eliminate or reduce the risk sufficiently. Clearly, there are many different solutions (strategies and equipment) to these three basic considerations. The options for solutions are often formulated into what is commonly described as a hierarchy, which helps you consider and select different control options for equipment to make working at height safe. We will now look at these in more detail.

Avoid

Work at height should be avoided when possible. If this is not possible, consideration should be given to using extendable equipment and assembling any materials or equipment at ground level.

Although the scope for this is often quite limited, examples include the following:

- Window cleaners often use long-reach cleaning equipment instead of working from ladders. Some high-rise buildings are equipped with automatic window cleaning equipment rather than using cradles.
- In some workplaces, lighting systems are designed so that they can be lowered to ground level for maintenance rather than using ladders or other temporary means of access.

Prevent

When you cannot avoid working at height, prevention involves working from an existing workplace/area that is known to be safe.

Examples include:

- working from solid surfaces or roofs with permanent fixed guardrails;
- the use of work restraint systems that prevent the worker from reaching an area where they would be at risk of falling; and
- temporary platforms such as MEWPs or scaffolding.

Minimise

Where there remains a risk of a fall, the employer must provide work equipment that minimises the distance and consequences of a fall. They also need to provide additional training and instruction to help prevent any person falling a distance liable to cause injury.

Examples include:

- use of harnesses and fall-arrest equipment (personal measures); and
- safety nets, airbags and soft-landing systems (collective measures).

There may be cases when personal and collective measures will both be used. This will depend on the work activity, and the measures required should be identified as part of a risk assessment. We will now look at both these measures in more detail.

Personal measures

In circumstances where collective restraint methods are impractical, then personal worker restraint or rope systems may need to be considered. Personal systems consist mainly of harnesses that are attached to an anchorage point by a lanyard. Lanyards can incorporate energy absorbers to help minimise the risk of injury from the harness. These systems use inertia systems that operate if the rope attached to the worker's harness is suddenly pulled into tension.

Personal measures include:

- Fall-arrest systems: these do not prevent falls but limit the distance of the fall and should prevent impact with the ground. They are often used in conjunction with other protection systems.
- Work restraint systems: these prevent workers from getting into a position where they might fall. In the main, they consist of a rope or lanyard that attaches the worker to an anchorage point. The length of the lanyard must be less than the distance from the anchor to the unprotected edge. In this way, the user is physically prevented from reaching the danger zone.
- Work positioning systems: these enable the user to work in tension or suspension to prevent or limit a fall. Examples are a boatswain's chair and a linesman's pole strap. A pole strap is used by electricity supply workers and frees a linesman's hands at the top of a pole. It consists of a strap that is passed around the pole and attached to a padded belt and harness worn by the worker, allowing the linesman to lean backwards.
- Rope access systems: these use two ropes, a working rope and a safety rope, each secured to a reliable anchor. The harness is attached to both ropes in such a way that the worker can get to and from the work area but the risk of falling is prevented or limited. This type of system can be used to access the side of a tall building where a cradle cannot be used. Workers can also use abseiling techniques to reach the work position.

Collective measures

These are usually the measures that will reduce the impact of fall from height and will protect more than one person. They include specific pieces of work equipment such as scaffolding and MEWPs; they do not rely on the intervention of the user to work once they have been correctly installed. Equipment such as nets and airbags placed beneath the work area are good examples of this.

Figure 1: Nets being used as a collective fall-arrest measure

CASE STUDY 1
Solar panel installation

A solar panel company and its director were sentenced after a worker's fatal fall from height.

The court heard that during installation of solar panels on the roof, the worker fell approximately 7 metres through a fragile roof ridge panel to the ground below, suffering fatal injuries.

The investigation found that no measures were in place to prevent falls from the roof or through the roof.

The investigating officer said: "This death could easily have been prevented if the company and director had acted to identify and manage the risks involved, and to put a safe system of work in place".[7]

8.2.3 Main precautions to prevent falls and falling materials

As well as the hierarchy of control measures for working at height discussed earlier, there are more general points to consider when planning work at height activities. Some of these we have already talked about, such as not working in bad weather.

It is essential that all work at height activities are planned, including carrying out a risk assessment of the activity to decide on the appropriate control measures for the work. A risk assessment will normally identify:

- whether working at height could be avoided;
- which workers are at risk of exposure;
- what work activities are causing the work at height hazard;
- if control measures should be implemented and what kind; and
- the effectiveness of existing control measures.

Work must also be supervised at all times by an appropriate competent person. The supervision should include checks that the place where work at height is to be carried out is safe. The work area must be checked before each use.

Preventing falling objects/materials

Falling objects can do as much damage to people below as an actual fall from height. It is therefore essential to try to reduce the likelihood of this happening as much as possible. We will now look at some of the methods that could be used.

One way of preventing falling objects while people are working at height (toeboards) will be discussed later. Other means include:

- Brick guards: these are made up of metal or strong plastic mesh where there is a risk that materials could be stacked at a height above the toeboards.
- Complete netting or sheeting of a scaffold or the use of extending 'fans': this does not prevent items falling but catches anything that does fall to stop people below being injured.
- Making items secure: items such as scaffold boards and pieces of equipment can be physically lashed to the access equipment. Tools can be kept secure by the worker wearing a toolbelt or tools/equipment can be anchored by lanyards to the access equipment.
- Good housekeeping procedures: tools, rubbish and other objects should not be left in positions where they could fall.
- Nothing should be thrown or tipped from scaffolds or other workplaces at height: all equipment and materials should be lowered by a rope system using a hoist. Waste materials (eg rubble) should be lowered in similar ways or transferred to ground level or into skips via waste chutes.

An important way of preventing injury is by excluding people from areas where things might fall on them. These areas should be clearly indicated and access prevented. They may be cordoned off by fences, cones, bunting or similar means, and signs should be used to indicate where the exclusion area applies.

Figure 2: Skip at the bottom of a waste chute for collecting waste

Where exclusion is not possible (eg when people need to gain access to a building where work at height is taking place or to use a footpath alongside it), then it may be necessary to erect covered walkways or similar protective structures. Vehicles working in areas where there is a risk of falling materials must be fitted with 'falling object protective structures' (FOPS) to protect the operator.

Even with all the collective measures in place, it may still be necessary for individuals to wear personal protection in the form of hard hats. Employers must also maintain and replace head protection as necessary and provide suitable storage for it.

In practice, most employers require that head protection be worn at all times on site, regardless of the precise circumstances regarding risk of head injury. Workers on a roof where there is no crane operating, for instance, may not be at risk from falling objects or from bumping their heads, but they would nevertheless normally be required to use head protection. Rules of this type are simply easier to understand and to enforce. Hard hats may sometimes be necessary in non-construction situations, such as warehouses or other storage facilities.

8.2.4 Emergency rescue

Work at height activities need to consider emergency and rescue arrangements. This is particularly relevant to fall-arrest systems because life-threatening 'suspension trauma' can occur unless there is prompt rescue. Emergency procedures must be tested to ensure that they are effective. Workers must be provided with suitable and adequate information, instruction and training in relation to the emergency procedures.

When developing emergency procedures, the different types of emergency and rescue scenario that might arise must be considered and then incorporated into the work at height risk assessment.

> **DEFINITION**
>
> *Suspension trauma* or *orthostatic shock* *occurs after a worker has fallen into a fall-arrest harness and is suspended in a hanging position until rescue arrives. When hanging in a harness, the leg straps support the body's weight. During this time, the leg straps crush the femoral arteries on the inside of the legs, cutting off blood circulation and causing risk of serious, life-threatening injury.*

8.2.5 Provision of training and instruction

Training usually forms part of a route to competence. Competent workers are essential for work at height. For more specialised access work, training is often accredited by trade bodies. The training usually lasts for a fixed period, after which the worker will need to requalify.

The same is true for MEWPs and other powered equipment. There are different qualifications for variations of work at height equipment (for example, mobile scissor lifts and mobile boom lifts). Training often includes legislation, accident control and prevention, and the use of personal protective equipment (PPE).

The situation is different for everyday equipment such as ladders and steps. Employers need to make sure that people have sufficient skills, knowledge and experience to perform the task, or, if they are being trained, that they work under the supervision of somebody competent to do so.

In the case of low-risk, short-duration tasks involving ladders, competence requirements may be no more complex than making sure employees receive instruction on how to use the equipment safely (for example, how to tie a ladder properly). This sort of training usually takes place on the job.

When a more technical level of competence is required, for example drawing up a plan for assembling a complex scaffold, formal training and certification schemes are the more usual ways to demonstrate competence.

Workers will also need to be given sufficient instruction regarding any equipment that is being used by someone who is competent to do so. Workers should understand how to safely use the work equipment. The instruction would usually form part of the training.

Workers will also require information on the hazards associated with both the work at height activity being carried out and the equipment that will be used during the activity.

8.2.6 General precautions when using common forms of working at height equipment

Ladders and stepladders

Ladders and stepladders are a common form of access equipment but account for a significant number of falls from height.

Ladders may be used as a means of accessing a permanent or temporary work platform (such as a roof or scaffold) or they may sometimes be used as the only means of access to a work at height position (ie where work is done from the ladder). Any work done from a ladder needs serious consideration.

There are different types of ladder:

- Fixed ladders: these are normally vertical and made up of a series of rungs or individual treads set into a wall (eg for access to the cab of a tower crane or a telecommunications mast).
- Leaning ladders: these are usually constructed of either wood or aluminium and made up of two vertical stiles joined by a series of horizontal rungs. Wooden ladders usually incorporate a metal reinforcing rod under each rung. Leaning ladders may have more than one section to allow the ladder to be extended.
- Combination ladders: these can operate either as a leaning ladder or a stepladder.
- Roof ladders: these are used as a means of climbing up sloping roofs, where the top end of the ladder hooks over the ridge of the roof.
- Stepladders: these are constructed in two sections, hinged together at the top. They can be constructed of wood, aluminium or fibreglass.

> **CASE STUDY 2**
>
> **Installing satellite equipment**
>
> A satellite-television installation company, which is no longer trading, was fined following the death of a worker who fell from the roof of a four-storey house.
>
> The company, which had a contract to carry out repairs on satellite TV and satellite faults, sent the worker to a property in London to fix two faulty satellite dishes.
>
> The worker was fixing a dish on the property's roof apex, which he had accessed via a dormer window. Before his fall he was also seen working on another satellite dish located on a flat roof. There were no witnesses but it is thought that he slipped while walking across the sloping part of the roof and fell 13.5 metres, landing on a side patio. He was pronounced dead at the scene.
>
> The HSE's investigation found that no risk assessment had been carried out before the work commenced and the worker had not been issued with a harness. The investigating HSE inspector said that the company should have accompanied the worker on a site visit to carry out a site-specific risk assessment once it was established that the building had more than two storeys.
>
> The inspector also revealed that the company had not asked for any references from the worker when it first employed him, nor did it ask him to present any training certificates. It also failed to assess his competence.[8]

Ladders should only be used for work at height in low-risk situations for short durations or when the site cannot be altered to accommodate a more suitable means of working.

To ensure the safety of workers using ladders, they must be:

- suitable for the job;
- of appropriate height;
- of adequate strength; and
- constructed of appropriate material, for example metal ladders should not be used when working with electricity.

In addition to this, ladders must not be damaged or painted, as defects could be hidden. Before use they should be cleaned and free of grease, and be structurally sound; for example, the ladder feet firmly attached to the frame, no missing rungs.

Siting of the ladder is also important. It should be placed, on a firm, uncluttered and non-slippery surface, positioned at a 75° angle (see Figure 3) and the top of it should rest on a strong upper resting point and not on a window or plastic guttering. It should be prevented from moving at the top and at the bottom; for example, it should be tied at the top and/or secured at the bottom.

Figure 3: Correctly angled ladder position

Other considerations include having adequate overlap between sections on extended ladders and the ladder should project at least one metre above the landing stage and both stiles when used for access. The ladder should be integrated whenever possible; if it is not possible, the ladder must be tied at the point where it meets the landing stage.

Figure 4: Incorrect ladder and step use

> **TIP**
> - Ladders should not be used for more than 30 minutes in one position.
> - Three points of contact (hands, feet and body) must be maintained.
> - Devices (toolbelts, hooks, etc) for supporting equipment or materials should be used when working off ladders.

Scaffolding

Scaffolding can be used to provide a safe working platform where there is no permanent structure in place. The two main types of scaffold are 'independent tied' and 'mobile tower'. An independent tied scaffold is free-standing of the structure to which it provides access. The structure bears none of the weight of the scaffold but it is secured (or 'tied') to the structure for stability.

> **DEFINITIONS**
> *There are several ways of 'tying' a scaffold to a structure. They include the following.*
>
> *A **through tie** makes use of windows or similar openings in walls. A vertical tube crossing the opening on the inside is attached to the scaffold.*
>
> *A **reveal tie** also uses openings in buildings but this time the tube is wedged in the opening.*
>
> ***Standards or uprights** are the vertical tubes that transfer the entire weight of the structure to the ground, where they rest on a square base plate to spread the load.*
>
> ***Ledgers** are horizontal tubes that connect between the standards.*
>
> ***Transoms** are horizontally placed from back to front, next to the standards. They hold the standards in place and provide support for boards.*
>
> ***Intermediate transoms** are added to provide extra support for boards.*

Scaffolding must be designed and erected by competent people (whether internal or an outside contractor) and according to recognised standards (see the *Preventing falls in scaffolding operations* guidance).[9] This must take account of the weight to be placed on the scaffold (eg by materials and equipment) as well as other forces that may act on it (eg wind).

Certain things need to be considered to ensure independent tied scaffolding is stable and safe.

Components must be sufficient in number, in good condition and erected as expected; for example, standards must be installed vertically.

When erected, there must be sufficient bracing and the scaffolding must be footed on firm ground using square base plates placed centrally on (and usually pinned to) sole boards.

The scaffolding must be tied to prevent the scaffold coming away from the building. The number of ties required will depend on the forces encountered; for example, if the building is on top of a hill and may be subject to strong winds.

General workplace issues

Checks will need to be made to ensure that working platforms are sufficiently wide to permit the safe passage of workers and safe use of plant or materials and to provide a safe working area. This means they will need to be closely and fully boarded, with no gaps, and have toeboards in place on the outside edge of the working platform. Toeboards will stop a worker's foot from sliding off the platform and will help to stop objects from falling from the platform.

Finally, the working platform must be provided with sufficient guardrails positioned so there is no gap for a worker to fall through. There should be no break in the protective guardrail except where needed for access.

In addition, scaffolds must be protected against vehicle collision and against causing injury to people. Measures may include the use of barriers, timber baulks or concrete blocks, warning signs and markings and lighting. Projecting parts of the scaffold need to be protected so that high-sided vehicles are not damaged and pedestrians do not sustain head injuries.

Partially erected scaffolds should be properly marked as such and their use must be prevented until their construction has been completed and inspected for integrity and safety. Signage systems are available that indicate the inspection status of the scaffold.

A scaffold should be inspected by a competent person before first use, following circumstances that may have affected its integrity and at suitable intervals thereafter. Work platforms involving the risk of falls should be inspected at least once every seven days.

A mobile tower scaffold or mobile access tower (see Figure 5) must possess many of the same features as fixed scaffolding in terms of guardrails, toeboards, bracing and a fully boarded work platform. In addition:

- The tower should be made secure by placing it on firm, level ground and, where necessary, by the use of outriggers and/or tying it to the building.
- Wheels should be locked when the tower is in position.
- The tower should never be moved while there is any equipment or person on it.
- Towers should be reduced to four metres or less in height before moving and checks should be made for overhead obstructions.
- The access ladder should preferably be integral to the tower (either internal or external) and provided with a hinged bar, gate or hatch door for access.
- Overloading should be avoided and, unless it has been specifically designed for the purpose, a mobile tower should never be sheeted, exposed to strong winds, loaded with heavy equipment or used to hoist materials or support rubbish chutes.

The inspection requirements for mobile towers are similar to other types of scaffolding but the requirements become more demanding when the mobile tower gets higher (usually when over two metres).

8.2 Working at height

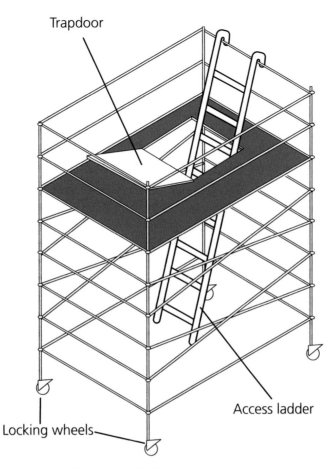

Figure 5: Mobile tower scaffold

Mobile elevating work platforms (MEWPs)

MEWPs are also commonly known as 'cherry pickers' or 'scissor lifts' and come in various types. They can be vehicle-mounted or independent and normally use a boom (articulated or telescopic) or scissor mechanism to lift the work platform.

MEWPs may be self-propelled or static. Many self-propelled MEWPs can be driven in the elevated position. Static MEWPs are usually transported by being towed behind a vehicle and can be manoeuvred into position manually. As with other work equipment, you will need to consider various issues before use.

To start with, the MEWP selected should be suitable for the particular task and should be well maintained and kept in good repair.

When positioning the MEWP, it must be on firm (and, where possible, even) ground. The outriggers should also be positioned on firm ground and the platform base should be level. When setting up, try to ensure this is not over weak areas of ground, for example where drains are located, and the MEWP should be positioned so that workers do not need to overstretch. Overhead obstructions such as power lines should also be identified and avoided when positioning the MEWP.

Before work commences, all workers must be trained in the safe use of the MEWP. Additionally, workers at ground level should be trained in emergency procedures, including emergency descent of the platform.

The equipment should be inspected before use to ensure it is working correctly. The inspection should check the platform has protection that will help to reduce the likelihood of a fall from height; things such as guardrails and toeboards should be in place. There should also be a means of attaching fall-arrest harnesses.

When in use, brakes should be applied, outriggers used and the MEWP should be protected at the base (with barriers, signs, etc) against collision and to ensure the safety of those on the ground. It is important that the platform is not overloaded (by people, equipment or materials); overloading could lead to failure of the MEWP. Workers must be appropriately supervised while the MEWP is being used.

Finally, MEWPs should not be used outdoors in adverse weather conditions for the reasons given earlier.

Trestles and staging platforms

Trestles are rigid frames that can support staging platforms; these are boards placed between two trestles, although more trestles can be used to extend the platform. Trestles are usually an 'A' frame and very robust. They can be made from a variety of materials, such as wood, steel, aluminium or synthetic materials. They are only meant for light, short-duration work; for example, decorating. They are usually supplied without edge protection, which could be an issue for some activities and workplaces.

Leading edge protection

All walkways and workplaces that involve people being at height should be suitably protected with guardrails. This includes balconies, staging, platforms and the edges of mezzanine floors. The height of guardrails from the floor should be sufficient to prevent falls. Intermediate guardrails should be provided unless children are likely to be present, who might use them for climbing. In such cases, the space between the top guardrail and the floor should preferably be filled by a solid feature, such as toughened glass. UK legislation and standards describe the dimensions of these safety features of buildings.

Figure 6: Edge protection with guardrail and toughened glass

Maintenance operations, such as elevator maintenance, should be carried out according to strict procedures to ensure that people are not put at risk by unprotected entry to lift shafts. Similarly, excavation work alongside pedestrian traffic routes should be suitably protected.

> **CASE STUDY 3**
> **Unprotected new stairway**
>
> A company pleaded guilty to a total of four health and safety offences that resulted in the death of a worker at a house building site.
>
> The prosecution arose as a result of an investigation into the incident, which took place when a worker fell approximately 2.8 metres through an opening on the first floor where a staircase was to be built. The worker died later as a result of his injuries.
>
> Speaking after sentencing, the investigating inspector said: "Working at height incidents are avoidable. All contractors must ensure that robust measures are put in place to prevent falls which can all too often result in serious injury or death".
>
> The investigation found that the company had earlier placed trestles and barrier tape to mark out and prevent access to the worker or any other person from falling.
>
> The inspector added: "Internal openings represent a serious hazard but simple measures such as fitting guardrails or secure boarding may be all that is required to prevent a tragedy".[10]

8.2.7 Prevention of falling materials through safe stacking and storage

Falling objects

In warehousing in particular, objects falling from height are a hazard to forklift truck operators and to pedestrians. Many of the causes of these accidents are due to unstable stacking of products when initially stored. The problem arises when the product is removed (sometimes called de-stacking).

Steps should be taken to prevent people being injured by falling objects. If there are areas or specific activities in the warehouse with a risk of material or an object striking someone, ensure that the area is clearly indicated and that unauthorised people don't enter it.

Mechanical handling

Mechanical handling equipment (eg a forklift truck or a reach truck) should be suitable for the job it is used for. All industrial truck operating areas should be suitably designed and properly maintained.

Figure 7: Stacking operations involving a potentially insecure load

Many goods are now transported on pallets. Shrink or stretch wrapping palleted loads usually provides greatest security, minimising the possibility of goods falling from the pallet when it is moved. Figure 7 shows a load that is potentially insecure and that could spill.

Safe stacking procedures may require pedestrians to wear safety helmets in warehousing areas.

This topic overlaps with 8.6: Safe movement of people and vehicles in the workplace in respect of materials moved by vehicles and the use of racking systems.

KEY POINTS

- Work at height is the cause of many workplace injuries and fatalities. A hierarchy of controls exists based on avoid, prevent and minimise.
- 'Avoid' consists of finding alternative methods of carrying out the work without the risk of a fall being present.
- 'Prevent' is where a safe means of access is found and can include the use of temporary work platforms (for example, tied and mobile tower scaffolds), MEWPs, ladders and steps.
- Fall protection systems include restraint systems, work positioning, rope access and fall-arrest systems.
- Falling objects need to be considered (for example, by using toeboards and brick guards).
- If objects are stored at height, then consideration needs to be given to the prevention of anything falling when the items are stacked or de-stacked.
- Where work at height risks require it, emergency rescue procedures need to be in place.

References

[1] HSE, *Working at height: A brief guide* (INDG401, 2nd revision, 2014) (www.hse.gov.uk)

[2] HSE, *Health and safety in roof work* (HSG33, 5th edition, 2020) (www.hse.gov.uk)

[3] HSE, *Fragile roofs: Safe working practices* (GEIS5, 2012) (www.hse.gov.uk)

[4] HSE, *Warehousing and storage: Keep it safe* (INDG412, 2007) (www.hse.gov.uk)

[5] HSE and the Ladder Association, *Using ladders and stepladders safely: A brief guide* (LA455, 2021) (https://ladderassocation.org.uk)

[6] National Access and Scaffolding Federation, *Preventing falls in scaffolding operations* (SG4:15, 2015) (https://nasc.org.uk)

[7] Adapted from *Roofing Today* (4 December 2019) (www.roofingtoday.co.uk)

[8] Adapted from *Safety and Health Practitioner* (18 August 2011) (www.shponline.co.uk)

[9] See note 6

[10] Adapted from *Safety and Health Practitioner* (11 June 2011) (www.shponline.co.uk)

8.3: Safe working in confined spaces

> **Syllabus outline**
>
> In this section, you will develop an awareness of the following:
> - Types of confined spaces and why they are dangerous
> - The main hazards associated with working within a confined space
> - What should be considered when assessing risks from a confined space
> - The precautions to be included in a safe system of work for confined spaces
> - When a permit-to-work for confined spaces would not be required

Confined spaces are found in many different industries. They can be the cause of serious accidents and sensible control measures need to be applied to ensure worker safety. This can include both technical measures (such as monitoring, or use of specialist access equipment) as well as procedural ones (such as permit-to-work systems, and assessments of training and competence).

What we will cover here closely relates to other areas of the syllabus, in particular permit-to-work systems, emergency procedures and hazardous substances.

> **DEFINITIONS**
>
> *The UK's Confined Spaces Regulations 1997 define a* **confined space** *as "any place, including any chamber, tank, vat, silo, pit, trench, pipe, sewer, flue, well or other similar space in which, by virtue of its enclosed nature, there arises a reasonably foreseeable specified risk".*
>
> *The Regulations define* **a specified risk** *as a risk to anyone working in the area from:*
> - *fire or explosion that causes serious injury;*
> - *the loss of consciousness arising from an increase in body temperature;*
> - *the loss of consciousness or asphyxiation arising from gas, fume, vapour or the lack of oxygen;*
> - *drowning due to an increase in the level of a liquid; or*
> - *asphyxiation due to the ingress of a free flowing solid or the inability to reach a respirable environment due to entrapment by a free flowing solid.*
>
> *The Health and Safety Executive (HSE) describes a confined space as "a place which is substantially enclosed (though not always entirely), and where serious injury can occur from hazardous substances or conditions within the space or nearby (such as lack of oxygen)".*[1]

Some organisations also refer to 'restricted spaces'.

> **DEFINITION**
>
> *A* **restricted space** *is an area where access and or egress is difficult or there is restricted working space that presents a hazard, or other risks or reasons that require controlled access.*

8.3 Safe working in confined spaces

Examples include small loft spaces, basements or cellars with low ceilings and difficult escape routes, plant rooms and substations.

These do not meet the full definition of a confined space but even so, confined space control measures may be invaluable.[2]

Introduction

Confined spaces can be deadly.

A number of people are killed or seriously injured in confined spaces each year in the UK. This happens in a wide range of industries, from those involving complex plant to simple storage vessels. Those killed include people working in the confined space and those who try to rescue them without proper training and equipment.[3]

> **TIP**
>
> The primary UK legislation dealing with confined spaces are the Confined Spaces Regulations 1997 and the Confined Spaces Regulations (Northern Ireland) 1999. According to the Explanatory Note, the Regulations:
>
> (a) prohibit the entry into a confined space for the purpose of carrying out work where it is reasonably practicable to carry out the work by other means;
> (b) require work in a confined space to be carried out only in accordance with a safe system of work;
> (c) impose requirements with regard to the preparation and implementation of adequate arrangements for the rescue of any person at work in a confined space in the event of an emergency;
> (d) provide that the Health and Safety Executive may grant exemptions from any requirement or prohibition of the Regulations in specified circumstances;
> (e) provide a defence in proceedings in respect of the duty to implement emergency arrangements.
>
> The Regulations are supported by HSE practical guidance in *Safe work in confined spaces. Confined Spaces Regulations 1997. Approved Code of Practice, Regulations and guidance* (L101).[4]

8.3.1 Types of confined spaces

Some confined spaces are fairly easy to identify, such as enclosures with limited openings.

Other examples are:

- storage tanks;
- silos;
- reaction vessels;
- enclosed drains; and
- sewers.

Less obvious examples are:

- open-topped chambers and inspection pits;
- vats;

Figure 1: Entry into a confined space

- combustion chambers in furnaces;
- ductwork;
- covered inspection pits;
- unventilated or poorly ventilated rooms; and
- temporary structures, including tents and marquees, for example when petrol generators are used inside the tent to provide electrical power for lighting or musical equipment.

8.3.2 The main hazards associated with working in confined spaces

Of itself, an enclosed place is not a confined space: it needs to have the capacity to cause a serious injury. Some of the hazards when working in confined spaces include the following.

Oxygen deficiency

This can occur:

- in grain silos and grain stores where oxygen is absorbed by the grains;
- where there is a reaction between some soils and the oxygen in the atmosphere;
- in ships' holds, freight containers, lorries, etc, as a result of the cargo reacting with oxygen inside the space; and
- inside steel tanks and vessels when rust forms, so slowly 'burning up' oxygen.

Toxic gases

These can occur:

- in pits where there is a build-up and displacement of oxygen due to the carbon monoxide exhaust gases of petrol engines;
- due to the presence of contaminants in the atmosphere caused by disturbing decomposed organic material in a bin, letting out toxic substances; and
- as a result of the build-up and release of gases like ammonia, methane, carbon dioxide and hydrogen sulphide in manure pits.

Suffocation or drowning

This can be caused by free-flowing solids, such as grain, sand or fertiliser.

Fire and explosion

Fire and explosion are common occurrences in confined spaces. The reasons include:

- flammable vapours that are heavier than air move and collect in pits and voids;
- flammable residues left on the internal surfaces of tanks such as solvents evaporate, giving off gas, fume or vapour. Raised temperatures can increase the rate of evaporation; and
- dust at high concentrations, for example in flour silos.

Heat and fatigue

These may be due to thermal conditions, leading to a dangerous increase or decrease in body temperature.

Fumes and gases

Exhaust fumes are dangerous. Equipment such as generators or compressors with petrol or diesel engines should not be sited where the fumes may collect near to pits or excavations. Similarly, vehicles in the area should be prevented from idling for long periods.

> **CASE STUDY 1**
> **Pump fumes kill two people in cellar**
>
> A man and his son died while pumping out water from the cellar of their rugby club following floods.
>
> The men were among a dozen people who were inspecting damage to the club. They stayed on to clear the cellar of floodwater and their bodies were found the following morning. It was thought they were overcome by fumes from the petrol pump they were using.
>
> A fire officer urged people to think about safety before using petrol-operated pumps.[5]

The build-up and release of gases like ammonia, methane, carbon monoxide and hydrogen sulphide are known to be hazardous. This is a known problem in sewers, where workers walk through sludge, disturbing pockets of methane and hydrogen sulphide.

Oxygen deficiency and oxygen enrichment

These two are often closely related. The normal atmospheric concentration of oxygen is about 21%. If the oxygen level falls below 17%, workers will rapidly suffer problems such as fatigue or unconsciousness, which can lead to coma and/or death.

> **CASE STUDY 2**
> **Trawler fatality**
>
> An engineer working on board a trawler was found collapsed inside a refrigerated saltwater tank. When he was found, three of his crewmates went into the tank to help him; they all suffered breathing difficulties and one also collapsed. Although the engineer was rescued from the tank, he could not be resuscitated. Two other crew members then put on breathing apparatus and rescued their struggling crewmates.[6]

Oxygen deficiency may occur 'naturally' as a result of organic decay processes using up oxygen and leaving an excess of carbon dioxide. Oxygen deficiency may also occur as a consequence of an area being deliberately purged by the use of an inert gas such as nitrogen. This is used to 'flush out' carbon dioxide or explosive gases such as methane.

> **CASE STUDY 3**
>
> **Farmworkers suffocated in container**
>
> Two workers died after being told to hold their breath while working in a nitrogen-filled store at a fruit farm.
>
> The workers suffocated in a container of apples when trying to retrieve fruit for an agricultural competition.
>
> Apples are generally placed in controlled-atmosphere stores for longer-term preservation, while refrigerated stores are used in the short term.
>
> Preservation in this case was achieved by reducing the oxygen level (the oxygen level was 1%) and increasing the nitrogen levels; this resulted in higher levels of carbon dioxide.
>
> The farm estate pleaded guilty to three health and safety offences and was given a substantial fine and ordered to pay costs.
>
> The farm manager was jailed for two and a half years for the manslaughter of the workers.[7]

Excess of oxygen (known as 'oxygen enrichment') occurs when oxygen concentrations are greater than 21%. This can often occur when flame-cutting equipment and oxygen cylinders are brought into confined spaces. The air quality in the confined space will need monitoring to ensure that it remains safe.

8.3.3 What should be considered when assessing risks from a confined space

An assessment of risks should of course be carried out to help determine the measures necessary for safety. The assessment should consider the hazards that were discussed earlier as well as:

- the task, including duration, activities being carried out, etc;
- the working environment, such as the type of confined space;
- working materials and tools, such as the use of intrinsically safe tools;
- the suitability of those carrying out the task, for example competence as well as any existing health conditions (asthma or other respiratory diseases) that may affect a worker's ability to carry out the activity; and
- arrangements for emergency rescue.

Sites/locations that have a number of confined spaces should record these in a confined spaces register. This will be one of the main sources of management information for confined spaces assessments.

The confined spaces register should contain a list of the confined spaces associated with each site and a site plan showing the reference number and location of the confined spaces.

Hierarchy of control

In terms of an approach to controlling the risks arising from confined spaces, the following general hierarchy applies:

- Avoid entry into confined spaces.
- If entry is unavoidable, a safe system of work must be followed.
- Adequate emergency procedures must be in place before work starts.

This approach is broadly in line with the hierarchy of controls used more generally for risk control measures. Figure 2 gives an example of a flowchart for controlling work in confined spaces.

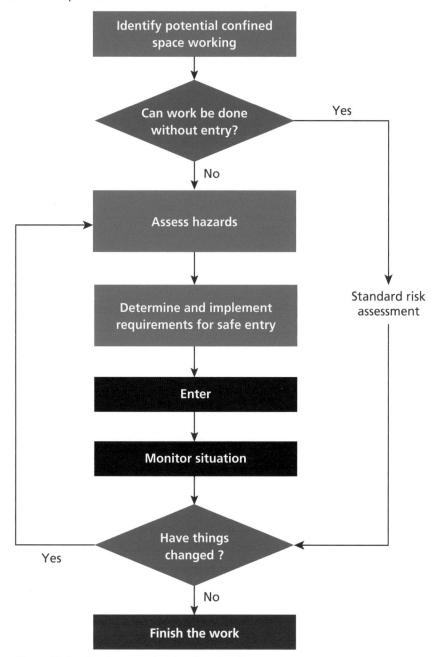

Figure 2: Confined spaces decision flowchart[8]

Avoiding entry might be achieved by using remotely controlled equipment, for example, which may be practical when inspecting and cleaning sewers and other confined spaces. This can include the use of drones and fibre-optic cameras for inspection purposes.

An alternative might be to 'unconfine' the space through the removal of wall panels and so on. In the case of shipbuilding activities, it may be possible, by careful planning, to undertake some of the fitting out before spaces become confined.

8.3.4 Precautions to be included in a safe system of work for confined spaces

Safe systems of work for general work activities are covered in detail elsewhere in the syllabus. It is important to say that confined spaces invariably require additional work activities that possess their own hazards above and beyond those of the confined space itself (such as moving, painting, sealing, repairing, inspecting).

A safe system of work is one where the necessary controls are identified and then documented in a formal way. For confined spaces, typical elements of a safe system of work may include the following.

> **DEFINITION**
>
> *The term **top worker** is used in the UK and refers to the worker controlling the access in and out of the confined space.*

Appointment of a supervisor

Supervisors have an important role to play and should be given responsibility to make sure that the necessary precautions are taken and then check safety measures at each stage of the work. Typical supervisor duties are:

- ensuring that entrants and 'top workers' are briefed on the health and safety precautions to take prior to confined space entry and work;
- authorising entry into confined spaces;
- ensuring that entrants and top workers adhere to entry procedures;
- ensuring rescue equipment and appointed rescue workers are available when confined space work is to be carried out;
- implementing control and preventative measures to manage all identified hazards;
- applying the entry permit and having it endorsed by the authorised manager before commencement of confined space work; and
- terminating the entry permit after completion of work.

We will look at the permit-to-work in more detail later.

The 'top worker' is responsible for:

- monitoring entrants entering and working in a confined space;
- maintaining regular visual and/or verbal contact with the entrants in the confined space;
- informing entrants to evacuate the space should the need arise; and
- alerting rescue workers to activate a rescue operation in the event of an emergency.

The main roles of a confined space entrant are:

- following entry and work procedures when carrying out work in confined spaces;

- carrying a portable gas/vapour measuring instrument for continuous monitoring of the atmosphere in the confined space for the full duration of their work;
- informing the top worker of any unsafe atmospheric conditions or when any emergency situation arises; and
- informing co-workers to evacuate from the confined space should the portable gas/vapour instrument alarm be activated due to unsafe atmospheric conditions.

Isolation of equipment

Confined space work may require isolation of equipment; for example, isolating pumps in sewer networks to prevent workers being swept away by effluent while working. Formal lock-off systems may be needed for this.

> **DEFINITIONS**
>
> **Lock-out** *is the placement of a lock-out device on an energy-isolating device in accordance with an established procedure. A lock out device is a mechanical means of locking that uses an individually keyed lock to secure an energy-isolating device in a position that prevent energisation of a machine, equipment or a process.*
>
> **Tag out** *is a labelling process that is always used when lock-out is required. The process of tagging out a system involves attaching or using an information tag or indicator (usually a standardised label) that includes the following information:*
>
> - *why the lock out/tag out is required (repair, maintenance, etc);*
> - *time and date of application of the lock/tag; and*
> - *the name of the authorised person who attached the tag and lock to the system.*
>
> *Together these are known as* **Lock Out, Tag Out (LOTO)**. *This means that the equipment cannot be restarted until all locks have been removed; this is especially useful if more than one person is working on the same piece of equipment at the same time.*[9]

Figure 3: Locking devices stop equipment being restarted

Cleaning before entry

This may be necessary to ensure fumes do not develop from residues, etc while the work is done. For example, in sewers, disturbing the effluent can release toxic gases such as hydrogen sulphide and methane.

Size of the entrance

This needs to be big enough to allow workers wearing all the necessary equipment to climb in and out easily and to provide ready access and exit in an emergency. For example, the size of the opening may mean choosing alternatives to bulky self-contained breathing equipment that could restrict safe entrance and egress.

Provision of ventilation

The number of openings (such as windows) may need to be increased to improve ventilation. Mechanical ventilation may be needed to provide an adequate supply of fresh air. This is essential when portable gas cylinders and diesel-fuelled equipment are used inside the space because of the dangers from build-up of engine exhaust. Petrol-fuelled engines should never be used in confined spaces. This is to stop the levels of carbon monoxide (in the exhaust emissions) from building up over time.

Testing the air

It may be necessary to test the air before a worker enters a confined space to check that it is free from both toxic and flammable vapours and that it is fit to breathe. Environmental testing should be carried out by a competent person using a suitable gas detector that is correctly calibrated. Where conditions may change, or as a further precaution, continuous monitoring of the air may be needed.

Provision of special equipment and lighting

Non-sparking tools and specially protected lighting are essential where flammable or potentially explosive atmospheres can occur. In certain confined spaces (such as inside metal tanks) suitable precautions to prevent electric shock may be required. This can include the use of extra low voltage equipment (typically less than 25V) and, where necessary, residual current devices. See 10.2: Preventing fire and fire spread for more information on control ignition sources, including suitable electrical equipment in flammable atmospheres.

Provision of breathing apparatus

Breathing apparatus is essential if the air inside the space is not fit to breathe because of gas, fume or vapour present, or lack of oxygen. Adding oxygen into a confined space can greatly increase the risk of a fire or explosion.

Access

One of the most important considerations is access to the confined space. Some situations may require horizontal access, such as grain silos (see Figure 4) and some may require vertical access (see Figure 5).

8.3 Safe working in confined spaces

Figure 4: Horizontal entry into a tank

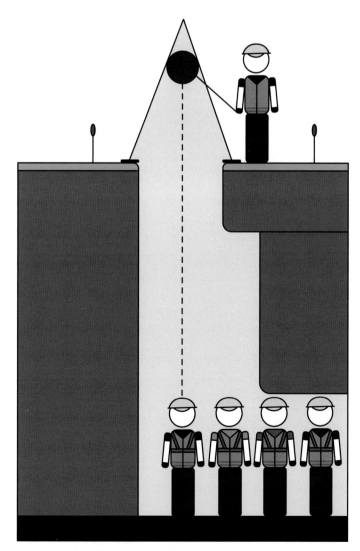

Figure 5: Vertical access to confined space[10]

General workplace issues

Suitable workers

Work in confined spaces is both physically and mentally demanding. Selection of workers forms an important part of the safe system. Considerations include physical size, fitness, underlying health issues, experience of working in confined areas, ability to wear breathing apparatus and not being claustrophobic. The selection process may involve both a regular medical examination and psychological assessment.

Physical abilities can reduce with age, so periodic reviews may be required; these should be more frequent if individual workers suffer specific health and mobility issues.

> **TIP**
>
> Medical screening could include:
>
> - health questionnaire, including questions about psychological issues such as claustrophobia;
> - height, weight, body mass index, waist and hip measurement;
> - blood pressure;
> - vision screening for near and distance;
> - standard urine test for protein and sugar;
> - lung function baseline;
> - hearing test baseline;
> - specific confined space related questions; and
> - respirator fit questions.

Training

Training on confined spaces is specified in the approved code of practice accompanying the Confined Spaces Regulations. In summary, training must cover:

(a) an awareness of the Confined Spaces Regulations (particularly the need to avoid entry to a confined space, unless it is not reasonably practicable to do so);

(b) an understanding of the work to be carried out, the hazards, and the necessary precautions;

(c) an understanding of safe systems of work, with particular reference to 'permits-to-work' where appropriate; and

(d) how emergencies arise, the need to follow prepared emergency arrangements and the dangers of not doing so.

Specific training for work in confined spaces will depend on an individual's previous experience and the type of work they will be doing.

Training courses may be categorised into working in low, medium and high risk environments. There are also additional courses for: top workers (entrant and non-entrant); emergency rescue and recovery of casualties from confined spaces; and managing work in confined spaces.

Confined space competence is achieved through a combination of training and practical experience.

Preparation of emergency arrangements

Examples of confined space emergencies include the following:

- A worker has an accident and is injured and they cannot get themselves out of the space without assistance.
- A worker becomes unconscious due to: exposure to hazardous fumes, gases or vapour; heat stress; or oxygen depletion.
- A worker falls ill and becomes immobile.
- The atmosphere changes in the space and workers need to get out quickly; for example, a gas detector alarm sounds.
- A fire or explosion occurs and workers need to get out of the space quickly.

General emergency procedures are covered in detail in 3.8: Emergency procedures. Confined space emergency arrangements will need to cover the necessary equipment, training and arrangements for practice drills. These should form an integral part of the planning and the arrangements should cover:

- communication methods, both with and between workers in the confined space and the 'top worker' at the entrance, and between the 'top worker' and the emergency services so that the alarm can be raised if needed;
- equipment for, and training of, the rescue team;
- emergency lighting, taking into consideration loss of power and the potentially explosive atmosphere that may be found. Emergency lighting can also include chemical light sticks and emergency distress strobes to help locate workers in a confined space;
- the need to shut down adjacent plant before attempting emergency rescue;
- provision of first aid;
- alerting emergency services; and
- practising emergency rescue drills.

Rescue equipment needs to be readily available. It should include ropes, resuscitation equipment, communication equipment, breathing apparatus, firefighting equipment and anything else considered essential for the confined space being worked in. The most sophisticated breathing apparatus will be useless if it is too large to allow entry to the confined space, so the suitability of all equipment must be established as part of the planning.

> **TIP**
>
> Workers who are not trained in proper rescue procedures should not carry out or be permitted to carry out rescue operations.

Rescue harnesses should be provided where necessary. Lifelines attached to harnesses should run back to a point outside the confined space.

8.3.5 When a permit-to-work would not be required

When an assessment identifies that the hazards to be faced are particularly serious, then a permit-to-work will be required. This normally applies to confined space working. Permit-to-work systems are covered in detail in 3.7: Permit-to-work systems.

For clarity, confined space working may require permits that control multiple hazards:

- hot work such as cutting and welding;
- work on pressurised systems;
- electrical work; and
- confined spaces.

Before a permit-to-work can be authorised, a permit to enter the confined space may be required, enabling inspections to take place prior to work being carried out.

In confined space working, a permit-to-work may have special checks and precautions. For example, continuous atmospheric testing may be required in deep excavations and tunnels and the permit-to-work should provide a specification of the required monitoring equipment.

Figure 6: Workplace identification of confined spaces

> **APPLICATION**
>
> Many workplaces keep a register of confined spaces. These are usually signed and often have a unique location number.
>
> In your workplace, find three examples of confined spaces. Use examples from another workplace if yours does not have any confined spaces.
>
> For each confined space, identify the hazards that could give rise to significant injury.

8.3 Safe working in confined spaces

> **KEY POINTS**
>
> - Entry into and work in confined spaces can be hazardous. The nature of the hazard can vary and may include oxygen deficiency, toxic gases, suffocation, fire and explosion and heat and fatigue.
> - Confined space hazards can give rise to specified risks, including loss of consciousness due to asphyxiation, drowning, loss of consciousness due to excess body temperature, serious injury related to equipment and serious injury related to falls, slips and trips.
> - A confined spaces hierarchy of control includes avoiding entry into confined spaces, following a safe system of work if entry is unavoidable and putting adequate emergency procedures in place before work starts.
> - Sensible control measures can reduce these risks, starting with a risk assessment that considers the task, the working environment, working materials and tools, the suitability of those carrying out the task and arrangements for emergency rescue.
> - A safe system of work for confined spaces may include considering the appointment of a supervisor, suitable workers, isolation of equipment, cleaning before entry, entrance size, ventilation, air testing, provision of special equipment, lighting and breathing apparatus, access, medical screening, training and emergency arrangements.

References

[1] HSE, 'Confined spaces' (www.hse.gov.uk)

[2] Water UK, *The Classification & Management of Confined Space Entries, Occasional Guidance Note. Water industry guidance* (3rd edition, 2019) (www.water.org.uk)

[3] HSE, *Confined spaces: A brief guide to working safely* (INDG258, 2013) (www.hse.gov.uk)

[4] HSE, *Safe work in confined spaces. Confined Spaces Regulations 1997. Approved Code of Practice, Regulations and guidance* (L101, 3rd edition, 2014) (www.hse.gov.uk)

[5] Adapted from BBC News (26 July 2007)

[6] Maritime and Coastguard Agency, 'Guidance: Enclosed spaces on sea-going vessels' (2022) (www.gov.uk/guidance)

[7] Adapted from 'Pump fumes kill two in flood town', BBC News (1 July 2015)

[8] Adapted from Health and Safety Authority (Ireland), *Code of Practice for Working in Confined Spaces* (2017)

[9] Canadian Centre for Occupational Health and Safety (www.ccohs.ca)

[10] Adapted from Water UK (see note 2)

8.4: Lone working

> **Syllabus outline**
>
> In this section, you will develop an awareness of the following:
> - What a lone worker is and typical examples of lone working
> - Particular hazards of lone working
> - Control measures for lone working
> - What should be considered when assessing risks of lone working

Introduction

Depending on the nature of the business, lone working may pose an organisational threat as well as a threat to individual workers. The cost and disruption from failing to manage lone working risks may be significant. For example, there is the potential for reputational damage, staff dissatisfaction and poor staff retention.

Lone workers who are young, pregnant or nursing mothers may be especially at risk.

Many trade and industry-specific supporting documents on lone working are available.[1]

There is no specific health and safety legislation that regulates lone working in the UK.[2] However, general duties under the Health and Safety at Work etc Act 1974 exist, as do duties to conduct risk assessments under the Management of Health and Safety at Work Regulations 1999 and Management of Health and Safety at Work Regulations (Northern Ireland) 2000 (see 1.3: The most important legal duties for employers and workers and 3.4: Assessing risk).[3]

8.4.1 What a lone worker is and typical examples of lone working

> **DEFINITION**
>
> *Health and Safety Executive (HSE) guidance defines **lone workers** as those who work by themselves without close or direct supervision.*
>
> *Anybody who works alone, including contractors and self-employed workers, is classed as a lone worker. Many aspects of travel for work can be considered as lone working.[4]*

Lone working is common in many modern workplaces. The HSE gives examples including:

- working alone at a fixed base, such as in shops, petrol stations, factories, warehouses or leisure centres;
- homeworking;
- working separately from other people on the same premises or outside normal working hours, such as security staff, cleaners, maintenance and repair staff; and
- working away from a fixed base, such as:
  - health, medical and social care workers visiting people's homes;

- workers involved in construction, maintenance and repair, including engineers, plant installation and cleaning workers;
- service workers, including postal staff and taxi drivers;
- engineers, estate agents and sales or service representatives visiting domestic and commercial premises;
- delivery drivers, including lorry drivers, van drivers and couriers;
- agricultural and forestry workers; and
- volunteers carrying out work on their own for charities or voluntary organisations (for example, fundraising).

CASE STUDY 1
Lone working and fall from height injury incident

A roofing company worker fell nearly 5 metres and landed on a concrete patio, suffering several major injuries. An investigation found the worker had been allowed to work alone on several occasions, without anyone monitoring what he was doing. The company failed to ensure the work was properly planned, adequately supervised and carried out in a safe manner.[5]

CASE STUDY 2
Lone working and exposure to hazardous substance incident

This incident occurred late at night and a laboratory worker was alone in the laboratory clearing up and placing items in a fridge. During the clear-up process, a chemical bottle broke, resulting in a chemical splash to the worker's eyes and face. This was primarily due to congested storage space in the fridge (housekeeping processes were inadequate). There was a lack of knowledge and awareness of the out-of-hours emergency procedures on site and the worker was unable to obtain assistance until a few hours later.[6]

CASE STUDY 3A
Lone worker missing presumed dead

Suzy Lamplugh was a 25-year-old estate agent who went missing and was presumed dead after apparently attending an appointment to show a house in Fulham, West London, to a man calling himself 'Mr Kipper' in July 1986.

Witnesses reported seeing a woman who resembled Suzy Lamplugh talking with a man near the property and then getting into a car.

Her company car was found outside a property for sale about half a mile away that evening. The handbrake was off, the car key was missing and her purse was found in a storage pocket in one of the car doors.

The Suzy Lamplugh case led to a campaign by a charitable trust that was set up in her name. Its purpose is to "reduce the risk of violence and aggression through campaigning, education and support". (See Case study 3B for examples of lone working safety advice produced by the trust.)[7]

It is important to stress that managers and supervisors can carry out lone working when they are working outside normal office hours, visiting remote sites or travelling for work.

8.4.2 Particular hazards of lone working

Often hazards of lone working may appear similar to those when working from a fixed place of work or when working alongside others. However, the ability to react and respond in lone working situations is often impaired. Another factor is that lone working often takes place in remote areas and in areas where the environment is harsh. The activities themselves can be hazardous (such as working at height) but working alone adds an additional hazard.

Lone working hazards can include:

- no prompt first aid or medical attention available in the event of an accident or emergency (for example, response times and distances to get help may be greater);
- no prompt medical attention in the event of sudden illnesses (for example, longer response times may affect the outcome of a medical incident);
- work-related stress due to isolation and not enough emotional and practical support (such as absence of physical company, loneliness);
- inadequate rest, hygiene and welfare facilities (for example, workers are unable to access facilities to change clothes, take breaks and make meals when compared with workers at a fixed workplace);
- aggravation of pre-existing physical and mental health conditions (for example, workers with diabetes may suffer because of inadequate facilities to take regular meals); and
- physical violence from members of the public and others, compared with workers who operate in pairs.

Workers' duties

The employer holds the main responsibility for protecting the safety, health and welfare of lone workers. However, lone workers – like other workers – have a responsibility to help their employer fulfil this responsibility. They must do the following:

- Take reasonable care to look after their own safety, health and welfare; for example, considering their own safety when selecting where they park their vehicle or where they choose to stay.
- Communicate promptly with their line manager and colleagues; for example, letting the employer know when they arrive and depart a job if that is the workplace procedure.
- Safeguard the safety and health of other people affected by their work. Extra vigilance may be required when working alone, compared with working with others or where there is additional help to check that other people affected by the work are safe.
- Co-operate with their employer's safety and health procedures; some workers can be reluctant to let their employer know where they are and to keep to agreed procedures for checking in.
- Use tools and other equipment properly, in accordance with any instructions and training they have been given. This can include properly using lone working devices and trackers (discussed later).
- Report all accidents, injuries, near misses and other dangerous occurrences. Being a lone worker still requires these incidents to be reported, in line with the workplace procedures.

Changing patterns of working

Patterns of work are changing, with greater use of zero-hours and casual contracts. Many people on these contracts are lone workers, working to deadlines and exposed to specific road risks for work-related journeys. Greater automation requires fewer workers to run computer-controlled production lines. Workers are also working until they are older. Disabled workers now have greater opportunities in the workplace. This means that employers need to think carefully when considering how to keep workers healthy and safe.

Lorry drivers are likely to experience long, unsociable hours, high physical and mental demands, and often long periods of sedentary work. Employers should monitor drivers' health regularly and adapt their work to accommodate any individual health needs.

Lone working can have adverse health consequences for workers, such as musculoskeletal disorders, stress, tiredness and fatigue, as well as issues associated with poor or irregular eating habits.

> **DEFINITION**
> A **zero-hours** or **casual contract** is usually put in place for intermittent work – workers are called on when they are needed. There is no guarantee of work from the employer and the worker does not need to accept the work if it is offered. Workers on these types of contract can work for several employers at the same time. However, the employer is still responsible for the health and safety of these workers while they are at work.

8.4.3 Control measures for lone working

When lone working cannot be avoided, a number of controls can be implemented:

- Manual methods of monitoring can be time consuming and unreliable and often include a large amount of paperwork. Significant advances in mobile technology have led many businesses to switch to app-based, lone worker solutions to help them monitor and protect their remote workers. Being able to monitor workers' whereabouts is extremely important in keeping them safe, as accidents can occur outside normal working hours. There is a balance to be struck between the privacy of the individual worker and the need for the employer to check that they are safe. This is especially true in relation to travel for work.
- When workers are travelling directly to and from their home, family and friends provide an opportunity for an early alert if the worker is late or has not been in contact with their home. Both parties need to know 'in case of emergency' (ICE) contact details.
- In the event of a serious incident, having these ICE details on a mobile phone or written down can be invaluable.
- Being able to identify someone's location is important in an emergency. Mobile phones and apps (such as what3words) can be used to do this, but some instruction is needed on how to do it.
- Regular communication should be maintained with lone working staff and procedures put in place so that workers can quickly communicate with their employer and raise the alarm if needed.
- Some industry sectors produce specific safety guidance on remaining safe; see the example relating to the food and drink sector (Case study 3b) and the tips on driving alone later on.

General workplace issues

> **CASE STUDY 3B**
>
> **Suzy Lamplugh Trust safety advice for hospitality staff**
>
> The Suzy Lamplugh Trust produced the following tips for lone workers in the hospitality sector.
>
> **Full list of tips for hospitality staff and businesses**
>
> - Consider exit points – can you get out safely in the event of an emergency?
> - Be aware of people's behaviour, and how it can change, for example with alcohol
> - Don't allow clients or guests into kitchen or staff-only spaces
> - Consider carrying a personal alarm
> - Open up/close the site in pairs where possible
> - If lone working must occur, use a buddy/tracing system so people aren't forgotten about or left alone for too long
> - Be aware of the personal information you give to clients and guests
>   - If you wear a name badge, consider just using a first name to reduce the potential of being tracked down by a client
>   - If you stay onsite, don't tell guests where you stay
> - Remember, violence and aggression are unacceptable, and must always be reported.[8]

Training as a control measure

Training is important where there is limited supervision to control, guide and help in situations of uncertainty. Training may be critical to avoid panic reactions in unusual situations. Lone workers need to be sufficiently experienced and to understand the risks and precautions fully. Employers should set the limits on what can and cannot be done while working alone. They should ensure workers are competent to deal with circumstances that are new, unusual or beyond the scope of training, such as when to stop work and seek advice from a supervisor and how to handle aggression.

Supervision as a control measure

Although lone workers cannot be subject to constant supervision, it is still an employer's duty to look after their safety and health at work. Supervision can help to ensure that workers understand the risks associated with their work and that the necessary safety precautions are carried out. Supervisors can also provide guidance in situations of uncertainty. Supervision of safety and health can often be carried out when checking the progress and quality of the work; it may take the form of periodic site visits combined with remote discussions in which health and safety issues are raised.

The extent of supervision required depends on the risks involved and the ability of the lone worker to identify and handle safety and health issues. Workers new to a job, undergoing training, doing a job that presents special risks or dealing with new situations may need to be accompanied at first. The level of supervision required is a management decision, which should be based on the findings of a risk assessment.

Technology – lone working safety devices

> **DEFINITION**
> A **lone working safety device** is a tool, app or service that allows for communication with workers and employers, or in more serious situations, emergency services. Lone working safety devices safeguard workers while they work or travel, giving them a quick and easy way to signal for help in an emergency. Devices can have a body camera ('bodycam') incorporated and may be connected to multiple mobile networks to help ensure a good signal. Some devices (known as 'man down' alarms) are designed to detect a fall or lack of movement by the wearer.

Technology is rapidly changing, and new devices and new software applications are becoming available that can detect falls and lack of movement, as well as being able to accurately locate where the user is. Many smartphones as well as smartwatches have these features.

Buddy systems

One very effective way of taking care of another co-worker is to have a buddy system.

> **DEFINITION**
> A **buddy system** is a system of organising workers into work groups so that each member of the group works collaboratively with at least one other worker in the group. The purpose of the buddy system is to identify at an early stage if a worker has failed to keep to their plan while lone working and to call for rapid assistance for their buddy in the event of an emergency.[9]

Workers on the same shift and who do similar tasks (even though not physically working alongside each other) can provide an informal support and monitoring system between themselves.

This works well as these buddy workers can often pick up early signs that something is wrong with their lone working buddy.

Home

Good communication between the employer and workers' contacts outside of work is also essential. If a worker is unexpectedly late home, there needs to be a clear line of communication to alert their workplace. If something happens at work, it is also important that the workplace has up-to-date contact details for the worker's next of kin.

8.4.4 What should be considered when assessing risks of lone working

Lone working is often associated with work activities that themselves have risks. Lone working can increase the risk to the individual (compared to the alternative of doing this work with someone else).

The requirement to conduct risk assessments is all-embracing and specific lone worker risk assessments are often contained in the broader risk assessment of the activity of which lone working is an important and integral part.

General workplace issues

In such cases, you need to especially consider what aspects could have a much greater effect on lone workers, for example:

- The nature of the work: some activities carry a higher risk than others, such as confined space entry or diving.
- Some activities carry a specific risk of violence at work, such as security work, handling cash and visiting potentially aggressive clients in their own homes.
- The individual: this could relate to, for example, their experience, level of training or vulnerabilities (such as disabilities or medical conditions).
- Environment: for example, if the work is in remote locations.

Lone working is part of a general risk assessment and often one of many hazards considered in workplace risk assessments. For example, Table 1 shows the lone worker entry for a sample office-based assessment. Other hazards considered in the assessment and not shown are slips and trips, manual handling and display screen equipment (among many others).

What are the hazards?	Who might be harmed and how?	What are you doing already to control the risks?	What further actions do you need to take to control the risks?	Who needs to carry out the action?	When is the action needed by?
Lone working	Workers could suffer injury or ill-health while out of office, eg visiting clients' offices, or while working alone in the office.	• Workers write visit details in office diary and give a contact number. • Workers not returning to the office after a visit call in to report this. • Security check all areas, including toilets, before locking up at night.	Whereabouts of workers 'out of the office' to be monitored by office-based workers.	Office admin team	From now on

Table 1: Lone working element of risk assessment for an office-based business[10]

Importantly, where lone working is present, the general five-step approach to risk assessment, discussed in 3.4: Assessing risk, should be applied and the following should be considered:

1. Identify hazards: the earlier list will be helpful for this.
2. Identify people at risk: this is a vital step for lone working as it is easy to overlook aspects of lone working for workers who are normally working in a group, such as an office worker who infrequently has to take cash to a bank. This is in addition to considering the safety and welfare of workers who predominantly work on their own.
3. Evaluate risk: this means taking account of what you already do and deciding if you need to do more. We have already considered many lone worker control measures, but this risk evaluation phase provides the opportunity to review the current measures in place and look at advances in lone worker devices. The range of control measures required is likely to be more extensive than when working as part of a team. There will be a need for reliable communications (mobile phones and radios for raising the alarm), means of evacuation in an emergency, additional welfare equipment to allow for long waits to get assistance, plus additional food and drink as necessary. These controls will be in addition to any measures required to control the risks from the primary work activities. It also may be helpful to consider the 'safe person concept' when dealing with travel for work.

8.4 Lone working

4 **Record significant findings**: in common with all risk assessments, the findings need to be recorded. Whether the lone worker assessment is separate or integrated into a more general one is a matter of choice for the employer.
5 **Reasons for review**: there is a need to encourage the reporting of lone worker incidents so that they are visible to managers. Workers may be unsure of what counts as incidents at work and this needs to be made clear.

We will now explore these elements of a lone worker risk assessment further.

When are you at work?

This question is often raised when workers have to travel as part of their job. It affects managers and supervisors but also delivery drivers and service engineers who are often home based and travel to and from their home address. So, at what point does a worker start and finish their work, and when does the employer's duty of care apply? In Europe the matter has been the subject of case law.

> **DEFINITION**
>
> *What does **at work** mean? There are a number of examples of modern work activities (specifically delivery drivers and service engineers) where the definitions of 'at work' have been unclear.*
>
> *The European Court of Justice (ECJ) ruled in* Federación de Servicios Privados del Sindicato Comisiones Obreras v Tyco Integrated Security SL *on the meaning of 'working time' for workers who do not have a fixed place of work.[11]*
>
> *The ECJ said that the time these workers spend travelling between their homes and the premises of the first and last customers can constitute 'working time'.*
>
> *The implication is that employers have duties towards the workers (including those relating to lone working) for the whole day, starting and finishing at their home location.*

Travel risks

Travelling for work raises many questions regarding lone working. Risk assessment is a key part of the preparation for travel (especially to another country). Some considerations are:

- people, places and activities being visited;
- the specific needs of individuals, such as medical conditions and disabilities;
- political, medical and security risks;
- infrastructure and contacts in the countries being visited;
- cultural awareness and training received by the individual;
- previous independent travelling experience;
- travel planning and vaccination schedules;
- personal safety and security training;
- communication arrangements, including whether a satellite phone is needed;
- physical environment (such as work temperatures, altitude);
- accommodation standards and availability;
- arrangements for travel within the country, including driving; and
- emergency arrangements, including local medical contacts, first-aid kits and emergency evacuation provisions.

General workplace issues

> **TIP**
>
> **Advice for workers driving alone in the UK**
>
> The following advice from a UK recovery service is useful for lone workers in the event of a vehicle breakdown.
>
> **What to do if you break down**
>
> Breaking down can be dangerous, particularly if you're on a fast road, motorway or dual carriageway. But remember that the 'hard shoulder' (safety lane) is only for emergencies, not for making calls, having a stretch or toilet stops.
>
> If your car has broken down, here's what to do to stay safe before you call for help and while it's on its way.
>
> 1 Make sure you're in a safe place
> 2 Put your hazard lights on
> 3 Stay well away from moving traffic
> 4 Wear a reflective jacket
> 5 Don't put a warning triangle on the hard shoulder
> 6 Call your roadside assistance company[12]

International travel and lone working

Assessing the risks of travel for work can be challenging and can involve an interaction between individual responsibilities and employer responsibilities. The environment into which the worker is entering cannot be controlled in the same way as on a fixed site. Instead the focus is on making the worker safe through a combination of individual and organisational actions.

Employer actions

International travel for work is now commonplace. Inevitably, the workers travelling internationally reflect the diversity in society, including women, transgender workers, older workers, students on internships or assignments and individuals with disabilities or long-term health conditions.

General arrangements should include the following:

- Ensure clearly defined internal escalation paths to ensure rapid escalation to manage incidents.
- Identify the stakeholder who will need to manage the organisational response, with clearly identified roles and responsibilities.
- Workers and destinations must be individually risk assessed.
- Decide whether security risks can be avoided, reduced, transferred elsewhere or tolerated.
- Health risks are generally best avoided or reduced.
- Risks alter with time: current information is needed and risk assessments updated as events change the environment and conditions for the traveller.

Crisis management

A three-level crisis management framework for international travel is helpful, including:[13]

- pre-crisis planning, screening, training and sourcing of accommodation;
- in-crisis decision-making, process and logistics management associated with the crisis, formation of a crisis management team and inter-agency co-ordination, and

- post-crisis support for the individual as well as organisational learning that feeds back into crisis preparation.

External assistance may be needed from specialist companies to provide in-country medical, evacuation and emergency support.

Individual actions

A dynamic risk assessment supports generic risk assessments that have been completed in advance for the well-known and routine risks.[14] Dynamic risk assessments enable the individual to be able to recognise emerging risks and to be in a position to make a decision about a situation that is not covered in a generic assessment.

Dynamic risk assessment is a systematic process and encourages consideration of the options available in an evolving situation. It provides an approach that challenges jumping to a solution without thinking through the options.

> **TIP**
>
> **Important information for individuals**[15]
>
> Useful documentation to copy prior to travel could include:
>
> - Passport and visas
> - Proof-of-life documents for countries affected by kidnapping and extortion (questions that only the traveller could answer if they were alive)
> - Driving licence
> - Vehicle documents (if using own vehicle)
> - Travel tickets
> - The itinerary
> - Personal crisis plan
> - Health insurance/how to assess medical advice
> - Credit cards/or other means of accessing funds
> - Emergency phone numbers
> - Copy of any repeat prescription
> - Means of sourcing required medications
> - Vaccination history
> - Blood group
> - Special medical instructions
> - Address list, including GP and dentist

> **CASE LAW**
>
> **Possible consequences of not dealing with foreseeable risk**
>
> In *Durnford v Western Atlas International Inc*[16] a worker brought a claim against his employer when he suffered a slipped disc due to an inadequate minibus that was supplied by his employer to transport him to the third-party premises where he was working while abroad. The Court of Appeal found that the employer had caused the worker to travel in conditions that were so extreme that there was a foreseeable risk of any person of an ordinary level of physical robustness succumbing to an injury.

KEY POINTS

- Lone working is commonplace and managed as part of everyday business. However, where the risks justify it, more formal measures need to be put in place to safeguard lone workers.

- Lone working hazards include the potential for accidents or emergencies arising out of the work and inadequate provision of first aid, sudden illnesses, inadequate rest, hygiene and welfare facilities and physical violence from members of the public and others.

- Risk factors will vary on a case-by-case basis and will depend on the susceptibilities of the individual workers, the working environment and what resources are nearby to help in the event of an emergency.

- While technologies can and do help, they are not foolproof and dedicated lone worker devices may be needed.

- Travel for work can pose particular lone working hazards, and the interplay between organisational and individual responsibilities needs to be clear.

- When lone workers are travelling anywhere for work, the focus should be on making them safe through a combination of individual and organisational actions, including dynamic and generic risk assessments.

References

[1] HSE, 'Lone workers' (2021) (www.youtube.com)

[2] TUC, *Lone working: A guide for safety representatives* (2009) (www.tuc.org.uk)

[3] NHS Staff Council, *Improving the personal safety for lone workers: A guide for staff who work alone* (2018) (www.nhsemployers.org)

[4] HSE, *Protecting lone workers: How to manage the risks of working alone* (INDG73, 4th revision, 2020) (www.hse.gov.uk)

[5] HSE prosecution case 4262529 (www.hse.gov.uk)

[6] Queen Mary University of London Health and Safety Directorate, 'Lone working case studies' (www.hsd.qmul.ac.uk)

[7] The Suzy Lamplugh Trust (www.suzylamplugh.org)

[8] Adapted from *Big Hospitality* magazine (7 August 2016)

[9] Adapted from *Occupational Safety and Health Administration (OSHA) standard 1910.120* (www.osha.gov)

[10] HSE, 'Example risk assessment for an office-based business' (2019) (www.hse.gov.uk)

[11] *Federación de Servicios Privados del Sindicato Comisiones Obreras v Tyco Integrated Security SL* [2015] EUECJ C-266/14

[12] AA, 'Broken down? Here's how to stay safe' (www.theaa.com)

[13] Institution of Occupational Safety and Health and International SOS Foundation, *Managing the safety, health and security of mobile workers: an occupational safety and health practitioner's guide* (2016) (https://iosh.com)

[14] For a review, see National Fire Chiefs Council, 'National Operational Guidance' (www.ukfrs.com)

[15] See note 13

[16] *Durnford v Western Atlas International Inc* [2003] EWCA Civ 306

8.5: Slips and trips

Syllabus outline

In this section, you will develop an awareness of the following:
- Common causes of slips and trips, including: uneven or unsuitable surfaces, trailing cables, obstructions in walkways, unsuitable footwear
- Main control measures for slips and trips, including: non-slip surfaces, maintenance, housekeeping

Context

Slips and trips are usually the most common cause of workplace injury. In workplaces where there are also a large number of non-workers (eg hospitals, shops), slips and trips also account for a large number of injuries to members of the public. Not only do they cause, on average, over a third of all major injuries, but they can also lead to other types of accidents, such as falls from height.[1]

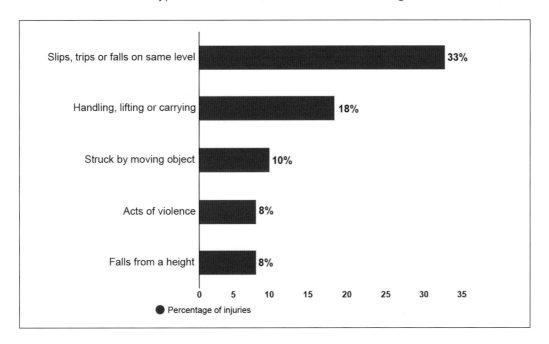

Figure 1: Non-fatal injuries reported by UK employers under RIDDOR by most common accident kinds (typical year)[2]

Of the accidents reportable under the Reporting of Injuries, Diseases and Dangerous Occurrences Regulations 2013 (RIDDOR), slips and trips accounted for:[3]

- 20% of reportable injuries to workers;
- two fatalities per year;
- 50% of all reported accidents to members of the public that happen in workplaces;
- cost to employers of £512 million per year (lost production and other costs);
- cost to health services of £133 million per year; and
- more major injuries in manufacturing and in the service sectors than any other cause.

General workplace issues

> **TIP**
>
> **What does the law say?**
>
> In England, Wales and Scotland:
>
> - The Health and Safety at Work etc Act 1974 (HASAWA) requires employers to ensure the health and safety of all employees and anyone who may be affected by their work, so far as is reasonably practicable. This includes taking steps to control slip and trip risks.
> - Employees have a duty not to put themselves or others in danger and must use any safety equipment provided.
> - The Management of Health and Safety at Work Regulations 1999 require employers to assess risks (including slip and trip risks) and, where necessary, take action to address them.
> - The Workplace (Health, Safety and Welfare) Regulations 1992 require floors to be suitable, in good condition and free from obstructions. People should be able to move around safely.
>
> In Northern Ireland, the equivalent pieces of legislation are:[4]
>
> - Health and Safety at Work (Northern Ireland) Order 1978;
> - Management of Health and Safety at Work Regulations (Northern Ireland) 2000; and
> - Workplace (Health, Safety and Welfare) Regulations (Northern Ireland) 1993.

What are slips, trips and falls?

Slips and trips on the same level account for a significant proportion of workplace accidents. These lead to losses in terms of pain and suffering, time off work and insurance claims. Slips, trips and falls are often grouped together, as in combination they account for the majority of workplace accidents.

Falls from the same height (or same level), covered in 8.2: Working at height, are often difficult to distinguish from slips or trips. Figure 2 provides a general guide to the differences.

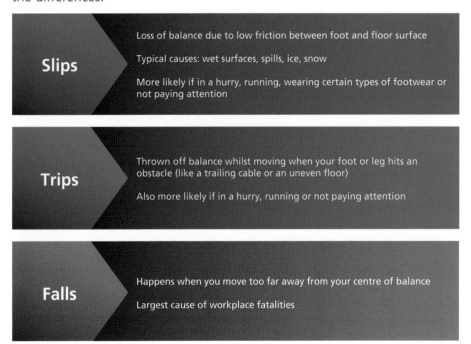

Figure 2: General guidance on slips, trips and falls[5]

8.5 Slips and trips

In terms of numbers of accidents in the UK, non-fatal slips, trips and falls on the same level are more than three times more frequent than non-fatal falls from height. That is, slips, trips and falls are the largest contributor to reported, non-fatal injuries at work.

Generic control measures

Good housekeeping is essential in preventing slips, trips and falls due to hazards such as spillages, trailing cables and other obstructions on floors. Workplace design issues are also important in avoiding trip hazards caused by changes in level, such as kerbs and steps.

Suitable floor surfaces, adequate lighting and suitable footwear are further generic factors to be considered in accident prevention.

8.5.1 Common causes of slips and trips

Slips

> **DEFINITION**
> *Slips are a loss of balance caused by too little friction between your feet and the surface you walk or work on.*

Slips can be caused by a range of factors, including wet surfaces, spills or weather hazards such as ice or snow. Slips are more likely to occur when workers hurry or run, wear the wrong kind of footwear or do not pay attention to where they are walking.

Causes of slips

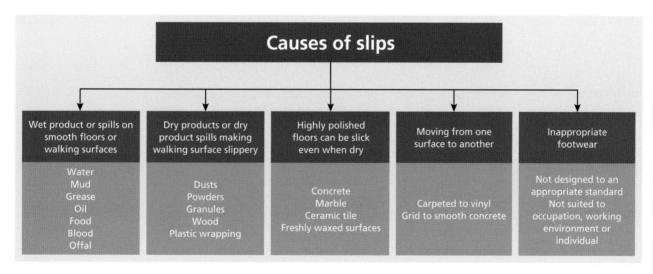

Figure 3: Causes of slips

Common injuries from slips

The most common injuries from slips are:

- sprains and strains;
- bruises and contusions;
- fractures; and
- abrasions and cuts.

The most common body parts affected are the:

- knee, ankle and foot;
- wrist and elbow;
- back;
- shoulder;
- hip; and
- head.

> **APPLICATION**
>
> In your own workplace, identify five areas where slips might occur.
>
> What practical measures are in place to mitigate against these risks?

> **CASE STUDY 1**
>
> **Leaking chiller cabinet causes supermarket slip risk**
>
> A health and safety officer visited a large retail organisation and noticed two large areas of wet flooring near to some chiller cabinets. An attempt to mark the areas and exclude people had been made using yellow cones, wire handbaskets and plastic trays.
>
> These measures were unsuccessful as people were still walking through the water. The officer decided that a formal approach was necessary and returned later that afternoon to investigate further.
>
> On arrival it was clear that the situation had not improved and the chiller cabinet condensate drain had been leaking onto the floor all morning.
>
> Advice was given to the manager about practical measures to keep the floor dry until the leak was repaired, using absorbent materials such as paper hand towels to soak up/contain the water.
>
> By the time the officer left the store the floor was dry and the situation was being effectively managed.
>
> The company had a spillage procedure. This clearly set out the ways in which the manager should have managed the leak and the wet floor but had not done so.[6]

Trips

> **DEFINITION**
>
> *Trip*
>
> - *Foot or lower leg hits object and upper body continues moving, resulting in loss of balance; or*
> - *Stepping down to lower surface and losing balance.*[7]

Causes of trips

Figure 4: Trips caused by trailing cables

There are many potential causes of trips. These may include:

- uncovered hoses, cables, wires or extension cords across aisles or walkways;
- clutter or obstacles in aisles, walkways and work areas;
- open cabinet, file or desk drawers and doors;
- changes in elevation or levels;
- unmarked steps or ramps;
- rumpled or rolled-up carpets/mats or carpets with curled edges;
- uneven walking surfaces such as raised paving stones and poorly constructed surfaces (see Figure 5), potholes on tracks and walkways and slow-growing tree roots lifting paving slabs;

Figure 5: Uneven surfaces

- thresholds or gaps, for example thresholds between doorways where there may be different surfaces. A good example of gaps that could cause trips/falls is that when some trains pull in at a platform, there is often a gap between the platform and the carriage step;
- missing or uneven floor tiles or bricks or other uneven surfaces;
- debris and accumulated waste materials if stored incorrectly (for example, if stored on or near a main walkway);
- trailing cables, pallets and tools in gangways/walkways;
- objects protruding from walking surface (for example, tree roots breaking the surface of a pavement);
- pavement/kerb drops;
- speed bumps/traffic calming measures;
- wheelchair ramps and kerbs; and
- improper cleaning methods and products causing surfaces to become slippery.

Environmental conditions that can contribute to trips include:

- poor lighting;
- glare;
- shadows;
- excess noise or temperature;
- fog or misty conditions; and
- poor housekeeping such as blocked walkways.

Other things that may aggravate the risk of trips include:

- carrying or moving cumbersome objects or simply too many objects at one time;
- not paying attention to surroundings or walking while distracted;
- taking shortcuts;
- being in a hurry and rushing; and
- bulky personal protective equipment, including improper footwear.

APPLICATION

In your own workplace, identify three areas where trips might occur.

What practical measures are in place to mitigate these risks?

Fall on the level

DEFINITION

*A **fall on the level** is a slip or trip that results in a worker hitting an object or the floor at the same level.*

The distinction between falls on the level and falls from height (as discussed in 8.2: Working at height) may seem a subtle one but it is meant to capture falls that have a slip or trip as their origin.

Figure 6: Trip with potential for a fall on the level

8.5.2 Main control measures for slips and trips

We have outlined the hazards associated with slips, trips and falls on the level; now we will look at the main control measures that can be applied to managing the risks. After that, we will give examples from three specific industry sectors: health and social care, public transport and kitchens.

Generic control measures

Clean up spills
Clean up spilled liquids, powders, etc immediately. If a liquid is greasy, make sure a suitable degreasing agent is used. After cleaning, dry the floor as much as possible. Use appropriate barriers and signs to tell people the floor is still wet and arrange alternative bypass routes. If cleaning is done once a day, it may be possible to do it last thing at night, so the floor is dry for the start of the next shift.

Avoid cables trailing across walkways
Use cable covers to securely fix cables to surfaces. If this is not possible, use high-visibility cables so that they can be seen easily by pedestrians. When carrying out repairs, alterations, etc, consider the use of battery-operated tools to avoid cables trailing across the floor. You should ensure that contractors' work is effectively managed so they follow the same rules.

Good housekeeping
Try to ensure work materials and equipment do not obstruct pedestrian traffic routes. Keep areas clear at all times through good standards of housekeeping and supervision. Remove rubbish and do not allow it to build up.

Fix floor coverings
Stop these becoming a hazard. This includes carpet tiles, vinyl coverings and mats. These need to be securely fixed and must not have curling edges. Preferably mats should be fixed onto the floor.

Non-slip surfaces
These should be installed where needed. Where that is not possible, consider keeping floors dry if wetness causes a problem. Use appropriate cleaning methods and avoid polishing if that is possible.

Suitable footwear
Make sure appropriate footwear is available and worn.

Warnings
Warn of risks of slips and trips by using signs, staff emails or noticeboards.

Doormats
Use absorbent doormats when there are wet or icy conditions forecast. These can absorb water and can help reduce spreading moisture onto the floor areas beyond doorways.

Steps
Provide adequate lighting on steps and add high-visibility tread edges to highlight the edges (for example, white/yellow painted or reflective edge to step). Provide handrails and use floor markings and signage. Signpost workers and visitors to lifts if they are carrying goods or luggage.

Figure 7: Stairs with handrails, painted stair edges and drainage

Pedestrian traffic routes

Pedestrian traffic routes in workplaces should be properly marked and procedures should be in place to prevent materials and other items from infringing into marked areas.

Maintenance and cleaning operations are particularly prone to causing slip and trip hazards. Strict procedures should be in place to exclude non-maintenance personnel from risk areas and/or suitable precautions should be taken to help reduce risk. Ice on paths and entrance steps can present a particular problem in winter.

Floors and surfaces of traffic routes should be of suitable construction. In particular, they should have no dangerous holes or slopes, they should not be uneven or slippery, and they should have effective means of drainage. Floors and traffic routes must be kept free of obstructions and from any article or substance that could cause a person to slip or trip.

Industry examples

Health and social care

Health and social care is the treatment of ill-health and medical conditions in hospitals, health centres and the community.

A particular concern in these environments is that patients or residents have fallen down a flight of stairs, resulting in serious injury or death. There are a number of elements that are specifically relevant to patients and residents in these environments and should be considered in the individual's care plan.

> **CASE STUDY 2**
> **Individual patient assessment on using stairs**
>
> A patient was identified as being at high risk of falls. The risk assessment identified that the stairs between the ground and first floor presented a real risk of harm. However, the individual had the capacity to understand the risk of using the stairs. The patient was mobile and wished to use the stairs. As a result of the assessment, improvements were made to provide handrails at the appropriate height on both sides and adequate lighting. The steps and floor coverings were in good condition and did not create a slip or trip risk. The assessment concluded that allowing the stairs to be used allowed for dignified living in the care setting, but the individual's condition, capacity to understand the risk and the support of their family were regularly reviewed.[8]

Public transport

All public transport (buses, trains, trams and planes) poses challenges when passengers move in great numbers. Additional hazards include moving walkways and escalators. Messages to the public through posters, public address systems, floor signage and other signage are key.

Distractions also contribute to accidents. On public transport these can include using mobile phones, eating and drinking on the move, looking for travel information, running and rushing with luggage.

General workplace issues

Kitchens

People working in kitchens and food services are quite likely to be injured through slips and trips. The pace of work makes accidents more likely to occur. Simple measures can reduce kitchen slip and trip accidents and injuries dramatically:

- Most slip injuries happen on wet floors or floors contaminated with food debris or fluids such as oil.
- Most trips are due to poor housekeeping.
- Planning ahead can help to avoid these problems during busy periods, when the pace of work increases.

Wet and contaminated floors are often the cause of slips. These can be stopped by:

- maintaining equipment to prevent any leaks of oil, water, etc;
- having a system for promptly reporting and using splashguards or edged work surfaces to contain spillages;
- using lids and covers for pans and containers, especially when they are being carried;
- having a well-maintained extraction and ventilation system to remove steam and grease before it can be deposited;
- using drainage channels and drip trays to carry water, stream drips and waste away from tilting kettles, pans and other equipment;
- positioning any 'messy' operations away from walkways and thoroughfares; and
- preventing water being walked into the kitchen or service area on people's shoes from outdoors or indoors by providing suitable floor mats.

CASE STUDY 3
Fast food restaurant accident

Workers were cleaning up after a busy night at a fast food restaurant. The cook was walking back to the kitchen from the pot-wash area, over a floor that had just been wet mopped. As he walked past the deep fat fryer, the cook suddenly slipped on the still damp floor. He instinctively reached out to try to break his fall, pulling over the electric deep fat fryer.

The fryer toppled over, spilling its entire contents, 35 litres of boiling hot oil, onto the cook and the floor. As the oil came into contact with the water residue on the floor, thick black smoke was produced, which set off the smoke alarms, making the situation worse.

Surrounded by hot oil, the cook couldn't get up from the floor. Each time he tried he slipped back. Eventually, a co-worker succeeded in sliding him out of the spilt oil, burning his own hands in the process. The cook suffered extensive burns to his ankles, legs, buttocks and chest and needed skin grafts. Another worker in the vicinity of the spilt oil also received severe burns to her right leg and ankle and also needed skin grafts. The company was prosecuted and fined after pleading guilty to two health and safety offences.[9]

8.5 Slips and trips

APPLICATION

Test your knowledge by identifying the slip, trip and fall hazards in this diagram.

A suggested answer appears at the end of 8.5 Slips and trips.

KEY POINTS

- Slip and trips (and falls on the level) are a common cause of accidents in the workplace.
- Broad categories of slip hazards are: wet product or spills on smooth floors or walking surfaces; dry products or dry product spills making walking surface slippery; highly polished floors that are slick even when dry; and moving from one surface to another.
- Some of the causes of trips are: uncovered hoses, cables, wires or extension cords across aisles or walkways; clutter and obstacles in aisles, walkways and work areas; unmarked steps or ramps; rumpled or rolled-up carpets/mats or carpets with curled edges; and irregularities in walking surfaces.
- Aggravating issues common to both slips and trips are: carrying or moving large and heavy objects or simply too many objects at one time; not paying attention to surroundings or walking while distracted; taking shortcuts; and being in a hurry and rushing.
- Good original design, efficient housekeeping and promptly cleaning up spills are all effective control measures.

APPLICATION ANSWER

The slip, trip and fall hazards you identified could include:

- Fall into sewer
- Fall from ladder
- Slip on cables lying on the ground
- Slip on surface due to inappropriate footwear
- Slip on ladder rungs
- Slip on wet surface around maintenance area

General workplace issues

- Trip on cables
- Trip on cone
- Trip on sewer cover
- Trip on cordless drill
- Trip on box
- Walk into open window, leading to a fall

References

[1] HSE, 'Non-fatal injuries at work in Great Britain' (www.hse.gov.uk)

[2] See note 1

[3] HSE, 'Slips and trips: Why does it matter?' (www.hse.gov.uk)

[4] Explanatory notes to the legislation state that these items are equivalent to those in Great Britain

[5] Adapted from Occupational Safety and Health Administration (OSHA), 'Slips, trips and falls' (www.osha.gov)

[6] Adapted from HSE, 'Getting to grips with slips and trips' (www.hse.gov.uk)

[7] Source: OSHA (www.osha.gov)

[8] HSE, 'Reducing the risk of falls on stairs' (www.hse.gov.uk)

[9] Adapted from HSE, 'Serious burn injuries lead to serious fine for fast food restaurant' (www.hse.gov.uk)

8.6: Safe movement of people and vehicles in the workplace

> **Syllabus outline**
>
> In this section, you will develop an awareness of the following:
> - Hazards to pedestrians:
>   - being struck by moving, flying or falling objects
>   - collisions with moving vehicles
>   - striking against fixed or stationary objects
> - Hazards from workplace transport operations (vehicle movement, non-movement)
> - Control measures to manage workplace transport:
>   - safe site (design and activity)
>   - suitability of traffic routes (including site access and egress pedestrian-only zones and crossing points)
>   - spillage control
>   - management of vehicle movements
>   - environmental considerations: visibility/lighting, gradients, changes of level, surface conditions (use of non-slip coatings)
>   - segregating pedestrians and vehicles and measures to be taken when segregation is not practicable
>   - protective measures for people and structures (barriers, marking signs, warnings of vehicle approach and reversing)
>   - site rules (including speed limits)
>   - safe vehicles
>   - suitable vehicles
>   - maintenance/repair of vehicles
>   - visibility from vehicles/reversing aids
>   - driver protection and restraint systems
>   - safe drivers
>   - selection and training of drivers
>   - banksman (reversing assistant)
>   - management systems for assuring driver competence, including local codes of practice

Workplace transport is a significant cause of accidents at work. The hazards to pedestrians and to drivers are clear: being struck by a moving, flying or falling object; collisions; and striking a fixed or stationary object. There are a wide range of vehicles used in the workplace, including lift trucks, mobile cranes, goods vehicles, car transporters, agricultural tractors and self-propelled mowers, among many others.

General workplace issues

And yet the control measures fall into three universal categories: safe workplace design, safe vehicle and safe driver. While the everyday measures are well known (one-way systems, speed ramps and training, to name a few), the challenge is to first identify that they are appropriate and then properly and consistently implement them.

As we mentioned, a wide range of vehicles operate in different types of workplace:

- Lorries and vans are frequently encountered in the wholesale, retail and logistics industries (as well as on any premises where goods are delivered or collected by road transport).
- Dumper trucks, mobile cranes and excavators are encountered on construction sites.
- Tractors and other agricultural vehicles are found on farms and some construction sites.
- Forklift trucks can be found in factories and warehouses and on construction sites.
- Cars used for business are found in all types of workplace.

Counterbalanced forklift trucks (and other mobile lifting equipment) pose particular hazards, and there are special training requirements for operators of these vehicles.

Wherever vehicles operate in a workplace, the safety of both drivers and pedestrians needs to be considered.

Every year, there are serious accidents involving vehicles used in the workplace, some of which result in people being killed. In a typical year, nearly a quarter of workplace fatalities are due to being struck by a vehicle.

> **TIP**
>
> In addition to the general duties that employers have under the Health and Safety at Work etc Act 1974 to both their employees and those not in their employment, other specific regulations include requirements about the safe movement of people and vehicles in the workplace. These are as follows.
>
> **Construction (Design and Management) Regulations 2015**
>
> Construction sites have particular hazards due to their temporary nature and the use of works vehicles in close proximity to construction workers. The Regulations cover hazards specifically relating to traffic routes and vehicles, summarised here.
>
> Traffic routes
>
> - Construction sites must be organised in such a way that pedestrians and vehicles can move without risks to health or safety.
> - Traffic routes must be suitable for the persons or vehicles using them, sufficient in number, in suitable positions and of sufficient size.
> - Traffic routes must be indicated by suitable signs where necessary, regularly checked and properly maintained.
> - No vehicle is to be driven on a traffic route unless that traffic route is free from obstruction and permits sufficient clearance.
>
> Vehicles
>
> - Employers must prevent or control the unintended movement of any vehicle.
> - No person must ride on any vehicle other than in a safe place provided for that purpose (a proper seat or workstation).

- No one shall be permitted to remain on a vehicle during the loading or unloading of any loose material (unless a safe place of work is provided for that purpose).
- Suitable measures must be taken to prevent a vehicle from falling into any excavation or pit, or into water, or overrunning the edge of any embankment or earthwork.

Workplace (Health, Safety and Welfare) Regulations 1992

In addition to the requirements specific to construction sites, there are requirements that apply to all workplaces for maintaining a safe working environment around vehicles.

These Regulations cover the condition of floors and traffic routes:

- Every floor in a workplace and the surface of every traffic route in a workplace shall be of a construction that is suitable for purpose.
- The floor, or surface of the traffic route, shall have no hole or slope, or be uneven or slippery so as to expose any person to a risk to their health or safety.
- Every floor shall have effective means of drainage where necessary.
- Every floor in a workplace and the surface of every traffic route in a workplace shall be kept free from obstructions and from any article or substance that may cause a person to slip, trip or fall.
- Suitable and sufficient handrails and, if appropriate, guards shall be provided on all traffic routes that are staircases, except in circumstances in which a handrail cannot be provided without obstructing the traffic route.

8.6.1 Hazards to pedestrians

Before turning to the specific hazards faced by people moving around in a workplace, there is one particular term that needs to be defined.

> **DEFINITION**
> *Traffic route* means a route for pedestrian traffic, vehicles or both, and includes any stairs, staircase, fixed ladder, doorway, gateway, loading bay or ramp.

'Traffic' is normally associated with vehicles on the road. However, the term 'traffic route' in the context of health and safety applies equally to pedestrian traffic as to vehicular traffic.

We will discuss traffic routes that are used solely by pedestrians at various points. Traffic routes used by vehicles will be addressed in 8.7: Work-related driving.

Moving, flying or falling objects

Pedestrians can be struck by moving, flying or falling objects. There are many examples, including pedestrians being struck by a moving vehicle or flying objects ejected from mobile work equipment, such as material from wood chippers or road planers. Examples would be steel coils falling off trailers or pallets splitting and falling off the forks of a forklift truck. In warehousing, unloading racking can disturb adjacent products that are stored at height that then fall onto pedestrians.

Collisions with moving vehicles

Pedestrians are also vulnerable when they come into contact with moving vehicles. The harm can occur from being struck by a vehicle. This includes incidents when the driver fails to see the pedestrian when manoeuvring, especially when reversing. It also includes those occasions when pedestrians intentionally approach a working vehicle to get the attention of the driver. Many large vehicles have a more restricted view of pedestrians the closer the pedestrian is to the vehicle.

Collisions can occur when a vehicle loses control, either by lack of driver attention (mounting a kerb) or when the vehicle loses grip; for example, skidding on a spillage or in icy conditions.

Striking against fixed or stationary objects

When pedestrians are moving about, there is always the chance that they might 'bump into things' due to inattention or distraction. Some situations, however, such as restricted headroom, may make such events more likely or more serious, and every attempt should be made to eliminate such hazards or bring them to people's attention. Signs may be required and, in some instances, surfaces may need to be padded. In certain work situations, workers should be required to wear 'bump caps'.

> **DEFINITION**
>
> ***Bump caps** are designed to provide a low level of protection. They protect the wearer against minor cuts and bruises when they are at risk of knocking their head against a stationary object (but they will not provide the same level of protection as a hard hat).*
>
>
>
> **Figure 1:** Bump cap
>
> *They are generally made from an inner plastic shell that is covered by material and usually take the form of a baseball cap. They are also, therefore, good at providing some protection from the sun.*

Where there is a chance that someone could walk into a window or a transparent or translucent surface in a wall, partition, door or gate, such surfaces should be made of safety material or protected against breakage. They should be appropriately marked or incorporate features to make their presence apparent.

Windows that open outwards at ground floor level should be kept closed whenever possible if they protrude onto a footpath. This will reduce the risk of people walking into them, especially if they are distracted and not looking where they are walking.

> **TIP**
>
> Pedestrians may also be at risk from other workplace hazards such as noise, airborne contaminants, extremes of temperature and electricity. Risk assessments should consider all who might be affected by these hazards, including those who might simply be 'passing through' or who are visitors or contractors.

8.6.2 Hazards from workplace transport operations

These are divided into:

- vehicle movement: including loss of control leading to collisions or overturning of vehicles, coupling and uncoupling and unstable and dangerous loads; and
- non-vehicle movement: including loading and unloading and vehicle maintenance.

Vehicle movement

Loss of control

Loss of control of a vehicle can lead to instability (causing a partial or complete overturn of the vehicle) or collision. A number of possible reasons may account for loss of control:

- Defective brakes, which could be due to lack of maintenance and/or mechanical failure.
- Defective steering, which could be due to lack of maintenance and/or mechanical failure.
- Tyre failure or tyres at the wrong pressure: underinflated or overinflated tyre pressures can cause vehicles to sit or tilt unevenly and so add to instability.
- Driving too fast, particularly when cornering.
- Harsh braking, which can cause a vehicle and/or its load to become unstable.
- Driving on steep gradients, either across the gradient (causing lateral instability) or with the gradient (causing longitudinal instability).
- Stability when working on gradients may be affected by the load's centre of gravity significantly shifting while being carried. For instance, a loaded trailer may have a higher centre of gravity than an unloaded one, reducing the slope that may be traversed safely. Particular techniques may be required when dealing with slopes and ramps. For example, a forklift truck carrying a load should reverse down a slope, whereas an unloaded one should drive forward.
- A forklift truck should carry its load so that the forks are just off the ground. This is both to keep the centre of gravity at a low level and to prevent the forks being at head height in the event of a collision with a pedestrian.
- Changes in level can cause instability, for example when a forklift truck inadvertently drives off the edge of a loading bay. Potholes or rough terrain can cause similar instability problems.

Accidents involving reversing assistants, crushed between a reversing tractor unit and a trailer, are also common.

Coupling/uncoupling

Incorrect procedures for the coupling and uncoupling of tractor units to and from trailers on large articulated vehicles have resulted in serious accidents. Falls from height while attaching or detaching pneumatic and electrical connections have already been mentioned. Other accidents have involved vehicles moving unexpectedly or trailers collapsing when the landing legs have not been properly locked in position.

Coupling and uncoupling of lorries and trailers should take place on firm, level ground (sufficient to support the weight of the trailer's landing legs). The whole area, especially the rear of the tractor unit, should be well lit. Drivers should be suitably trained in the correct procedures and properly supervised.

Unstable/dangerous loads

Linked to unsafe 'loading/unloading' is driving with an unstable or dangerous load ('live' loads). For example, bulk bags or other materials swinging from raised forks can be particularly hazardous. Pallets that are not secured or bound together with steel bands or with shrink-wrap can shift during transit. Long lengths of pipe on the forks and carried over uneven ground can destabilise lift trucks.

Figure 2: Building materials swinging from uplifted forks could become an unstable load if the lift is not planned correctly

Non-vehicle movement

Loading/unloading

One of the main hazards associated with the loading and unloading of vehicles is workers falling from the vehicle, particularly flatbed lorries and trailers. Many workers are seriously injured or killed each year after falling from vehicles (cabs or trailers) during either loading or unloading activities (this includes while loads are being secured). The resulting health effects can either be acute, such as broken arms and legs and minor back sprains, or more chronic and/or life-changing, such as severe back injuries and cognitive disabilities resulting from head injuries.

Other issues specific to loading include the following:

- Uneven loading can affect stability if, say, the load is all on one side of the trailer.
- Overloading could shift the centre of gravity in a trailer, which could cause the vehicle to become unstable, especially going around corners. It also puts excessive pressure on tyres and brakes; this could affect the driver's ability to brake in time or steer correctly. In addition to this, overloaded vehicles will cause excessive wear to infrastructure such as roads and bridges.

8.6 Safe movement of people and vehicles in the workplace

> **TIP**
>
> **Checklist**
> - Are loading/unloading operations carried out in an area away from passing traffic, pedestrians and others not involved in the loading/unloading operation?
> - Are the load(s), the delivery vehicle(s) and the handling vehicle(s) compatible with each other?
> - Are loading/unloading activities carried out on ground that is flat, firm and free from potholes?
> - Are parking brakes always used on trailers and tractive units to prevent unwanted movement, for example when coupling vehicles?

Vehicle maintenance

Vehicle maintenance could fall under both vehicle and non-vehicle movement. For example, if during maintenance an engine is switched on, the vehicle could move if the brakes and/or wheel chocks are not applied. However, the majority of the maintenance issues relate to 'non-movement' of the vehicle; we will now look at this in more detail.

Figure 3: Wheel chocks can help to stop vehicle movement during maintenance activities

Here are some of the main hazards associated with vehicle maintenance:

- Vehicle falling from stands or jacks, especially if an incorrect piece of lifting equipment has been selected. You should always check that any piece of equipment used for lifting vehicles can support the weight of the vehicle and has been designed for this activity. Vehicles could also fall from lifting platforms if they have been incorrectly positioned; it is vital, therefore, that the vehicle is stable before lifting begins.
- Draining or repairing fuel tanks: there is a risk that this activity could lead to a fire and/or an explosion. You should, therefore, try to ensure that there is a suitable area available for this activity and never carry out this activity in or near to vehicle repair pits. You should also follow any safe systems of work for this activity.

- Exposure to hazardous substances, such as isocyanates in spray paints, dusts from sanding operations, asbestos (some older cars may still have parts such as clutch pads that contain asbestos), mineral oils (engine oils), battery acid. It is essential that the correct personal protective equipment (PPE) is worn when carrying out these activities. For example, when:
  - spray painting, use air-fed respiratory equipment;
  - handling batteries, wear gloves, safety glasses/goggles and aprons to protect the body; and
  - sanding, use a face mask that will offer protection from the dust being produced.

  As well as wearing PPE, it is essential to ensure that workers have good hygiene standards, such as washing hands after handling oils and regular cleaning of overalls.

- Being hit by exploding tyres during the inflation process; it is always a good idea to use a tyre cage when inflating commercial vehicle tyres to protect workers if an explosion does occur.
- Asphyxiation/loss of consciousness could happen when engines are left running/idling while maintenance activities are carried out. These types of activity should always take place in well-ventilated areas.
- Falls from height, especially when working on large commercial vehicles such as buses and trucks. When carrying out these activities, you should try to ensure that safe systems of work are in place and are followed.

8.6.3 Workplace transport control measures

Workplace transport safety can be seen as consisting of three elements: safe site, safe vehicle and safe driver.

Safe site

The design of traffic routes is a particularly effective control measure. Traffic routes should be sufficient in number and suitable in terms of dimensions and surfaces used for the number and type of vehicles that will be using them. Changes in level and obstructions should be avoided or clearly marked, steep slopes should be avoided if possible and speed reduction (traffic calming) measures should be incorporated where appropriate.

Pedestrians and vehicles should be separated as much as possible, ideally with physical barriers between vehicle and pedestrian traffic routes.

Where permanent physical barriers are impracticable, then traffic routes should be clearly delineated by road markings. All traffic routes should be well lit, particularly where they coincide, such as at crossing points. In this respect, crossing points should be made clear to drivers and pedestrians by road markings, speed humps and/or signs. Exit doors in use by pedestrians should not open directly onto a vehicle route.

> **CASE STUDY 1**
> **Worker killed by moving farm vehicle**
>
> A Somerset farm was fined more than £60,000 after a worker was killed in an incident involving a moving vehicle. On the day of the accident, a self-employed farm worker was assisting with jobs including animal welfare checks and mucking out a large pig shed close to a telehandler (telescopic materials handler).
>
> As the telehandler, fitted with a bucket, was scraping muck from the floor using short manoeuvres, the worker was struck and killed as the vehicle reversed.
>
> Speaking after the case, the investigating inspector said: "Being struck by a moving vehicle has been the biggest cause of workplace fatalities on farms for several years. Farmers should properly assess their workplace transport risks and separate people and vehicles where reasonably practicable. HSE will not hesitate to take appropriate enforcement action against those that fall below the required standards".[1]

Other measures include:

- Using one-way systems wherever possible.
- Eliminating the need for vehicles to reverse (for example, by providing large turning circles) or providing designated reversing areas that are kept free of pedestrians.
- Segregating vehicle and pedestrian routes by physical barriers as much as possible or by clearly marking routes with painted lines and signs.
- Designing traffic routes so that drivers and pedestrians have a clear view, particularly at crossing points. Barriers should be installed at places where pedestrians might be tempted to cross but where it could be unsafe to do so. Where vehicles such as forklift trucks exit or enter buildings, it may be possible to incorporate transparent plastic doors so that drivers can see what is on the other side of the doors and pedestrians can see when vehicles are about to emerge or enter.
- Providing adequate lighting on traffic routes and at crossing points in particular.
- Installing mirrors on traffic routes to allow drivers (or sometimes pedestrians) to see around blind corners and other obstructions.
- Incorporating features such as 'pedestrian refuges' in loading bays to prevent a person in the bay being crushed by a vehicle reversing into it.
- Ensuring that there are suitable warnings on traffic routes (such as signs, road markings and flashing lights at crossing points) to alert drivers and pedestrians to the presence of each other. Signs should also be used to advise of speed limits, road priorities, restricted access, one-way systems, restricted height or width and other dangers. Note that it is good practice for workplace road signs to conform to the same design and standards (the types of road sign seen on public roads should be used on roadways in a workplace).
- Maintaining traffic routes is required to keep them in good condition. Potholes should be filled as quickly as possible; these could affect the stability of transport using the route. Similarly, eliminating slippery surfaces, for example due to ice and oil, should also be carried out as soon as possible after the condition has been noted. Having procedures in place to manage spills, potholes and icy conditions should help to ensure that the required actions are taken as soon as is possible.
- Having a traffic management plan on large sites. This needs to consider delivery times and waiting areas for incoming vehicles, together with the order and timing of offloading and the onward travel times.

General workplace issues

> **APPLICATION**
>
> A large warehouse is being built that has 15 loading bays at the rear and a worker/public car park (with spaces for 200 cars) to the front and side of the building. Inside the warehouse will be an office area, a storage/packing area with floor-to-ceiling racking and a worker canteen with seating for 50 people.
>
> What are some of the issues that should be considered when designing traffic routes?

Safe vehicles

The use of vehicles designed to allow the maximum possible vision around the vehicle is crucial. This is shown most clearly in Figures 4 and 5, where good and bad design are shown to have a significant effect on the driver's sightlines. Extra equipment such as mirrors can help to see down each side of a vehicle and sometimes to the rear. Convex mirrors also allow a wider field of vision and can eliminate the 'blind spot'. However, drivers of large vehicles may still have a restricted visual field either low down at the front of the vehicle or to the rear of the vehicle.

Vehicles are often used both in the workplace and on the public highway. For example, in London the Direct Vision Standard sets down minimum requirements for reducing the number of accidents involving pedestrians and cyclists. These design standards improve standards for lorries in the workplace as well as on the public roads. They also specify side guards to prevent cyclists and pedestrians falling under the moving wheels of vehicles.

Figure 4: Good direct vision[2]

Figure 5: Poor direct vision[3]

438 Managing Health and Safety in the UK

> **CASE STUDY 2**
>
> **Restricted visibility for driver proves fatal**
>
> A worker was fatally injured when he was trapped between a JCB and a Volvo dump truck. The employer instructed the JCB driver to reverse his vehicle but failed to direct the reversing manoeuvre despite being in a position to do so. As the driver reversed down a steep slope into the shed, he felt a bump and realised that he had struck the rear of the dump truck. As he eased forwards, the victim's body fell to the ground between the vehicles.
>
> **Advice**
>
> When vehicles are reversing it is important that aids such as mirrors are fitted and maintained to allow the driver to see down the sides of the vehicle. To aid visibility at the rear of the vehicle, CCTV cameras can be fitted or a banksman can be used to direct the vehicle's manoeuvre.[4]

Rear-facing CCTV cameras with a screen in the cab are used in many vehicles to allow the driver to see if there is anyone behind the vehicle before it reverses. Similarly, vehicles may be fitted with detection systems to alert them to the presence and proximity of an object at the rear of the vehicle while reversing.

Here are a few other examples of how vehicle movement could be made safer:

- Use warning devices on vehicles that alert pedestrians to their presence, such as reversing alarms, horns and flashing beacons.
- Employ traffic marshals in areas where vehicles operate to oversee vehicle activities and to ensure that pedestrians who need to be in the same area are not put at risk.
- Drivers and pedestrians wear high-visibility clothing in areas where vehicles might be operating.
- Install driver protection measures, such as lap belts, seat belts and airbags, which can also have a significant effect on reducing the seriousness of any injury following a collision.
- Introduce speed limits.

The speed of a vehicle colliding with an object or person is a determining factor in the degree of damage or injury. Therefore, speed limits and their enforcement are an important control measure. This can be through warning signs, flashing speed signs and speed bumps.

Vulnerable objects, such as storage racking, should be suitably protected from vehicles; barriers may be required to protect other structures, particularly where a collision with racking or fuel tanks is possible.

Older buildings are often supported by cast-iron pillars, which are strong but brittle on impact. A forklift truck or other vehicle colliding with these structural features can have disastrous consequences if the pillars are damaged or weakened.

Maintenance of vehicles

This is also important to reduce the risk of the failure of brakes, steering, tyres, etc. Vehicles that are used on public roads should be subjected to annual roadworthiness tests. Vehicle servicing should take place according to the manufacturer's recommendations and, additionally, regular inspections should be made to ensure that the vehicle can be driven safely.

Pre-operational checks

For most vehicles, these should be made daily or at the start of each shift. A pre-use checklist can be useful for ensuring that all items are included and have been checked. In the case of a forklift truck, the vehicle should be examined more thoroughly by an authorised person periodically. If any item is found to be defective and the defect cannot be rectified immediately, the fault should be reported to a supervisor and the vehicle taken out of commission until it has been corrected.

> **TIP**
>
> It is good practice for each work vehicle to have an individual log where inspections and remedial measures are recorded. No operator or other person should make any repair or adjustment unless specifically trained and authorised to do so. Other work vehicles should undergo similar pre-use checks and regular examination, the period of which will be determined by manufacturer recommendations, level of duty, exposure to abnormal conditions and so on.

Many work vehicles are fitted with 'roll-over protective structures' (ROPS) and/or 'falling object protective structures' (FOPS). ROPS (see Figure 7) are a form of anti-roll structure that may protect a driver in the event of an overturn. They need to work alongside restraint systems, such as seat belts, to prevent a driver attempting to jump clear of the vehicle only to be trapped between the ROPS and the ground.

FOPS (see Figure 6) are to protect drivers by preventing objects dropping from height onto them. They are usually found on vehicles that lift and handle goods above the height of the operator's seat, such as forklift trucks, telescopic materials handlers and cranes.

Figure 6: FOPS protects driver from falling objects

8.6 Safe movement of people and vehicles in the workplace

Figure 7: Tractor with integral roll-over protection and cab

Safe drivers

Key requirements for safe drivers are:

- Training and competence assessment should be carried out for operators of lift trucks.
- The competence of lift truck operators should be reassessed at regular intervals or as required by legislation, or when new equipment is supplied or new risks arise, such as changes to working practices.
- Lift trucks quite often have special training requirements due to the hazardous nature of the work and because no equivalent highways test exists.
- Training drivers of vehicles other than lift trucks (and not specifically covered by legislation, such as agricultural tractors) should also be considered and to a similar standard.
- Supervise all drivers (including those visiting the site).
- Conduct medical checks as required by legislation, guidance or company policy.

Selection and training

To guarantee the competence of operators, suitable arrangements for selection, training, monitoring and supervision should be in place. For vehicles that are used on public roads, drivers should, of course, hold a current licence for the particular category of vehicle being used. When selecting new drivers, an employer might want to take into account their previous driving experience, driving record, references from previous employers, general attitude and so on.

In particular, forklift trucks (including telehandlers) and most vehicles required to lift and move materials have a counterbalance. Because of the hazards of overturning or tipping, specialist training is required. The HSE's approved code of practice L117 for rider-operated lift trucks defines the three stages of operator training (outlined in Definitions box).[5]

> **DEFINITIONS**
>
> **Basic training** – *this is a training course followed by a test.*
>
> **Specific job training** – *this covers knowledge of the workplace and handling attachments.*
>
> **Familiarisation training** – *this involves working on the job under close supervision.*

Other vehicle training (such as for diggers, dumper trucks or soil scrapers) is often based on these same three principles. Although there is no legal requirement for renewing the training certificates after a set time, awarding bodies often grant certificates for a fixed period, normally 3 years.

> **ADDITIONAL INFORMATION**
>
> Basic training for operating a forklift truck may cover:
>
> - the responsibilities of operators to themselves and others;
> - the basic construction and main components of the lift truck, including how it operates and its load-handling capabilities and capacities;
> - the handling attachments that may be used with the lift truck;
> - the purpose of the controls and instruments, and how to use them;
> - the various forms of load, and the procedures for their stacking, destacking and separation;
> - assessing the weight, and, where relevant, the load centre of a load and deciding if the load is in the truck's actual capacity (safe working load); and
> - the factors that affect machine stability, such as turning, load security and integrity, and speed and smoothness of operation.

Fitness to drive

UK health and safety legislation does not define fitness to drive for workplace driving (off-highway) but HSE directs businesses to the guidance given by the Drivers Medical Unit of the Driver and Vehicle Licensing Agency (DVLA).[6] There are separate medical standards for:

- Group 1: holders of ordinary driving licences; and
- Group 2: heavy goods vehicle (HGV) and public service vehicle (PSV) licence holders.

The requirements of the driving task need to match the fitness and abilities of the driver. For most work, a standard equivalent to Group 1 will be appropriate. In some cases, a more stringent (Group 2) standard may be required. For example, when:

- moving highly toxic or explosive materials;
- working in a particularly demanding environment;
- working at night; or
- operating large, heavy vehicles.

8.6 Safe movement of people and vehicles in the workplace

> **TIP**
>
> Fitness to drive on the highway in the UK is defined in the following legislation:
> - Road Traffic Act 1988; and
> - Motor Vehicles (Driving Licences) Regulations 1999 (as amended).

> **TIP**
>
> Safety should form an integral part of operator/driver training. Records should be kept of all training given to individual operators/drivers, including any conversion and refresher training carried out. Training of workers should include instruction on site rules; visiting drivers should be similarly instructed on the site rules in force and the need to adhere to them.

In addition to the DVLA requirements, it is sensible for operators and potential operators to be medically screened before employment and at regular intervals thereafter. With regard to driver fitness, the following points should be considered:

- General:
  - Drivers should have full movement of their trunk, neck and limbs and be capable of normal agility.
  - They should have a steady and reliable temperament.
  - Those dependent on alcohol or non-prescribed drugs should not be employed as drivers.
- Vision and hearing:
  - Good judgement of space and distance is required.
  - If vision is corrected by glasses, these must be worn.
  - Ability to hear instructions and warning signals is important.
- Medical conditions:
  - Conditions such as epilepsy and diabetes do not necessarily bar someone from operating vehicles but expert medical advice should be sought. Workers with a restricted driving licence may be excluded from driving certain categories of vehicle on the road.

Reassessment

HSE recommends screening all existing and potential workplace transport operators for fitness before employment and at 5-yearly intervals from age 45. Group 2 licences are renewable every 5 years from age 45 and, when an individual is both a workplace transport operator and holds a Group 2 licence, these assessments can be made at the same time. A workplace transport operator who continues after age 65 should have annual assessments for fitness.

> **CASE STUDY 3**
>
> **Gamekeeper found dead following accident on an all-terrain vehicle/quad bike**
>
> The trustees of a country estate were fined after admitting a health and safety breach in connection with a gamekeeper's death.

General workplace issues

> The 53-year-old, who was employed as a temporary stand-in gamekeeper, sustained serious injuries to his pelvis when the quad bike he was driving overturned on a slope. However, his absence was not detected until 52 hours later, at which point a search was initiated. His body was found some 200 yards away from the scene of the accident, in a separate field.
>
> He had no means of raising the alarm, although there was a mobile phone signal and the normal gamekeeper (who was undergoing surgery) had been issued with a phone.
>
> It appeared the injured gamekeeper had attempted to reach a nearby farmhouse to seek help and had opened a farm gate in order to get there. The trustees were prosecuted because the injured gamekeeper clearly did not die immediately and if he had a means of communication then he would have had an opportunity to summon help.[7]

KEY POINTS

- Safe movement of people can be assured by a systematic, risk-based review of the hazards that vehicles pose in the workplace.
- Safe workplace design may include one-way systems, avoiding or limiting the need for reversing, condition (width, stability and drainage) of roadways, segregation of pedestrians, crossing points, general lighting, traffic signals and site mirrors.
- Safe vehicle considerations may include seating position, visibility lines for seeing pedestrians and other road users, engineering aids such as proximity sensors, cameras, vehicle lighting, mirrors and warning klaxons.
- Safe driver checks may include approved training courses for particular vehicle types, for example lift trucks (including telescopic materials handlers), industry (not mandatory) schemes for minimum requirements for driving construction vehicles (such as dumper trucks, rollers and tarmacking machines) and refresher/requalification intervals as determined on an industry basis.

References

[1] Adapted from Ben Bloch, 'Somerset farmworker dies in forklift tragedy – owner fined huge amount' *Somerset Live* (16 December 2021) (www.somersetlive.co.uk)

[2] Adapted from Transport for London, 'Guidance for the HGV Safety Permit' (https://tfl.gov.uk)

[3] See note 2

[4] Adapted from: HSE, agriculture case studies 'Poor visibility for driver proves fatal' (www.hse.gov.uk)

[5] HSE, *Rider-operated lift trucks. Operator training and safe use. Approved Code of Practice and guidance* (L117, 3rd edition, 2013) (www.hse.gov.uk)

[6] DVLA, *Assessing fitness to drive: a guide for medical professionals* (2016, updated 2018) (www.gov.uk)

[7] Adapted from HSE, agriculture case studies 'Lone working – gamekeeper found dead following accident on an ATV' (www.hse.gov.uk)

8.7: Work-related driving

Syllabus outline

In this section, you will develop an awareness of the following:
- Managing work-related driving:
  - plan
    - assess the risks
    - policy
    - work-related driving taken account of by top management
    - roles and responsibilities
  - do
    - co-operation between departments (where relevant)
    - adequate systems in place, including maintenance strategies
    - communication and consultation with the workforce
    - provision of adequate instruction and training
  - check
    - monitor performance (ensures the policy is working correctly)
    - ensure all workers report work-related road incidents or near misses
  - act
    - review performance and learn from experience
    - regularly update the policy
- Work-related driving control measures:
  - safe driver (competence – checks on level of skill/experience, validity of driving licence; provision of instruction; fitness to drive)
  - safe vehicle (vehicles fit for purpose for which they are being used; maintained in a safe condition – checks on MOT/service history; adequate safety devices; maximum load weight not exceeded; adequate restraints for securing goods)
  - safe journey (planning of routes; realistic work schedule – enough time to complete the journey safely, allowing for driving breaks; consideration of weather conditions; consideration of legal driving hours where relevant)
- Hazards associated with the use of electric and hybrid vehicles:
  - silent operation/pedestrians not being aware of vehicles manoeuvring
  - availability and location of charging points
  - electric shock for high voltage components and cabling
  - retained electrical charge in components even when the vehicle is switched off
  - unexpected movement of the vehicle or engine components due to the motor's magnetic forces
  - potential for the release of explosive gases and harmful liquids from damaged batteries

Introduction

The risk of workers driving on public roads needs to be managed like any other risk at work. Here we will consider work-related driving risks at a management level and around the continual improvement cycle of Plan-Do-Check-Act.

We will deal with the necessary considerations for managing work-related driving, including risk assessment and control measures (driver/vehicle/journey). Finally, we will look at the hazards of electric and hybrid cars.

> **TIP**
>
> Standards and guidance for employers on work-related driving can come from a number of sources. Here are some:
>
> - Royal Society for the Prevention of Accidents (RoSPA), *An introduction to managing occupational road risk*.[1] The syllabus closely follows this document;
> - RoSPA, *Driving for work using own vehicles*;[2]
> - RoSPA, *Driving for work: Safer journey planner*;[3]
> - Road Traffic Act 1988;
> - Driver and Vehicle Standards Agency (DVSA), *Rules on drivers' hours and tachographs for vehicles used for the carriage of goods in Great Britain and Europe*;[4]
> - DVSA, *Rules on drivers' hours and tachographs for passenger vehicles in Great Britain and Europe*;[5]
> - Health and Safety Executive (HSE), 'Driving and riding safely for work';[6]
> - HSE, 'Electric and hybrid vehicles';[7]
> - Brake, a registered charity campaigning for road safety, sustainable transport and helping victims of road crashes;[8]
> - European Agency for Safety and Health at Work, *Review of successful occupational safety and health benchmarking initiatives*;[9]
> - HM Government, *Tachographs: Rules for drivers and operators*;[10] and
> - HSE, *Rider-operated lift trucks. Operator training and safe use. Approved code of practice and guidance (L117)*.[11]
>
> Supplementary reading that goes beyond the syllabus requirements can also be found at:
>
> - Rail Safety and Standards Board (RSSB) produces a range of guides and research reports on driver fatigue that are not just for rail but for road as well;[12] and
> - International Standards Organization, ISO 39001:2012 *Road traffic safety (RTS) management systems — Requirements with guidance for use*.[13]

> **DEFINITIONS**
>
> **Drivers** are those who drive a vehicle for work and tend to fall into two different categories:
>
> - professional drivers of goods, passenger vehicles, private hire vehicles or anyone whose main job activity is driving; and
> - people who drive frequently to occasionally as part of their job, but driving is not their main activity (for example, salespeople, managers covering several sites, office workers driving to a meeting location).
>
> **Journeys:** anyone who travels during their working hours makes a journey no matter what the distance. This also includes travelling from home to a location that is not their normal place of work, for example a worker travelling from home to a work-related training course.
>
> **Vehicles** fall into several categories:
>
> - specific purpose-built (fleet) vehicles provided by the employer – these can range from large trucks to company cars and vans;
> - vehicles leased or hired for work-related purposes; and
> - 'grey fleet' vehicles owned by individual drivers that could also be used for work-related driving.

Management approach to work-related driving

Workers driving on public roads is an occupational risk. It needs to be managed in terms of setting policy, allocating responsibilities, assessing the risks, implementing appropriate control measures, monitoring their effectiveness and reviewing the situation at suitable intervals. The syllabus aligns with the Plan-Do-Check-Act (PDCA) cycle (see Figure 1) and our discussions are arranged similarly.

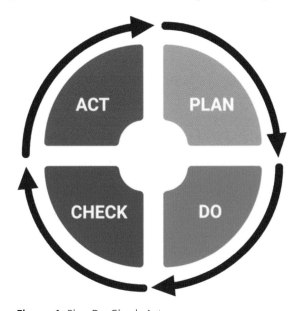

Figure 1: Plan-Do-Check-Act

General workplace issues

Road injury statistics

The statistics on road-related injuries are stark. According to the charity Brake, in a typical year, approximately 1,500 people are killed on UK roads.

The trend relating to serious injuries is more difficult to analyse, as data collecting methods have changed. However, the total figure is still in excess of 22,000 serious injuries a year, or an average of around 60 a day.

Car occupants comprise the highest number of road deaths, which is not surprising given that car traffic comprises around 80% of all travel on British roads.

The approximate proportion of road deaths by transport type in a typical year are:

- car occupants = 42%;
- pedestrians = 24%;
- motorcyclists = 20%;
- pedal cyclists = 10%; and
- other = 5%.

These figures relate to all road users and not just to those who have accidents while at work. This is simply because work-related accidents are often not recorded separately.

8.7.1 Managing work-related driving

Managing work-related driving can be broken down into the PDCA model. We will now take a look at each stage in turn.

Plan

Collect data and information about work-related driving

This will include an inventory of drivers and the vehicles they are qualified to drive or ride, the journeys they make and the business mileage they drive or ride annually (this can help to identify broad categories of risk based on exposure). An inventory of accidents and insurance claims will also be a useful source of information.

Assess the risks

The arrangements in place for managing work-related road risks should, like other occupational risks, be based on an assessment of the risks. An assessment of road risk should follow a recognised procedure, such as the HSE 'five steps' approach as discussed in 3.4: Assessing risk.

When identifying the hazards, reference should be made to risks associated with driving for work, which can be conveniently categorised as those related to:

- the driver (and the standard of driving);
- the vehicle; and
- the journey.

While some of the risks may be out of the employer's control, many can be influenced by sensible precautions and proactive measures:

- Prioritise areas for action, including considering journeys, drivers and vehicles with the highest risks and potentially most severe consequences; for example, drivers who travel the highest number of business miles or who have to start and finish their journeys particularly early or late in the day.

- Risk assessments need to be checked and updated regularly and whenever there is a significant change in a relevant activity (new type of vehicle) or there is a serious incident or near miss.
- Consult your workers, especially drivers, riders and managers and their representatives, particularly on matters concerning the schedules and time constraints that they have to work to.
- Define roles and responsibilities in the policy and procedures for all staff in the organisation, as well as consultants and subcontractors who use the road when working for your organisation.

Driver-related risks

Behavioural factors account for many accidents, including drug/drink-driving, excessive speed and reckless driving. Inexperience and inadequate assessment of road risks are also important aspects. Research has shown that drivers under 25 years of age tend not to have attention capabilities, are unable to judge safe speeds and have underdeveloped hazard recognition and accident-avoidance skills. However, it is not simply a question of limited experience; it is also a question of maturity.

Aggression can play a part too. Aggressive drivers are more likely to drive fast, brake harshly, drive close to preceding vehicles, change direction without warning and put themselves in other positions of risk than better-natured drivers. The extent of aggressive behaviour on the road depends on someone's personality, such as their degree of competitiveness, concern for others and attitude to risk. It can also be affected temporarily by anger over a particular incident (whether or not road related), anxiety, stress and alcohol.

Vehicle and journey hazards

As part of assessing the risk, consideration needs to be given to the hazards involved and this includes the vehicle used and the objects it might collide with, together with anything that might cause loss of control of the vehicle, such as an icy surface. Also, the term 'road hazards' is often used and includes other risks, such as driver fatigue.

Identification of the hazards should take account of driver-related, vehicle-related and journey-related elements and might be achieved by consultation with relevant workers, particularly those who drive extensively.

The people who might be at risk are generally drivers, passengers and other road users (including pedestrians). Those groups of workers at particular risk include long-distance drivers and recently qualified drivers.

In evaluating the risks, the likelihood of harm should be considered and a decision taken on whether existing precautions are sufficient. Unless the road hazard can be eliminated altogether (for example, by preventing travel in certain circumstances), then some residual risk will remain.

As with other risk assessments, the significant findings must be recorded. The record should show clearly how the organisation intends to manage occupational road risk.

The final step is to review the assessment and revise it if necessary. A periodic review should check that the assessment is still valid, but a special review may be prompted by vehicle accident data suggesting that there might be a problem or by significant changes to vehicle specifications or the journeys carried out.

General workplace issues

Having assessed the risk, it should then be possible to start implementing the control measures that have been identified by the assessment as necessary to help reduce the risk to as low a level as is reasonably practicable.

Policy

A policy on work-related road risk should consider, among other matters:

- Reducing risk by limiting exposure; for instance, by restricting driving hours or by encouraging workers to use other means of travel (such as trains and taxis) instead of driving.
- The maximum number of driving hours that a worker should be allowed to undertake per day (9 hours) and the maximum number that a worker should be allowed to undertake without a break (4.5 hours).
- The organisation's arrangements for workers' use of private vehicles for work. This will include the need to determine that the vehicle is suitable, that the worker holds a current driving licence valid for the type of vehicle, that the vehicle is insured for business use and, as a minimum, is serviced on a regular basis. When there is any doubt over a vehicle's suitability, then arrangements should be in place to hire a replacement.
- An organisation's environmental policy may specify a preference for public transport over the use of private vehicles, and this may be incorporated into its policy on vehicle use.
- The organisation's rules regarding the use of mobile phones in vehicles: the use of hand-held phones while driving is dangerous (and illegal) and the policy should refer to this fact (perhaps with reference to disciplinary action where appropriate). However, the policy should also specify the organisation's rules on the use of hands-free kits.
- Elimination of the hazard; for instance, by arranging meetings that use telephone or video-conferencing facilities rather than everyone travelling to a single location (such an option will also have cost and environmental benefits).
- The organisation's rules regarding smoking in vehicles: in Great Britain, vehicles used for work are considered as a workplace and no smoking is allowed – irrespective of whether there is another person in the vehicle.

ADDITIONAL INFORMATION

Rules on drivers' hours are complex. In Great Britain, for goods vehicles, these can be summarised as:

Daily duty hours	Maximum of 11 hours in a day
Daily driving	Maximum of 10 hours in a day
Exemptions	- dealing with an emergency; for example, a major disruption to public services or danger to life; - using the vehicle for private driving and not for work; - driving off-road or on private roads during duty time; and - driving a vehicle used by the armed forces, police or fire brigade.[14]

More stringent rules apply to passenger-carrying vehicles.

8.7 Work-related driving

> **APPLICATION**
>
> Consider the driving risks in your organisation. What should the health and safety policy address?

Top management commitment
The policy on work-related road risk must be seen to apply to, and have the commitment of, all levels of the organisation, from senior management downwards. Therefore, the person given the responsibility for implementing the policy must have sufficient authority to exert influence and to apply the policy requirements. In organisations with widespread use of vehicles, the transport manager will often be the person given much of the responsibility.

Do

Co-operation
In large organisations, it is likely that different departments will share some of the responsibilities for work-related road risk. It is important in such cases that there is co-operation between these departments and a shared understanding of the organisational requirements.

Adequate systems of work
Systems of work cover the practical elements of managing work-related driving. Systems should include, for instance:

- Setting minimum vehicle standards by controlling the hazard using engineering means as much as possible (for example, reversing sensors, energy-absorbent crash protection systems and ergonomic design); providing in-vehicle personal protection, such as seat belts, airbags and head restraints.
- Carrying out checks on vehicle documents, including whether vehicles have the correct registration documents, insurance and vehicle tax, and that drivers have 'business use' insurance when using their own vehicles for work purposes.
- Completing statutory safety checks: these are the legal requirements that must be met for vehicles to be safe to operate, such as annual safety inspections.
- Vehicle familiarisation: provide vehicle briefing sessions for drivers with new vehicles for the first time.
- Tachographs and speed limiters: make sure that workers know and adhere to any rules about tachographs and speed limiters for any they use.

> **DEFINITION**
>
> ***Tachographs** (tachos) record information about driving time, speed and distance. They are used to make sure drivers and employers follow the health and safety rules on drivers' hours.*

Communication and consultation
The policy and procedures on work-related driving should be communicated to all workers on a regular basis, using the full range of internal communication methods.

A driver's handbook is needed to clearly set out the expectations, rules and procedures for work-related driving.

Introduce a journey planning procedure
The most effective way to reduce at-work road incidents is to reduce the amount of at-work driving. Journey planning should start by deciding whether a journey by road is necessary, or whether it can be made by rail or air travel or replaced with remote communications.

Emergency procedures
Vehicle breakdowns, medical incidents, major traffic disruption and bad weather can all give rise to emergencies while driving for work. Emergency procedures may include aspects of lone working (see 8.4: Lone working) and personal safety. There may also be a need for a business mobile phone that has good network reception in the areas where travel is likely.

There needs to be a system for drivers to report incidents, near misses and motoring offences.

Instruction and training
Managers, supervisors and team leaders need to be trained to manage work-related road safety as part of their health and safety responsibilities. This includes a system for approving journeys and, when appropriate, overnight accommodation if journey times are too long.

Managers should lead by personal example and follow the organisation's policy and procedures; for example, by complying with the rules on the use of mobile phones. They should also monitor that their staff are actually implementing the procedures properly and should provide regular opportunities for them to raise issues or concerns.

Check

Checking consists of a number of elements:

- Monitor implementation to help ensure that work-related driving policies and procedures are properly implemented. Examples include policies for checking driving licences; driver training and assessment; health surveillance; vehicle safety checks; staff appraisals that include work-related driving performance; road safety briefings; and systems for investigating driving incidents.
- Compliance with systems for reporting work-related driving incidents.
- Monitor trends in accident and near miss data.
- Investigate incidents to find common themes and/or root causes. Some organisations have used in-vehicle technologies such as dashcams to help investigate incidents. This is potentially challenging as it can be seen as a 'spy in the cab'.
- Other telemetry is available for heavy goods vehicles, and tachometers are commonly installed. Some other navigation systems are available for a wider range of vehicle types and these can give data about speed and acceleration, as well as times when the vehicle is stationary.
- Review claims data from insurance providers to see if there are any trends.

Also under this 'checking' category may be considerations about whether the risk assessment is fit for purpose.

Act

To assess whether the policy on work-related risk is effective, it is good practice to evaluate the business's policies and procedures against standards (such as ISO 39001:2012 *Road traffic safety (RTS) management systems — Requirements with guidance for use*) and benchmark them against what other businesses are doing to manage work-related driving risks.

> **DEFINITION**
> *Benchmarking is a planned process by which an organisation compares its health and safety processes and performance with others to learn how to reduce accidents and ill-health, improve compliance with health and safety law and/or cut compliance costs.*

Benchmarking can be a way of comparing numerical data as well as exchanging narrative accounts of what has and has not been effective in managing work-related driving.

Checking: active data

This can include records that show whether the policy requirements are being adhered to. Checks should be made to determine whether work vehicles are being maintained in accordance with agreed maintenance schedules and whether drivers are complying with rules relating to hours of driving and hours of work. Maintenance records will show the sorts of faults that are being found on vehicles and the criticality of such faults. A review of these records could show that maintenance schedules or interim inspection procedures need revising. Records of workers' driving histories will also show those who might be considered at greatest risk so that safe driving initiatives can be targeted appropriately.

Checking: reactive data

This will largely focus on reported road incidents in terms of both accidents and enforcement action. Workers should be made aware of the need to report every such incident that occurs while at work as well as some that occur outside work, such as offences that might lead to disqualification (such as drink-driving) or other driving convictions. The reporting system is only likely to be fully effective if workers can use it without fear of recrimination. Many refer to this as having a 'just culture'.

8.7.2 Work-related driving control measures

While many of these risk control measures are considered as part of the risk assessment process (when thinking about what more needs doing), the syllabus treats the control measures separately:

- Safe driver: competence, including checks on level of skill/experience, validity of driving licence, provision of instruction and fitness to drive.
- Safe vehicle: vehicles fit for purpose for which they are being used, including being maintained in a safe condition, adequate safety devices, maximum load weight not exceeded and adequate restraints for securing goods.
- Safe journey: planning of routes, including realistic work schedule, enough time to complete the journey safely, allowing for driving breaks, consideration of weather conditions and consideration of legal driving hours where relevant.

Safe driver

A key aspect is a driver's competence to drive. In many instances, it is considered insufficient to accept that a worker is competent to drive based simply on their possession of a valid driving licence (that is, a licence that is in date and valid for the type of vehicle being driven). This may have been obtained many years previously or, conversely, obtained recently when little driving experience has been gained. Driver competence needs to be tested initially, continually monitored and improved through training and instruction.

Fatigue

Driving, especially long-distance driving, can be tiring; this is made worse where the vehicle is not designed for long-distance journeys. The potential for driver fatigue (and falling asleep at the wheel) therefore increases, which in turn increases the accident risk. For example, research suggests that driver fatigue accounts for around 20% of serious accidents on fast roads in the UK, with 40% of sleep-related accidents being thought to involve commercial vehicles.

> **CASE STUDY 1**
> **Driver fatigue**
>
> Two men died in a road traffic accident as a result of their employer failing to ensure that they were sufficiently rested to work and travel safely.
>
> Both workers died when the driver fell asleep at the wheel of the work van and came off the road, crashing into a parked van, while driving back home after a night shift.
>
> The previous day the driver had left home at 4.30am and had driven to the site, arriving at 7.30am to carry out work on the site. The expected work did not take place, so, after waiting until midday, the driver started the drive back to the depot, arriving at 3pm.
>
> On his way to the depot he was asked to take on an overnight welding job in Stevenage and, with his colleague, set off from the depot at 7.18pm, arriving at the site at 9.47pm.
>
> The two men then undertook welding jobs from 11.15pm, leaving the site once they had finished at 3.40am. The crash occurred at around 5.30am as the driver was driving back home.
>
> Experts told the court that the driver may have fallen asleep at the wheel or experienced 'microsleeps,' which hugely increased the risk of a traffic accident.
>
> The company did not follow its own fatigue management procedures nor did it comply with the working time limits for safety-critical work, such as welding, which insist there should be a 'minimum rest period of 12 hours between booking off from a turn of duty to booking on for the next', and it did not conduct a sufficient and suitable risk assessment of the driver's fatigue.
>
> An approximate calculation is that the workers were driving for over 10 hours and working for over 9 hours, totalling a shift of more than 19 hours.[15]

A driver's biological clock (the circadian rhythms) has a major influence on tiredness and, as a result, sleep-related road accidents tend to peak during the periods between midnight and 6am and from 2pm to 4 pm, when alertness is naturally low.

Night shift workers have been found to be particularly vulnerable to sleep-related road accidents when driving home in the early morning on quiet, monotonous roads. The risk is particularly high after their first night on night shift and especially when the shift has lasted for 12 hours or more.

A large meal can induce sleepiness, and alcohol (even in small amounts and particularly if taken in the afternoon) and certain drugs, including prescribed medicines, may also have an effect. Certain medical conditions, in particular obstructive sleep apnoea and narcolepsy, can cause excessive sleepiness in sufferers.

Distraction

Distraction is a common cause of road accidents, whether it is because of something happening outside the vehicle (such as another accident) or inside the vehicle (such as adjusting the radio or satnav, smoking or using a mobile phone). Of these, the use of hand-held mobile phones has probably received most attention in recent years. The distraction caused by mobile phones is both physical and cognitive (mental).

Health

Eyesight is also an important physical characteristic for drivers and the Driver and Vehicle Licensing Agency (DVLA) prescribes 'standards of vision for driving'. These standards are higher for Group 2 licence holders. Limited vision is likely to affect many people who perhaps do not realise how their eyesight has changed over the years, particularly among older people and those who have never worn corrective lenses. It is, therefore, important that drivers take regular eye tests that will detect deterioration in eyesight at an early stage. Certain eyesight conditions must be notified to the DVLA.

As well as having good vision (with the aid of corrective lenses if necessary), it is important that drivers are generally fit and healthy. Drivers with certain conditions, such as epilepsy and diabetes, must notify these to the DVLA and may be refused a driving licence or have conditions placed on them, such as the need to renew their licence at regular intervals (subject to a satisfactory medical report).

Vehicle safety features

Many design improvements have been made to vehicles in recent years to make them safer, including anti-lock braking systems (ABS), electronic stability control systems, anti-jackknife devices, blind spot sensors, reversing sensors, rear view cameras, the provision of front and rear seat belts, airbags (front and side), laminated windscreens, side impact protection bars, crumple zones and pedestrian protection systems. Some of these are 'active' measures that are designed to help prevent an accident and some are 'passive' measures that help to reduce the extent of any injury.

Other factors, such as visibility and ergonomic considerations, can also affect safety (see 8.6: Safe movement of people and vehicles in the workplace). The safety features possessed by any particular vehicle can affect both the risk of an accident and the likelihood and severity of harm.

Suitability is a vital consideration when selecting vehicles for work. For example, rough terrain vehicles with high ground clearance, off-road tyres and weather protection for the operator are needed for agricultural and construction work.

Many work vehicles are used for carrying and moving loads. The maximum safe working load needs to be adhered to otherwise it can cause the vehicle to become unstable. Many vehicles have load indicators to show that the vehicle is working in its safe operating envelope.

General workplace issues

Having a secure load is important. For lorries it is essential that loads are properly secured with straps, chains or dunnage (or similar) before the vehicle drives on the public highway. The load carried by a vehicle can also affect the risk, either because it is inherently dangerous (flammable, caustic, toxic, etc) or because it is unstable or in some other way insecure.

Pallets and flexible intermediate bulk containers (FIBCs) (see Figure 2) are often used to package bulk goods (for example, stone, sand and other building materials) for easy loading onto, and unloading from, road vehicles and transport in workplaces (upon receipt or despatch).

Figure 2: Pallet and FIBC

Liquids are often moved in palletised liquid bulk containers (see Figure 3). For safety, the condition of the pallet needs to be sound, as does the container for the liquid or bulk materials, which should be properly secured before being transported.

Figure 3: Liquid bulk container

Service and maintenance

Vehicles that are not serviced effectively can present a serious risk to their occupants and to other road users. Replacement of vehicle parts to manufacturers' schedules can prevent serious failures of key mechanical parts (such as steering and brakes). For this reason, the UK mandates periodic testing (MOTs) by an independent authority once the vehicle has reached a certain age.

Safe journey

There are various aspects of a journey that can affect risk – the length of the journey (in distance and time), the type of road used, the physical condition of the road, environmental and weather conditions, work schedules and the actions of other road users.

Journey planning

The time spent driving inevitably affects driver fatigue (see Safe driver) but so does the length of the working day. If a person has an early morning start to attend an all-day meeting or other work commitment some distance away, it is likely that they will suffer tiredness to some extent on the journey home.

The actual extent may depend on other factors, such as the type of road, the general level of stress, dietary intake and the quality and quantity of sleep the worker had the previous night. For professional drivers, at least those driving certain vehicles, fatigue caused by driving is often recognised by national legal requirements to limit the number of hours spent on the road, both in total and in periods during the day. This may be recorded on a tachograph.

TIP

Some factors to consider (adapted from RoSPA guidance) when planning safe journeys are:

- Use safer alternatives. Can the meeting be held via video conferencing? If a face-to-face meeting is essential, is public transport safer, cheaper and more environmentally friendly?
- Limit distances/travel times. Have rules for limiting travel. This can be done by setting onsite meeting times, for example 10:00–15:00, to allow for safe journey times.
- Monitor cumulative driving times and driving distances over weekly and monthly periods.
- Have realistic work schedules that allow for foreseeable delays due to traffic congestion or bad weather. Permit overnight stays when travel times exceed agreed limits.
- Avoid driving in bad weather such as poor light, high winds and very high or very low temperatures.
- Specify safer routes, especially if there are regular destinations used by drivers. For example, refuse collections near to a school are best avoided at school opening and closing times.

Type of road

While there are far fewer accidents per mile on fast roads, they tend to be more serious due to higher speeds and the greater likelihood of multi-vehicle collisions. Nevertheless, such highways by their nature do not possess many of the hazards associated with other roads; including bends, steep gradients, brows of hills and dips, intersections, close proximity of oncoming traffic, variations in lane and road width, pedestrians, cyclists and frequently stationary or turning vehicles.

General workplace issues

As well as the type of road, its condition can also affect risk. Wet and icy roads make skidding more likely and can increase braking distances. A poor surface, particularly one containing potholes, can also make an accident more likely, especially for motorcycles. Materials used in road construction can affect factors such as tyre grip and the amount of water thrown up by tyres in wet weather.

Journey conditions

Environmental conditions, such as bright sunlight, heavy rain or fog, can have obvious effects on visibility and hence on the risk of an accident. The time of day (day or night, or the transition between the two) is also relevant in this respect. Adverse weather conditions, such as rain, fog, snow and cold temperatures, can cause the road surface to be wet or icy.

Journey schedules

The time of the journey may be dictated by strict work schedules (for example, the timing of a meeting or the need to meet a delivery deadline). The schedules should be realistic to allow sufficient time without the need to drive at excessive speeds.

Arrangements should be made to ensure that schedules are sufficiently flexible to allow drivers to behave safely on the road or that additional journey time is allocated where appropriate. In some of the prosecutions involving excessive hours, it was pointed out that drivers could not have completed the schedules set by their employer without working those hours.

Other road users

'At work' drivers continually face the dangers created by the actions of other road users, whether these are other drivers/riders, cyclists or pedestrians. The presence of each type of road user can be related to the type of road; for instance, pedestrians and cyclists are more likely to appear on urban roads. While the actions of other road users may appear to be out of the control of 'at-work' drivers and their employers, 'at-work' drivers should take any possible steps to predict or foresee such actions and to prevent or avoid the dangers.

APPLICATION

Imagine you are a buyer for a retail outlet and are planning an annual winter visit to one of the outlet's main suppliers. The supplier is based 250 miles away and the visits usually last for 6 hours (including lunch). Due to the location of the supplier, you must drive to the venue.

What should you consider when preparing for and carrying out this visit?

8.7.3 Hazards associated with the use of electric and hybrid vehicles

Electric vehicles

TIP

Terminology

Electric vehicles (EVs) are also referred to as battery electric vehicles (BEVs), hybrid electric vehicles (HEVs) and plug-in hybrid electric vehicles (PHEVs).

Electric vehicles use a large-capacity battery and electric motors to drive the vehicle. The battery needs to be charged from specialist charging points connected to the electricity supply. These can be at home or work or at public facilities.

Batteries provide a limited range of mileage, although some energy may be recovered during braking and driving downhill.

Hybrid vehicles

Hybrid vehicles have two sources of energy: an internal combustion engine (using either diesel or petrol for fuel) and a battery. Hybrid vehicles will use the two sources of power automatically and may use both simultaneously. The smaller petrol or diesel engine tops up the batteries. Some energy is also recovered from the vehicle braking systems.

A plug-in hybrid vehicle can have its battery charged directly from charging points.

Silent operation

Although some vehicles exhibit a 'whining' noise to alert pedestrians to their presence, road (tyre) noise is still present. Nevertheless, the lack of noise may lead pedestrians not to notice a vehicle's approach.

Availability of charging points

While charging points are becoming more numerous, the hazard for drivers of electric and hybrid vehicles is running out of energy. And while fast chargers are available, there is the prospect that a delay or a route deviation could lead to a worker becoming stranded for several hours.

So while environmentally beneficial, these vehicles do have workplace implications in terms of requiring more detailed journey planning (and considering realistic vehicle ranges) and having a recovery plan in the event of the vehicle running out of energy.

Electric shock from high voltage components and cabling

Unless a specific task requires the vehicle to be energised, it is necessary to isolate and/or disconnect the high voltage battery in accordance with manufacturer's instructions. Workers need to carry out a visual check for damage to high voltage cabling and electrical components before starting any work on the vehicle.

The locations of high voltage cables in EVs should be identified before carrying out hot work, such as panel replacement, cutting or welding.

Retained charge

Even when isolated, vehicle batteries and other components may still contain large amounts of energy and retain a high voltage. Only suitably trained workers using appropriate high voltage tools and test equipment should work on EVs.

Unexpected movement

Remote key fobs should be kept away from the vehicle to prevent accidental operation of electrical systems and accidental movement of the vehicle. Keys should be locked away with access controlled by the person working on the vehicle (in the same way as other electrical isolation methods). If the keys are required during the work, the worker should check that the vehicle is in a safe condition before the key is retrieved.

Potential for release of explosive gases

Traditional lead-acid batteries (as sometimes used in forklift trucks) are known to give off highly explosive gases (hydrogen and oxygen) during charging. Lithium-ion batteries do not do the same; however, there are other risks associated with damage to these batteries that could lead to 'thermal runaway' (possible fire and explosion risk).

The most common form of battery found in traditional electric vehicles (such as forklift trucks) is lead-acid. These batteries contain sulphuric acid (battery acid or electrolyte). Electrolyte is a general term used to describe a non-metallic substance such as sulphuric acid or salts that can conduct electricity when dissolved in water. The plates of the battery (anode and cathode) are made of lead.

Lead-acid batteries can produce explosive mixtures of hydrogen and oxygen gases when they are being charged. Batteries should be charged in well-ventilated battery-charging areas.

Lithium-ion (Li-ion) batteries are compact, lightweight batteries that hold considerable charge and work well under repeated discharge-recharge conditions.

The batteries are commonly found in EVs. Although accidents are rare, when they do occur it may be spectacular, resulting in an explosion or fire.

Like lead-acid batteries, a Li-ion battery consists of two electrodes separated by an electrolyte. One of the components (the lithium-based electrolyte) reacts vigorously with water. If a Li-ion battery is punctured or mechanically damaged, then the lithium in the contained parts of the battery cells can react violently with water in the air. This starts a chain reaction in the cells and is known as 'thermal runaway'.

This is why Li-ion batteries in EVs pose a specific fire hazard that is often difficult to extinguish and can subsequently reignite.

KEY POINTS

- Work-related driving can be managed in line with the Plan-Do-Check-Act cycle.
- Risk assessment is the main planning tool, with hazards grouped by driver (and the standard of driving), vehicle and journey.
- Control measures can be grouped as Do, Check and Act.
- Do covers eliminating the hazard (for example, by arranging remote meetings), reducing it (for example, by restricting driving hours) and controlling it by engineering means (such as reversing sensors) and personal protection (such as seat belts, airbags and head restraints).
- Check includes involving staff or their representatives, identifying all significant road travel hazards and risks, and identifying and prioritising reasonable control measures designed to improve driver, vehicle and journey safety.

- Act covers monitoring both active data (for example, records that show if policy requirements are being adhered to) and reactive data (such as reported road incidents).
- Hazards associated with electric and hybrid vehicles include their silent operation (pedestrians may be unaware of their approach); the number, location and variations in charging points (meaning journeys may need to be adjusted); electric shock (particularly when servicing and maintaining these vehicles); unexpected movement (due to stored electrical energy in batteries); and explosive gases and fires (resulting from both lead-acid and lithium-ion batteries.

References

[1] RoSPA, *An introduction to managing occupational road risk* (2016) (www.rospa.com)

[2] RoSPA, *Driving for work using own vehicles* (2018) (www.rospa.com)

[3] RoSPA, *Driving for work: Safer journey planner* (2011) (www.rospa.com)

[4] DVSA, *Rules on drivers' hours and tachographs for vehicles used for the carriage of goods in Great Britain and Europe* (2016, updated 2018) (www.gov.uk)

[5] DVSA, *Rules on drivers' hours and tachographs for passenger vehicles in Great Britain and Europe*, (2015, updated 2020) (www.gov.uk)

[6] HSE, 'Driving and riding safely for work' (www.hse.gov.uk/roadsafety)

[7] HSE, 'Electric and hybrid vehicles' (www.hse.gov.uk)

[8] Brake (www.brake.org)

[9] European Agency for Safety and Health at Work, *Review of successful occupational safety and health benchmarking initiatives* (2015) (https://osha.europa.eu)

[10] HM Government, *Tachographs: Rules for drivers and operators* (www.gov.uk)

[11] HSE, *Rider-operated lift trucks. Operator training and safe use. Approved Code of Practice and guidance* (L117, third edition, 2013) (www.hse.gov.uk)

[12] See www.rssb.co.uk

[13] International Standards Organization, ISO 39001:2012 *Road traffic safety (RTS) management systems – Requirements with guidance for use* (2012) (iso.org/standards)

[14] HM Government, 'Drivers' hours: GB domestic rules' (www.gov.uk)

[15] Adapted from Kellie Mundell, 'Exhausted welders killed in crash land employer in court' *IOSH Magazine* (15 May 2020) (https://www.ioshmagazine.com)

ELEMENT 9

WORK EQUIPMENT

9.1: General requirements

> **Syllabus outline**
>
> In this section, you will develop an awareness of the following:
> - Providing suitable equipment, including the requirement for CE marking within the UK and Europe
> - Preventing access to dangerous parts of machinery
> - When the use and maintenance of equipment with specific risks needs to be restricted
> - Providing information, instruction and training about specific risks to people at risk, including users, maintenance staff and managers
> - Why equipment should be maintained and maintenance conducted safely
> - Emergency operation controls, stability, lighting, markings and warnings, clear workspace

Introduction

There are many types of work equipment, including hand tools, machines, powered tools and unpowered tools.

We start by setting out general rules for all forms of work equipment; this covers statements of general principle that will be expanded on as necessary later. We will then cover the particular issues associated with the use of hand-held tools, whether powered or not, before moving on to deal with the wide range of machinery hazards that might be encountered in the workplace. Finally, we draw everything together by explaining the control measures that are used to reduce risks from machinery hazards.

> **TIP**
>
> The Provision and Use of Work Equipment Regulations 1998 (PUWER) is the main source of requirements for work equipment.
>
> PUWER covers a wider range of topics that we will draw on, the main ones being:
>
> - General requirements, including:
>   - suitability of work equipment;
>   - maintenance and inspection;
>   - specific risks;

9.1 General requirements

- information, instruction and training;
- dangerous parts of machinery;
- protection against specified hazards;
- high or very low temperature;
- control systems: start, stop, emergency stop, isolation from sources of energy;
- stability;
- lighting;
- maintenance operations; and
- markings and warnings.
- Mobile work equipment (MWE), including:
  - employees carried on MWE;
  - rolling over of MWE;
  - overturning of forklift trucks;
  - self-propelled work equipment;
  - remote-controlled self-propelled work equipment; and
  - drive shafts
- Power presses.

Work equipment is defined under PUWER as being any machinery, appliance, apparatus, tool or installation for use at work (whether exclusively or not). Examples are many and varied, and include:

- dumper truck;
- drone;
- lawnmower;
- crane;
- power press;
- 3D printer;
- photocopier;
- computer;
- hand saw;
- portable drill;
- meat cleaver; and
- mobile access equipment.

Examples of things that are not 'work equipment' in this context include:

- livestock;
- private cars (discussed in 8.7: Work-related driving);
- substances (such as acids, alkalis, slurry, cement, water – hazardous substances are covered by specific legislation); and
- structural items (walls, stairs, roofs – these are more appropriately covered by building regulations and standards).

Some examples of work equipment, such as certain types of mobile access equipment, have already been discussed in 8: General workplace issues.

APPLICATION 1

Make a note of things that you think would be defined as work equipment.

9.1.1 Providing suitable equipment

Any work equipment needs to be suitable, but suitable for what? The answer comes in two parts.

Firstly, work equipment must be suitable for the intended task – it must be robust enough to withstand the harder and more frequent use that it will get in the workplace. For example, a stepladder designed for home use need not be very strong because it does not get used often. However, it may be unsuitable for use in a workplace because it will be used frequently and will therefore be much more likely to wear out quickly and fail. The employer will need to provide industrially rated steps that are tough enough to withstand frequent use, not to mention a certain amount of misuse or rough treatment.

Secondly, work equipment must be suitable for the working environment where it will be used. It is very important to take into consideration any hazards created by the location, such as where electrical equipment is being used in wet conditions on a building site or in flammable atmospheres.

For example, in the UK and Europe, equipment for use at work must meet certain minimum standards. Manufacturers show that the standards have been met by affixing a 'CE' (*Conformité Européenne*) mark to the item and issuing the purchaser with a written document known as a 'Declaration of conformity'.

As a consequence of leaving the European Union (EU), the UKCA mark is replacing the CE mark for equipment sold in Great Britain (CE marking continues for goods sold in Northern Ireland). However, in nearly all important respects it is an identical system to CE marking. And, even if the UK had never been a member of the EU, British manufacturers would still be making use of the CE system for exports to the EU.

Figure 1: The CE mark shows equipment meets European standards

9.1.2 Preventing access to dangerous parts of machinery

We will talk more about different forms of guard and safety device later, but, as a general statement of principle, work equipment should be designed or constructed to prevent access to those parts that might cause harm. This cannot always be achieved for every piece of equipment but the employer is still under a duty to make sure that access is prevented where possible. Take, for example, a grinding wheel; although it is clearly not possible to fence the entire wheel, a large part of the wheel and all of the associated working mechanism should be securely guarded, which will significantly reduce the risk associated with using such a tool.

Figure 2: Grinding wheels should be guarded

9.1.3 Restriction of use and maintenance of equipment with specific risks

Many, if not most, pieces of work equipment present specific risks to health or safety. Think, for example, of the hazards presented by the following:

- woodworking chisel;
- light fitting;
- soldering iron; and
- lathe.

9.1 General requirements

Legislation does not, and cannot, spell out the hazard potential for every piece of work equipment – that is the purpose of a risk assessment. However, it is possible to identify a number of specified hazards, such as those that can cause equipment to catch fire or overheat, ejection of parts or substances, or the disintegration of the machine or its parts (such as a burst grinding wheel, for example) that create higher risks. Where such hazards exist, it stands to reason that higher levels of worker competence will be required.

When work equipment presents a specific risk of injury, its use and maintenance should be restricted to those who are competent. The consequences of someone who does not have the appropriate training and experience using or attempting to maintain such equipment can be very serious, not only for the person concerned, who will in all probability be badly injured, but also for the organisation that allows them to use the equipment.

Maintenance activities often require the removal of safeguards to gain access to parts of machinery that need lubricating or adjusting. Maintenance workers will therefore need specific training and skills to enable them to carry out their tasks safely.

It may be that the employer will only allow the use of work equipment by those that it authorises. To become authorised, a worker will not only need to prove their competence to use the equipment but will also need to have the employer's permission to use it. This leads to the next point – providing information, instruction and training.

9.1.4 Providing information, instruction and training on specific risks

We have already pointed out that the range of items that can be considered work equipment is huge. That being the case, it is also true that the amount of information, instruction and training will vary greatly. For example, there would be more training for the use of a pillar drill than for a hand drill. Use of an office shredder requires minimal training, whereas much more will be needed for the operation and maintenance of an industrial shredder in a paper mill or recycling plant.

> **DEFINITIONS**
>
> *Information*: in this context we mean one-way communication for the benefit of the user. Examples include signs or warnings. This might be suitable as a reminder of safe working practice or hazards.
>
> *Instruction*: this is another form of one-way communication that can include written material from manufacturers and suppliers as well as safe systems of work from the employer. It may be difficult to assess whether instructions have been fully understood.
>
> *Training*: this allows the worker the opportunity to practise a skill under supervision. In this way, an assessment of their understanding, skills and competence can be made and any problems or misunderstandings can be corrected before the worker becomes 'authorised' and is allowed to use the equipment alone.

Limited information is all that may be needed for some items of work equipment, such as simple hand tools, whereas a prolonged period of training may be needed for a machine operator. As with so many other aspects of health and safety management, 'proportionality' should be considered when deciding on the level of training required.

Work equipment

Health and safety issues should be woven into all aspects of training so that when the worker eventually starts to use a particular machine or control a particular process, they do so safely.

> **APPLICATION 2**
>
> In any work situation, training may be required to cover three aspects of work. Consider what these might be.
>
> In developing a training programme, what do you think are some of the factors that an employer will need to bear in mind when identifying training requirements?

The three aspects of work that will need to be covered in a training programme will be:

1 Day-to-day operation of (in this case) the equipment – starting, stopping, loading, unloading and so on.

2 Foreseeable abnormal situations and the actions to be taken if such a situation was to occur – stoppage of the line, a blockage, a drill bit shattering, a tyre puncturing and so on.

3 Conclusions to be drawn from experience in using the work equipment, meaning that, following training, the worker should have sufficient knowledge to be able to provide feedback about the equipment and its use in the workplace (for example, whether there are any faults with the machine or if it is being used unsafely by others).

When identifying training requirements for safe use of work equipment, it is necessary to consider the worker's current level of competence (knowledge, experience and skills). Any gaps in these areas should be filled by providing appropriate additional training.

> **APPLICATION 3**
>
> It is important to remember that an employer should train not only those who use work equipment but also the worker's supervisors and managers.
>
> Why do you think that supervisors and managers should receive training? Should it be the same level of training as an operator gets?

Supervisors may have started out as machine operators, so may already be aware of many of the issues, but this is not always the case. In any event, they will need to be aware of the hazards associated with the work equipment and, crucially, have knowledge of safe (and unsafe) working practice so that they can effectively manage their workers.

Lastly, the training programme should also take account of the circumstances in which the work is carried out (for example, if working alone or under close supervision of a competent person).

9.1.5 Maintaining equipment safely

Equipment should be inspected and maintained to check that it is kept in efficient working order and in good repair and any safety devices are functioning correctly. The extent and complexity of maintenance will vary enormously, from simple checks on hand tools to a substantial maintenance programme for a complex system.

Possible risks to those carrying out maintenance

As we mentioned earlier, during maintenance, safeguards may have to be removed to access parts of machines, and this creates an obvious risk of injury. However, there are other problems associated with maintenance activities that increase the risk of injury to maintainers.

> **APPLICATION 4**
>
> Take a few moments to consider a range of points that you think may be relevant when maintaining equipment. Compare your list with the following points.

As well as being exposed to working parts following the removal of guards, maintainers will also need to contend with some or all of the following:

- exposed live electrical connections;
- awkward posture or position;
- difficulties in access, such as the need to work at height;
- stored energy, such as flywheel energy, pneumatic or hydraulic energy, or capacitance;
- hazards associated with other tools, for example welding hazards;
- chemicals such as solvents, oils and grease; and
- heavy components that need to be removed and replaced.

You may have identified other issues that could be added to this list. Seeing this range of issues should make clear why maintenance activities account for a large number of serious injuries and fatalities.

Machine safety checklists

Machine safety checklists can be used when carrying out maintenance to ensure that all essential tasks have been carried out. These are filled in by the supervisor or some other responsible person.

Machine safety interlocks and guards				JV Mouldings Ltd	
Details					
Machine type and location				Number	
Safety interlocks and guards checked					
Date	Time	Signature	Comments		

Table 1: Example of a machine safety checklist

Competence

Maintenance work should only be carried out by competent workers. It is acceptable, indeed often desirable, that in-house workers should be used whenever possible. PUWER has specific duties relating to the inspection of equipment that presents special hazards, such as manufacturing machinery, steam boilers, lifts, hoists, power presses, etc. In such cases, it will almost certainly be necessary to bring in outside contractors or insurance assessors to undertake these specialist inspections.

There are many different types of maintenance:

- Breakdown maintenance is carried out after a failure, when the equipment has broken down.
- Condition-based maintenance involves monitoring the condition of critical parts and providing maintenance if required. A good example of this is looking at the tyres on your car – if they look flat, then you would pump them up; if the tread was almost gone, then you would buy new tyres.
- Routine maintenance includes periodic inspection and testing, perhaps based on the recommendations of the manufacturer, and should take into account any legal requirements. A car provides another good example of this; a driver would check fluid levels periodically.
- Planned preventative maintenance has the primary aim of preventing failures occurring while the equipment is in use. This is best achieved through a written system of work and instructions that trigger inspections, testing and, perhaps more importantly, periodic replacement of components and equipment before they reach the stage where they will fail. Defective parts should be replaced immediately. A motor car's servicing schedule is a good example of planned preventative maintenance.

> **TIP**
>
> Full records of all testing and inspection should be kept for examination at any time by enforcement authorities or insurers that may be investigating an accident.
>
> It is recommended that, whatever the work environment, a maintenance log should be kept, recording all work carried out and aiding future planning. This could be done via a written log by using general computer software such as spreadsheets. Alternatively, there are many commercially available apps and computer packages capable of storing, retrieving and printing out the schedules that will be required for a maintenance log.

9.1.6 Safe use of work equipment and emergency controls

We will now look at issues that are commonly associated with more complex or larger equipment (including machinery) as opposed to simple hand tools. For instance, start, stop and emergency controls will not generally be found on work equipment where the risk of injury is low (for example, desktop computers that have no exposed moving parts).

On equipment where the risk of injury is higher, it should not be possible for a machine to start without the application of a control; that is, there must be a positive action required to start the machine. This is particularly important in a situation where the machine has been stopped automatically by a safety device, such as someone treading on a pressure-sensitive mat around a robot; the robot should not automatically restart as soon as the person withdraws from the pressure-sensitive area – it should be necessary to reset the machine first.

Modifying the movement of any work equipment, such as by slowing or redirecting it, will require control of the equipment and associated components' energy of motion; for example, the rotary motion of a grinding wheel, the forward or backward linear motion of a tractor or other vehicles. Some situations may be complex as regards energy, such as a travelling crane carrying a ladle containing molten metal; you might like to think of the challenges involved in safely controlling the motion of such a set-up. Some electrical systems such as transformers may involve very large amounts of electromagnetic energy; again, this energy will have to be controlled as the system 'slows down'. (As any electrical engineer will tell you, 'breaking' an electrical circuit is usually more demanding than 'making' the circuit.)

Emergency controls

An emergency stop system (such as an e-stop 'mushroom' or a pull cord along a conveyor) should be readily available for an operator to activate. As with normal stopping, energy will still have to be released before the machine comes to a full halt. In an emergency, it may be acceptable for the system to come to a halt more rapidly than for a conventional stop; for example, many pieces of rotating electrical equipment can be slowed very rapidly by employing electrical braking.

Figure 3: A common emergency stop device

Emergency stops are intended to produce a rapid response to a potentially dangerous situation and they should never be used as functional stops during normal operation.

All controls and emergency stops must be clearly visible, instantly identifiable and appropriately positioned. In the case of start and stop controls, 'appropriately positioned' means that the operator should not have to reach across a danger zone to operate the machine or need to travel a distance to do so. An emergency stop control should be positioned where it can easily be operated in an emergency – either right next to the danger zone or as a master emergency stop that will isolate power to several machines.

Stability

Larger items of work equipment or machinery will need to be stable when in use. For example, this might result in limitations being placed on the height of a mobile tower scaffold, or it may require equipment to be clamped in place or bolted to the floor to prevent movement. Another example would be the use of stabilisers (outriggers) on mobile cranes.

Figure 4: Use of outriggers to stabilise the equipment while in use

Lighting

It is essential that, wherever work equipment is used, there is suitable and sufficient lighting. In some cases where the task involves perception of detail, for example when using certain measuring devices, it may be necessary to provide additional lighting.

A problem associated with rotating machines is the 'stroboscopic effect'. If machines are operated in areas where lights flicker, then the rotating part may appear to stand still. This can have serious consequences for a worker who then touches the machine thinking that it is stationary. In such cases, additional task lighting will help to reduce the strobing effect.

It should also be remembered that the best option for lighting is almost always natural light, since this is what our eyes have evolved to cope with and that allows us to see all colours in the spectrum. Some low-quality artificial light may not allow users to see colours clearly. This can cause problems when colour differentiation is critical, such as in electrical works or when reading colour codes on pipes, etc. In such cases, the artificial light will need to be improved or another source of better quality light added.

Marking and warnings

When necessary, all work equipment should be clearly marked in a manner that conveys information that is clear and concise, for example:

- maximum rotation speed of an abrasive wheel;
- safe working load of a lifting device; or
- maximum operating pressure of an air vessel.

A warning normally takes the form of a notice or a warning device (for example, 'Hard hats to be worn') or an audible signal, such as a ringing bell to indicate that a crane is about to move, or a flashing light on a forklift truck.

Safety helmet must be worn

Figure 5: Common warning sign – hard hats to be worn

Workspace

Designing an effective, safe and efficient workplace is a fascinating, challenging and vitally important subject. A workplace is complex; its daily operation will include the movement of workers, machines, raw materials, etc, which should be made as efficient as possible.

9.1 General requirements

The workplace should be designed so that workers and visitors passing through do not need to be too close to operating machinery. Noisy equipment should be located so that as little noise as possible is released to other work areas; this may necessitate the designation of areas where hearing protection is required.

APPLICATION 5

Spend a few minutes thinking about a work area with which you are familiar.

- How well organised is it?
- How might things be improved?
- Could the introduction of new equipment be accomplished effectively?

ADDITIONAL INFORMATION

Responsibilities of users

Users of work equipment have duties under general UK legislation (see 1.3: The most important legal duties for employers and workers). Specifically, these are to:

- take care of their own and others' safety, for example by ensuring machinery guarding is used that will protect the worker and others from things like flying debris from the machinery operation;
- comply with instructions – especially important when operating any type of equipment;
- use safety devices and protective equipment correctly and not deliberately tamper with them so that they will not work; for example, bypassing an interlocking guard (see 9.4: Control measures for machinery); and
- report any faults with work equipment/machinery to an appropriate person; this will ensure that other users of the equipment will not be harmed when operating it.

Any of these duties can apply to users of work equipment. Take a few minutes to think of some examples of your own.

KEY POINTS

- Equipment provided in any workplace must be suitable for the work requirements.
- Access to dangerous parts of machinery should be prevented in the normal course of work and steps taken when the equipment requires maintenance and/or repair; equipment should be maintained safely and the reasons for this understood.
- There will be situations in which the use and maintenance of equipment with specific risks needs to be restricted.
- There should be appropriate provision of information, instruction and training to anyone who could be at risk, including users, maintenance staff, managers and visitors.
- Actions required in emergency situations include stabilising equipment, emergency lighting and clearing the workspace.

9.2: Hand-held tools

> **Syllabus outline**
>
> In this section, you will develop an awareness of the following:
> - General considerations for selecting hand-held tools (whether powered or manual):
>   - requirements for safe use
>   - condition and fitness for use
>   - suitability for purpose
>   - location to be used in (including flammable atmospheres)
> - Hazards of a range of hand-held tools (whether powered or manual) and how these hazards are controlled

In 9.1: General requirements we discussed work equipment in general and the requirements of the Provision and Use of Work Equipment Regulations 1998 (PUWER). These requirements apply equally to hand-held tools. You will come across many hand-held tools, both unpowered and powered.

Here we are concerned with the requirements for selecting hand-held tools, whether or not they are powered, and the general hazards they present.

9.2.1 General considerations for the selection of hand-held tools

Hand-held tools can either be non-powered, such as hammers, files, chisels, saws and screwdrivers, or powered, such as electric drills, angle grinders, sanders, welding machines, circular saws and strimmers. The most obvious difference between the two types is the addition of a power source. Powered hand-held tools generate much more energy than non-powered tools and therefore have the potential to cause much more serious harm.

Whichever type you use, there are some general considerations to think about when selecting tools.

Requirements for safe use

Fuel sources such as petrol introduce further complications that will need to be taken into account when conducting risk assessments on the storage, refuelling and use of such equipment. Workers will need to be trained in appropriate steps for refuelling, how to avoid spillages, appropriate ventilation, overfilling, ignition sources such as static electricity, etc.

Personal protective equipment (PPE) will be required for many, or perhaps most, hand-held tools. A range of protection measures may be needed. These may include:

- anti-vibration gloves on equipment with significant vibration properties (such gloves may not always be particularly effective in reducing exposure to vibration and will usually need to be used with other control measures);

- barrier creams/thin gloves to protect against agents that could cause dermatitis;
- eye protection; and
- hearing protection.

> **APPLICATION 1**
>
> Can you think of any other PPE that is likely to be required for the use of hand-held equipment?

However, PPE may be just one aspect of the safe use of a piece of equipment in a particular situation. It is more important that workers are fully trained on how the equipment works; for example, adjusting/reversing the speed of a drill, using the appropriate adaptor when reaching into an enclosed area with a paint sprayer, how to 'set' the cutting tool when wood turning.

It may also be important that workers receive appropriate supervision while using the equipment, especially if they are new users. They should also receive relevant information on the use of the tool. For example, they should be aware of all the hazards associated with it.

There may also be specific hazards (such as noise and vibration) that should be identified and controls required for these (covered in more detail elsewhere). As with other hazards, a risk assessment must be used to assess all foreseeable hazards from using hand-held tools.

Condition and fitness for use

You first need to consider the condition of the tool. For example, checks must be made on the equipment to identify any faults, such as cracked casings, frayed cables, damaged plugs, or broken or blunt accessories such as blades for saws. If the tool is not in a good condition, it should not be used, and the condition should be reported to the appropriate person.

You also need to consider the fitness of the tool. This ties in closely with suitability for purpose, which we will discuss next.

Fitness for use is whether the tool selected will be able to cope with the job that is being carried out. So, for example, you would use a hammer when nailing two pieces of wood together, not a wrench.

> **APPLICATION 2**
>
> Thinking of your own experience, can you identify an occasion when you may have used a hand-held tool that was not quite fit for the job?
>
> Remember, hand-held work equipment covers a whole range of items, from knives to hammers and from electronic tablets to pneumatic drills.

Suitability for purpose

'Suitability for purpose' is one of the main considerations when selecting hand-held equipment.

As discussed earlier, you need to ensure that the tool you select is fit for the job to be done. For example, it is very unlikely that a hand-held saw would be suitable for cutting up a felled tree; a chainsaw would be a better fit for this job as it will be

Work equipment

quicker and take away some of the effort of doing the job. There may be less risk of musculoskeletal injuries as the 'sawing motion' and the effort that goes with this will not be required.

However, you must also realise that using a suitable tool, like the chainsaw, will come with its own set of hazards, such as refuelling and manual handling issues.

Figure 1: Equipment being used should be suitable for the job being done

In our example, a chainsaw will be much quicker but there is a higher risk of serious injuries if the equipment is used incorrectly or without the correct PPE to accompany it.

It is important to remember that 'suitability for purpose' must encompass the 'life cycle' of the work in question. This would include: taking the clean, dry, well-maintained piece of equipment to the place of work; putting it safely aside during breaks in the work; changing the battery; and cleaning and storage after work.

Where tools will be used

It is important to consider the environment in which work is taking place, so that the appropriate tools are used. In particular, intrinsically safe devices which greatly reduce the risk of sparks, for example drills, torches, radios and laptops, should be available for use in any areas where flammable atmospheres may be present, such as confined spaces (see also 10.2: Preventing fire and fire spread).

You will also need to consider the amount of space available to move around in. For example, when working in confined spaces you are unlikely to use a pneumatic drill due to its bulkiness (as well as the other hazards that this would cause).

Indoor or outdoor use is also an important consideration. Obviously, any equipment that will be used outdoors will need to be robust, as the equipment will generally be used in harsher environments. If it is likely to be wet (either because of the weather or the location of the work, such as next to a river), it is important to ensure that equipment selected is waterproof and powered by battery, if relevant.

You may also need to consider lighting. For instance, will additional lighting be required to supplement natural lighting to allow the equipment to be operated safely?

9.2.2 Hazards associated with hand-held tools

Later we will be dealing with the main mechanical and other hazards associated with machinery. In that context, we will be focusing on large, fixed machines. However, similar hazards often exist with portable power tools, so here we offer an introduction to the subject. There is some overlap with other elements in the syllabus, as indicated.

In no particular order, the hazards associated with portable power tools include:

- use of incorrect tool for the job being done (as discussed earlier);
- repetitive movements;
- faulty or poorly maintained equipment, such as loose hammer heads, blunt chisels, faulty plugs, frayed cables;
- parts rotating at high speed, such as drill chucks;
- electricity (see 11: Electricity);
- handling and storage of fuels such as petrol, diesel and biofuels (these are used in many construction and agricultural tools);
- ejection of materials;
- excessive noise levels (see 5.1: Noise);
- vibration (see 5.2: Vibration);
- creation of dust from cutting activities (see 7: Chemical and biological agents);
- sparks;
- hot surfaces, especially from those tools that use combustion engines as a power source; and
- trailing cables.

TIP

Some risks associated with the misuse of non-powered hand tools include:

- strained muscles, etc following prolonged use, repetitive movements or through having to use excessive force, for example when using a blunt chisel;
- being struck by flying materials, such as when a screwdriver breaks when being used as a lever to open a paint can;
- handles coming loose from hammers and striking the user or others in the vicinity; and
- cuts and abrasions sustained if the tool slips when in use (possibly because of excessive force or inattention on the part of the user).

Figure 2: Old and worn hammers can break when in use

Work equipment

Although there are risks from non-powered hand tools (see Tip box), it has to be said that these are relatively low and that most people use such tools without incident. However, there are a number of simple control measures that can be taken to further reduce the risks.

Control measures for hand-held tool hazards

Some of the hazards mentioned earlier will require specific control measures (for example, noise and vibration) and these are covered in detail elsewhere.

But some common control measures that can be applied to hand-held tools include the following:

- Use the right tool for the job, such as a proper lever to open a paint can or an ergonomic screwdriver when inserting a lot of screws.
- Maintain tools in good condition, such as keeping cutting tools sharp to minimise the force required.
- Carry out pre-use checks that the tool is not showing obvious signs of damage or wear that could increase the risk of injury.
- Store tools correctly when not in use to limit the potential for damage.
- Use toolbelts or lanyards to reduce the risk of tools falling from height.
- Ensure that high-speed rotating parts, such as the disc on an angle grinder or disc cutter or the blade of a circular saw, are adequately guarded.
- Use two-handed controls, such as may be found on an electric hedge trimmer. These may require the operator to simultaneously depress up to three separate switches, meaning that the user's hands cannot be near the blade (of course, this will offer no protection to any nearby individual).
- Test portable electrical appliances and make sure only intrinsically safe electrical equipment is used in flammable atmospheres (see 10.2: Preventing fire and fire spread).
- Use double-insulated tools or a reduced low voltage system (see 11: Electricity).
- Provide safe storage of fuels (such as petrol) in well-ventilated areas away from ignition sources and have clear systems of work for decanting and refilling operations, plus emergency procedures in case of spillage.
- Use eye protection to guard against flying debris, for example when using a strimmer or chainsaw.
- Select equipment with low noise and vibration characteristics and use hearing protection to guard against noise induced hearing loss.
- Wet down to reduce dust levels or wear respiratory protective equipment.
- Take care over placement of trailing cables; consider use of battery-operated tools as an alternative.
- Maintain a clean and tidy workplace with adequate lighting levels to minimise the chances of slips, trips and falls.
- Provide training in the proper use of tools; this is likely to be predominantly about developing high levels of skill, but safety should be built into this process.
- Use competent workers for high-risk hazards, such as storage and handling of fuels.
- Provide information on the hazards associated with the tool.
- Introduce administrative controls, such as job rotation, limiting the time spent on activities.
- Provide any other PPE that may be required.

9.2 Hand-held tools

Figure 3: PPE being used while grass strimming

Since there is a greater range of more serious hazards associated with powered tools, it should come as no surprise that more sophisticated control measures are needed for these.

To begin with, as with non-powered tools, the worker should choose the right tool for the job. In addition, equipment should be maintained in a state of good repair and efficient working order and should be periodically inspected by a competent person. The exact frequency of such inspections will depend on the amount of use and the chance of the equipment being damaged. Pre-use checks should also be performed.

Workers must be competent to use the equipment, familiar with any safe systems of work and know what to do if the equipment fails in use or if defects are identified.

CASE STUDY
Worker injured by unguarded farm machinery

Summary
A casual worker was injured when his arm was drawn into a potato harvester. The accident happened on the worker's second day at work on the machine. At the end of the day he reached across a processing section of the machine that should have been guarded to retrieve the last remaining potatoes from the machine as it was running clear. His arm was drawn up to the elbow in an in-running nip, breaking his fingers and forearm bones and removing muscle from the top of his forearm. The worker was airlifted to hospital.

Action
The farmer was prosecuted for failing to take effective measures to prevent access to dangerous parts of the machine. He pleaded guilty and was fined £5000 plus prosecution costs of £1561.

Work equipment

Advice

This accident was entirely preventable. The farmer had failed to guard dangerous parts on the second-hand machine that had been brought back into use after standing idle for a number of years. Unguarded or inadequately guarded machines continue to be a source of numerous serious accidents to adults (and children) on farms. All machines must be effectively guarded to help prevent any contact with the dangerous parts.

Employers have a duty (PUWER, reg 11 Dangerous parts of machinery) to ensure that equipment for use at work is effectively guarded before being used. A simple system for checking over equipment before use would have identified a missing guard.[1]

APPLICATION

Consider that you are going to use an electric saw to cut up some planks of wood for a project. This is likely to take you around 2 hours of continual work.

What control measures would you put in place before starting this task?

KEY POINTS

- General issues to consider when selecting powered or manual hand-held tools include the need for training and PPE.
- The condition of hand-held tools should be checked and whether they are able to cope with the job should be considered.
- The tool's suitability for purpose must cover the 'life cycle' of the job, including its cleaning and storage.
- It is important to think about the location in which the equipment will be used, including potential hazards such as a flammable atmosphere, inadequate ventilation and confined space.
- The specific hazards associated with various hand-held tools, both powered and manual, include excessive noise and vibration, creation of dust and sparks, trailing cables and ejection of materials.

Reference

[1] Adapted from HSE, 'Check equipment before use' (www.hse.gov.uk)

9.3: Machinery hazards

> **Syllabus outline**
>
> In this section, you will develop an awareness of the following:
> - Potential consequences as a result of contact with, or exposure to, mechanical or other hazards (see ISO 12100:2010 (Table B.1))
> - Hazards of a range of equipment:
>   - manufacturing/maintenance machinery (including bench-top grinder, pedestal drill)
>   - agricultural/horticultural machinery (including cylinder mower, strimmer/brush cutter, chainsaw)
>   - retail machinery (including compactor)
>   - construction machinery (including cement mixer, bench-mounted circular saw)
>   - emerging technologies (including drones, driver-less vehicles)

> **TIP**
>
> The main piece of guidance that we will consider is ISO 12100:2010 *Safety of machinery – General principles for design – Risk assessment and risk reduction*.[1]

Main mechanical and other hazards

ISO 12100:2010 is an internationally recognised standard relating to machinery safety. It looks at specific machinery hazards and the potential consequences if the hazard is realised. In addition, some machinery may present non-mechanical hazards, which may not at first sight be so obvious. It is important to understand that many of the consequences will not happen in isolation but could happen in combination. For example, should a car fall from lifting equipment during maintenance it could crush and trap anyone who is underneath.

> **DEFINITION**
>
> *A machine or machinery* can be defined as an "assembly, fitted with or intended to be fitted with, a drive system consisting of linked parts or components, at least one of which moves, and which are joined together for a specific application".[2]

We will first look at three types of motion that can be used to describe all possible machinery movements and then various hazards presented by the various movements of machinery.

Movement of machinery parts can be classified as one or more of the following:

- rotary motion;
- sliding (linear) motion; and
- reciprocating motion – forwards and backwards (can be classified as a form of linear motion).

These forms of motion will generate the mechanical hazards as defined in ISO 12100:2010. We will now look at each of the consequences of mechanical hazards from ISO 12100:2010.

9.3.1 Potential consequences of mechanical hazards

Being run over

This is one of the less common consequences but it still needs to be considered. This is most likely to happen when a worker is operating moving machinery. For example, paving machines laying asphalt are slow moving and may need adjusting during the operation. If the activity is not done correctly, there is a risk that a worker could be run over during the process. This is especially relevant where the machine moves forwards and backwards to complete an operation.

> **CASE STUDY**
>
> A nine-person crew was working with an asphalt road-widening machine. A member of the crew, who had only been on the job four days, was tasked with adjusting the machine's side-mounted spreader arm by walking alongside and to the rear of the paver. In this particular application, the paver laid down two layers of asphalt. The first layer was paved with the machine going forward. Then the operator reversed the machine to the starting point and put down the second layer while the first was still hot. After the paver had made a long, forward pass, the new worker decided to hitch a ride on the machine as it was backing up for the second pass. When they jumped up on the machine, their foot slipped, which resulted in them being run over by the right front tyre.
>
> The paver weighed 40 tons and amputated the victim's left leg and crushed their pelvis, with substantial damage to the thigh and groin areas. The worker later died in hospital.
>
> Accident investigators speculated that fresh, wet asphalt on the bottom of his boots may have caused the slip. The machine operator had told the new worker not to ride on the side of the machine, but to climb on top if he wanted a ride. Also, co-workers suggested that the victim might have been fatigued or suffering from heat stress.[3]

Being thrown

This involves a person being thrown from a machine. For example, any rider-operated equipment could result in the driver and/or passenger being thrown during acceleration or deceleration if they are not properly restrained.

Crushing

Figure 1: An example of a crushing hazard

Crushing occurs when one part of machinery moves against a part of the body that is in some way trapped against a fixed object, such as a wall or another part of the machine.

Other examples of crushing hazards include:

- the ram of a forging hammer;
- the tools of power presses;
- the callipers of spot welding machines;
- the closing 'nip' between two plate motions;
- the table of a machine tool and a fixed structure such as a nearby wall;
- reversing vehicles in a loading bay; and
- industrial robots.

Figure 2: Workers could be crushed between the industrial robots and the fences/barriers on either side

Work equipment

Cutting or severing

Cutting hazards can be presented by the wide range of machines used to cut and shape every type of material, whether wood, metal, stone, plastic, etc. Cutting or severing is caused by one blade, rather than two (which would be shearing, as described later). Examples of machines that produce this type of consequence include:

- power saws;
- industrial paper guillotines;
- grinding machines;
- planer machines;
- laser cutting and forming equipment;
- handsaw blades;
- rotary knives; and
- water jet cutting.

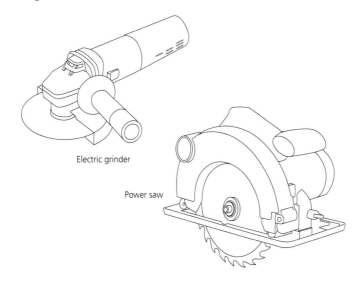

Figure 3: Electric grinder and power saw

Any device similar to those pictured in Figure 3 is capable of cutting due to its own speed of movement when it comes into contact with the body. Cutting or severing effects may be made even more serious if the body part is somehow trapped and unable to move away.

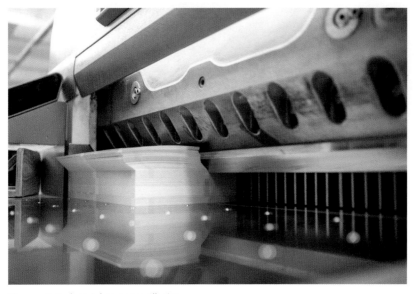

Figure 4: Industrial paper guillotine

Drawing in or trapping

Later we will look at 'in-running nips' as part of entanglement consequences. Here we will look at drawing-in/trapping consequences. These tend to occur when something is drawn in between gear wheels or rollers. It is usually a worker's hand that will get drawn in and become trapped.

Figure 5 shows some more illustrations of common 'drawing-in' hazards:

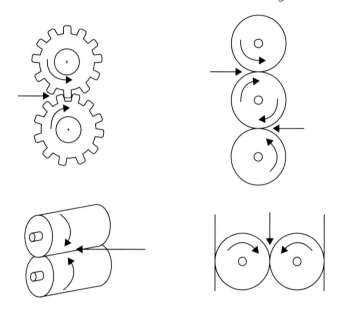

Figure 5: Examples of common 'drawing-in' hazards

Entanglement

Entanglement may occur if a person or their clothing, hair, jewellery, etc comes into contact with moving parts of a machine. There are several ways in which this can happen.

Contact with a single rotating surface

These may be a source of danger, even when they are rotating slowly; probably the greatest hazard presented by a slowly rotating surface comes from possible entanglement. Examples of single rotating parts include:

- spindles (for example, on an abrasive wheel);
- chucks (for example, on a drill or a lathe); and
- rotating workpieces (pottery wheel, woodturning).

Catching on projections or in gaps

The entanglement hazards are increased if the rotating part has projections or gaps. The distinction between projections and gaps may not always be clear; for instance, gear wheels can be viewed either as a series of projections or a series of gaps between the teeth.

Either way, they can catch onto clothing and draw the person into the machine (as well as causing cuts or abrasions). Examples of rotating machine parts include:

- fan blades;
- spoked wheels and pulleys;
- gear wheels;
- flywheels;

- projecting pins and keys; and
- drive belts on machinery.

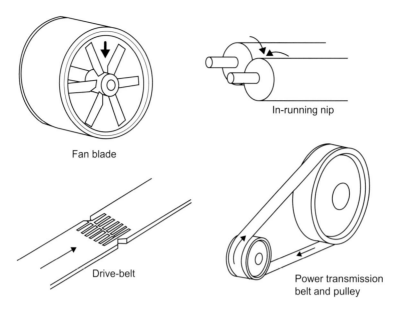

Figure 6: Fan blade, drive belt, power transmission belt and pulley

Catching between two parts

The entanglement hazards are even greater if the machine contains parts that rotate or move against each other to create in-running nip hazards. There are various ways this can happen:

- between counter-rotating parts, such as gear wheels, or material could be drawn between two rolls;
- between rotating and tangentially moving parts, such as a power transmission belt and its pulley; a chain and chain wheel; a rack and pinion; or metal, paper, rope, etc and a reeling drum or shaft; or
- between rotating and fixed parts, for example revolving mixing and mincing mechanisms, which are found in a wide variety of food processing equipment.

Some mechanisms may comprise both sliding and turning movements and therefore there is a much higher risk of injury. An example of this is a baling machine.

Figure 7: Baling machine

However, entanglement is not the only harm that in-running nips can cause. They can also injure the person directly through shearing or crushing.

> **TIP**
> In all the entanglement cases outlined here, the risk is increased by loose clothing, gloves, ties, jewellery, hair, use of cleaning brushes, rags, and so on.

Friction and abrasion

Friction burns can be caused even by relatively smooth parts if they operate at high speed, for example, the exposed rim of a laundry centrifuge. Abrasions usually occur when rough surfaces move against each other. Friction or abrasion hazards can damage a worker's skin and/or muscles.

Other examples of friction or abrasion hazards include:

- the periphery of an abrasive wheel;
- belt sanding machines; these come in various forms, such as stationary types of belt sanding machines in which the material to be sanded is 'taken to the sander' and hand-held types in which the sander is taken to the object (floor, furniture) to be sanded;
- material running onto a reel or shaft; and
- a conveyor belt and its drum or pulley.

Impact

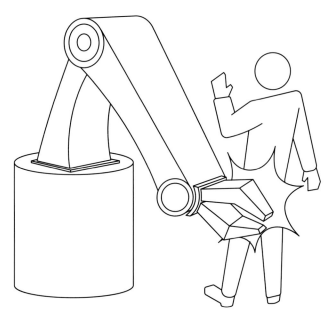

Figure 8: Being struck by a robotic arm could cause an impact injury

Various impact injuries may occur if a person is struck by, for example:

- the traversing motion of a machinery part, such as a blow from a robotic arm (see Figure 8). In addition to the common use of robotic systems in factories and workshops around the world, they are increasingly used outside the factory/workshop setting;
- any part of the equipment that projects in some way; either the 'active' part of the machine, which actually does the welding, for example, or the counterweights that keep the system balanced; and
- moving vehicles.

Impact injuries may be even more serious if the victim is crushed between the moving part and a solid object such as a wall.

Shear

Shear is when one or more surfaces move relative to each other in the same plane, creating a trapping point where they cross, just like a pair of scissors. Machine parts can move in such a way and with sufficient force to cause very serious damage to any parts of the body that come between them. Such shearing hazards can occur between machine parts such as:

- the moving parts of a metal planing machine and its bed;
- the blade and the fixed edge of a manual guillotine; and
- connecting rods and fixed parts.

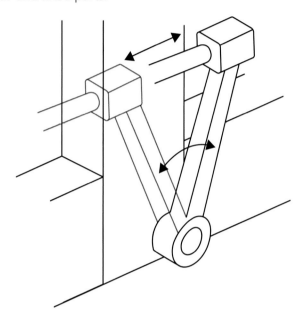

Figure 9: Example of a shearing hazard between two reciprocating parts

Slipping, tripping and falling

As with other slips/trips/falls covered elsewhere, they are also potential consequences associated with mechanical hazards; for example, falling from a piece of working-at-height equipment. This topic is discussed in more detail in 8.2: Working at height and 8.5: Slips and trips.

Stabbing and puncture

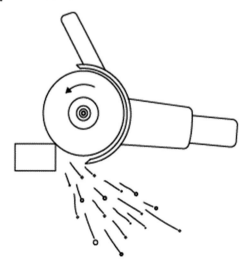

Figure 10: Particles being ejected from a grinder

Stabbing and puncture could be caused by a part of the machine itself, either in its normal course of motion or perhaps when parts of the machine are ejected when the machine fails in some way, for example a drive band on a conveyor system breaking. In addition to the stabbing and puncture hazards presented by the actual machine and its components, you also need to take account of any material hurled from the machine in its normal course of operation, such as wood chips from a woodturning device, thin flakes of hot metal from a grinding wheel and so on.

The body may be penetrated by the following:

- rapidly moving parts of machinery or pieces of material; for example, sewing and drilling machines; and
- flying objects ejected from machinery, which may be parts of the actual machinery; for example, the flying shuttle of a loom, a bursting abrasive wheel or other material, such as flying swarf or debris from grinders, rotary mowers and hedge cutters.

(Note: this discussion is not concerned with the hazards presented by medical 'sharps' even though this hazard is clearly 'stabbing'. Such biochemical hazards are discussed in 7: Chemical and biological agents.)

Suffocation

This is not a common consequence from mechanical hazards but should be considered during the machinery risk assessment where relevant. For example, the risk of suffocation may be high in grain storage and conveying systems.

Non-mechanical hazards

Most non-mechanical hazards associated with machinery are covered elsewhere. The non-mechanical hazards include:

- electrical hazards (see 11: Electricity);
- thermal hazards: some machinery will produce either extreme heat or cold and consequences from this include burns (from heat or cold), dehydration, discomfort, frostbite, scalding and injuries caused by radiating heat sources;
- noise hazards (see 5.1: Noise);
- vibration hazards (see 5.2: Vibration);
- radiation hazards (see 5.3: Radiation);
- material/substance hazards (see 7.1: Hazardous substances and 7.2: Assessment of health risks);
- ergonomic hazards (see 6.1: Work-related upper limb disorders and 6.2: Manual handling); and
- hazards associated with the environment in which the machine is used, such as temperature, wind, moisture and dust.

> **APPLICATION**
>
> Think of some examples of machinery with which some of these non-mechanical hazards are associated. Are there any examples in your own workplace?

Combination of hazards

Although this is not a specific hazard, it should be considered when carrying out a risk assessment. For example, many machines will have both noise and vibration issues; therefore, suitable control measures must be implemented for both these hazards.

One example is the use of a high-power laser-cutting machine, which presents both the 'obvious' hazard of the intense laser beam and also a 'material/substance hazard' because of the large amount of toxic fumes produced during the metal cutting process.

Other mixed hazard examples include high-frequency noise produced by machines used to cut tiles (that is, in addition to the cutting hazards) and welding equipment for which the electrical hazards may be overlooked.

9.3.2 Hazards associated with specific machines

Some of these machines have already been mentioned, but we include them again here for convenience and completeness:

- manufacturing/maintenance machinery – bench-top grinder, pedestal drill;
- agricultural/horticultural equipment – cylinder mower, strimmer/brush cutter, chainsaw;
- retail machinery – compactor, checkout conveyor; and
- construction machinery – cement mixer, bench-mounted circular saw.

Specific guarding requirements are given in 9.4: Control measures for machinery. Here we look at some of the common hazards and controls for these specific items of equipment. You will see that each piece of equipment has a combination of hazards that we looked at earlier.

Type and example of machinery	Hazards	Controls
Manufacturing/maintenance machinery Bench-top grinder	• Stabbing/puncture: ejection of flying materials • Drawing-in hazard: between tool rest and grinding wheel • Entanglement hazard: from rotating spindle • Abrasion hazard: contact with rotating wheel • Material/substance hazard: grinding dust, which may be a health hazard depending on what material is being ground • Electricity • Noise • Vibration	• Fixed guarding to sides and rear of the wheel • Properly adjusted tool rest • Wheels changed and dressed, and grinder used only by a competent person • Portable appliance testing (PAT) • Possible use of local exhaust ventilation • Eye/full-face protection • Users to ensure no loose clothing, hair, jewellery, etc

9.3 Machinery hazards

Type and example of machinery	Hazards	Controls
Manufacturing/maintenance machinery Pedestal drill	• Entanglement: in the rotating chuck • Stabbing/puncture: ejected material • Entanglement: nip points between drive belts and pulleys (older machines) • Material/substance hazard: metal cutting fluids (dermatitic agent) • Ergonomic hazard: some repetitive activities • Noise • Electricity • Vibration: blunt drill bits (could also cause overheating of material being drilled or cause work pieces to snag on the drill bit)	• Fixed guard over gearing, belts and pulleys • Movable guard over chuck • Pedestal securely fastened to floor or bench • Use of mini-vice to hold work piece • Competent users • Regular maintenance and inspection • Emergency stop control • Eye protection
Agricultural/horticultural equipment Cylinder mower	• Entanglement: with the rotating mower cylinder or its drive train • Shearing: sharp blades If powered: • Material/substance hazard: petrol • Noise • Thermal hazard: hot engine surfaces	• Fixed cover over top of blade cylinder, engine/motor • Safe storage of fuel, with procedures for filling • Wearing of eye protection • Hearing protection • Information, instruction and training for users
Agricultural/horticultural equipment Strimmer/brush cutter	• Stabbing/puncture: flying debris (stones, etc) • Cutting/severing: contact with cutting head • Entanglement: in rotating parts • Material/substance hazard: petrol • Material/substance hazard: engine fumes • Noise • Vibration • Ergonomic hazard: from repetitive movement • Material/substance hazard: biological agents (for example, from disturbed animal faeces or snakes) • Ultraviolet radiation: from the sun	• Cover fitted over strimming cable/cutting blades • Ergonomically designed handholds • Safe storage of fuel, with procedures for filling • Job rotation to limit effects of vibration and noise exposure • Use of harness to support the weight of the strimmer • Eye/full-face protection • Hearing protection • Gloves

Work equipment

Type and example of machinery	Hazards	Controls
Agricultural/horticultural equipment Chainsaw	• Cutting/severing: contact with chain • Environment hazard: uneven ground with the potential for trips and falls • Stabbing/puncture: ejected debris • Ergonomic hazard: 'kick back' when cutting • Material/substance hazard: petrol • Thermal hazard: burns from hot engine parts • Material/substance hazard: engine fumes • Noise • Vibration • Ergonomic hazard: from prolonged use and possible poor posture	• Guard between handhold and chain • Fixed guard over motor • Chainsaw designed to minimise vibration and direct exhaust away from operator • Competent users only; must also be fit and able-bodied • Job rotation • Safe storage of fuel, with procedures for filling • Regular maintenance and inspection • Full-face protection • Chainsaw jackets, boots, shin guards and gloves (Kevlar)
Retail machinery Compactor	• Trapping and/or shearing: between compactor ram and compactor body • Electricity • Impact hazard: from vehicles • High pressure: from hydraulic system	• Fixed guard around compactor and operating mechanism • Interlocked access doors • Maintenance and inspection • Information, instruction and training for operators • Markings and warning signs
Construction machinery Cement mixer	• Entanglement: in rotating parts • Drawing-in hazard: between motor pulley and belt drive • Electricity (if electrically powered) • Material/substance hazard: petrol (if petrol driven) • Material/substance hazard: cement dust • Ergonomic hazard: from loading and unloading the cement mixer	• Fixed guards over the motor and drive gear • Use by trained operators only • Safe fuel storage and filling procedures • PAT (if electrically powered) • PPE to reduce effects of noise, protect the hands and reduce inhalation of cement dust
Construction machinery Bench-mounted circular saw	• Cutting/severing: working near exposed blade • Stabbing/puncture: such as ejected dust or splinters • Drawing-in hazard: between belts and pulleys • Electricity • Noise • Vibration	• Fixed guard over motor and blade • Self-adjusting guard over section of blade used for cutting • Securely bolted to work bench • PAT • Use and blade change by competent persons only • Eye and hearing protection • Possible need for extraction ventilation

9.3 Machinery hazards

Type and example of machinery	Hazards	Controls
Emerging technologies Drones	• Entanglement: with propellers • Cutting/severing: from propellers • Electrical hazard: from charging • Ergonomic hazard: when carrying drone (effort required) • Impact: speed of drone could cause injury to people in the area or damage to property • Stabbing/puncture: fragments from propellers or body of drone if impact with ground or stationary object	• Move away from drone before starting the propellers • Ensure fully charged before flying – reduces risk of crash • Check environment before flight (stay away from overhead obstructions such as power lines) • To be flown by competent people • Maintenance and inspection (before use) • Ensure drone does not go out of communicating distance with controller (reduces risk of crashing) • Ensure GPS lock with drone before flying
Emerging technologies Driverless vehicles	• Impact: with other vehicles, pedestrians or buildings (loss of control or hijacking of system by third party) • Being run over: pedestrians, due to loss of control • Electricity: during charging • Radiation: from high levels of electromagnetic field radiation required to power the car • Material/substance hazard: lithium batteries highly combustible – could lead to fire. Could be thermal runaway if battery is damaged during accident	• Regular maintenance/servicing of vehicle • Regular checks on vehicle charging stations • Ensure software is latest version • Avoid stress and anxiety by ensuring passengers are happy to travel in vehicle

APPLICATION

What do you think might be the consequences of the mechanical hazards presented by an abrasive wheel?

While the abrasive wheel is not a common piece of work equipment in most workplaces, you can see that, even in one small machine, most of the mechanical hazards may be present.

Managing Health and Safety in the UK 491

Work equipment

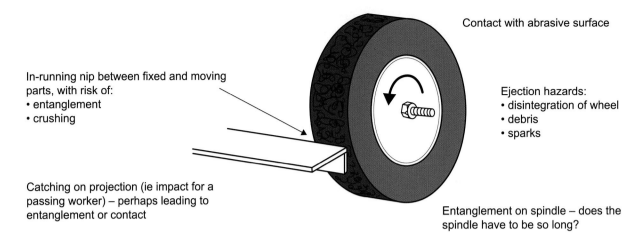

Figure 11: The hazards of an abrasive wheel

KEY POINTS

- Potential consequences as a result of contact with, or exposure to, mechanical or other hazards are described and defined in ISO 12100:2010.
- Hazards are associated with equipment that might be found in any workplace, including those relating to emerging technologies.
- Mechanical hazards can be generated by the rotary, sliding or reciprocating motion of machinery parts.
- Consequences of mechanical hazards can include being run over by or thrown from a machine; being crushed, cut or trapped; and getting friction or abrasion burns, impact injuries and stabbing or puncture wounds.
- Non-mechanical hazards associated with machinery include electrical, thermal, noise, vibration, substance and ergonomic hazards.
- Combinations of hazards should be considered in risk assessments.

References

[1] International Organization for Standardization, ISO 12100:2010 *Safety of machinery – General principles for design – Risk assessment and risk reduction* (2010) (www.iso.org)

[2] See note 1 (clause 3.1)

[3] Adapted from *Equipment World* (April 2016)

9.4: Control measures for machinery

Syllabus outline

In this section, you will develop an awareness of the following:

- The basic principles of operation, advantages and limitations of the following control methods:
  - guards: fixed, interlocking and adjustable/self-adjusting
  - protective devices: two-hand, hold-to-run, sensitive protective equipment (trip devices), emergency stop controls
  - jigs, holders, push-sticks
  - information, instruction, training and supervision
  - personal protective equipment
- Use of the above control methods for the range of equipment listed in 9.3
- Basic requirements for guards and safety devices:
  - compatibility with process
  - adequate strength, maintained
  - allow for maintenance without removal
  - do not increase risk or restrict view
  - are not easily bypassed

We are now going to look at the basic principles of operation, plus the advantages and limitations of a range of guards and protective devices commonly found on workplace machinery.

Before we examine machine guarding, however, we will first set out the definitions that we will use (some of these will already be familiar to you, particularly from 1: Why we should manage workplace health and safety).

DEFINITIONS

- **Machinery** *is apparatus for producing or applying power, having fixed or moving parts each with definite functions.*
- A **hazard** *is something that has the potential to cause harm, such as personal injury or ill-health.*
- **Risk** *is a measure of the likelihood that a hazard will result in harm together with the resulting severity.*
- A **safeguard** *is a guard or device designed to protect persons from danger.*
- A **guard** *is a physical barrier that prevents or reduces access to a danger point.*
- A **safety device** *is a device other than a guard that eliminates or reduces danger.*
- A **safe working practice** *or **safe system of work** is a method of working that eliminates or reduces the risk of injury.*

- *An **interlock** is a safety device that interconnects a guard with the control system or the power system of the machinery.*
- ***Failure to danger** is any failure of the machinery, its associated safeguards, its control circuits or its power supply that leaves the machinery in an unsafe condition.*
- ***Failure to minimal danger** (also known as **failure to safety**) is any failure of the machinery, its associated safeguards, control circuits or its power supply that leaves the machine in a safe condition.*
- ***Integrity** is the ability of the devices, systems and procedures to perform their function without failure or defeat.*

> **TIP**
>
> The main piece of legislation associated with this topic is the:
>
> - Provision and Use of Work Equipment Regulations 1998 (PUWER); or
> - Provision and Use of Work Equipment Regulations (Northern Ireland) 1999.
>
> The Health and Safety Executive (HSE) has also produced
>
> - *Safe use of work equipment. Provision and Use of Work Equipment Regulations 1998. Approved Code of Practice and guidance* (L22).[1]
>
> This approved code of practice has also been approved for use in Northern Ireland by the Health and Safety Executive for Northern Ireland (HSENI).

The principles, techniques and practices for the safeguarding of machinery are relevant to designers, manufacturers, suppliers, installers and users.

9.4.1 Principles of operation of machinery control measures

As a general statement of principle, employers need to take effective measures to prevent access to dangerous parts of machinery or stop their movement before any part of a person enters a danger zone.

The term 'dangerous part' can be taken to mean any piece of work equipment that could cause injury if it is being used in a foreseeable way.

Guards

As with many things in the world of health and safety management, we can apply a hierarchy to the choice and use of engineering controls.

The hierarchy F I A T is commonly used when discussing machine guarding.

- **F** fixed guards
- **I** interlock guards
- **A** automatic guards
- **T** trip devices

Fixed guards

A fixed enclosure guard will completely enclose the hazard at all times while, if necessary, allowing entry and exit for the raw materials being worked on.

9.4 Control measures for machinery

Note that we refer to 'completely enclose the hazard', not 'completely enclose the machine'; a classic example of this is a nip guard across the gap between the rollers on a conveyor system.

Fixed guards should be removable for machine maintenance but should require the use of a special tool to do so. Clips or similar are not adequate methods of fastening a fixed guard.

A limitation of a fixed guard is that it is not necessarily going to be connected to the machine's power supply, so it is possible to operate the machine without the guard in place.

Figure 1: Fixed guard over belt drive

The guard, therefore, needs to be 'conveniently removable' (perhaps with special tools) but does not need to be 'instantly and easily removable'. A guard that proves difficult to remove (perhaps the screws holding it have rusted) may end up not being replaced in the belief that this will save time when maintenance is next required.

Another form of fixed guard is the distance guard; this is usually a fence around the hazard, such as safety fencing around an automated welding line. Gates into the welding area will be interlocked with the welding mechanisms.

Another example of a fixed guard is a tunnel guard on a machine.

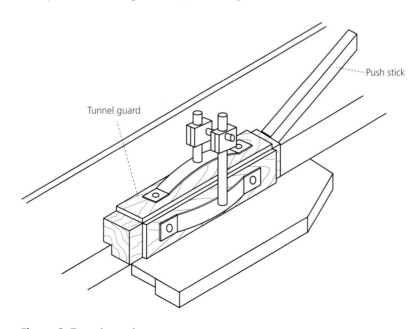

Figure 2: Tunnel guard

Self-closing guards

These are a form of fixed guard that operate on a spring-loaded mechanism, which ensures that the guard is kept in place while still allowing the equipment to be used. The self-closing guard on a circular saw is a good example of this. As the saw is used to make the cut, a portion of the guard will be pushed upwards, allowing the blade to come into contact with the work piece. At that point, it will be hard for the operator to get their fingers in the way. Once the cut is finished, the self-closing guard will spring back into place over the blade, thus preventing access to it.

Adjustable guards

An adjustable guard is yet another variation on the fixed guard, only instead of being permanently fixed in place or self-adjusting, it is manually adjusted.

An example of such a guard would be one over a bench-mounted circular saw. These provide a very effective means of guarding the top of the saw blade but have limitations in that they do not automatically conform to the thickness of the material as a self-adjusting guard would do, but must be set each time the saw is used. They also need a user with a good standard of skill and competence to use them effectively.

Interlocking guards and systems

When enclosure by a fixed guard is not possible, an interlocking system of protection may be practicable. An interlock system will ensure that, if the hazard is exposed while the machine is in operation, the power will be switched off. An interlock guard will prevent a machine from starting up at the wrong time or stop a machine operator from making an unsafe move.

Some examples of equipment using interlock guards include food mixers and washing machines. Interlock guards can be electrical, pneumatic (air pressure), hydraulic (liquid pressure) and mechanical interlock devices.

Mechanically interlocked gates can also be used. An example is a hydraulic press used in forming metal objects such as sinks. When the gate is open and material is being fed into the press, the interlock mechanism will be designed to prevent the press from operating. When the gate is closed, the interlock mechanism allows the press to be set in motion.

> **TIP**
>
> Interlocks are an effective method of ensuring that people are kept from coming into contact with a hazard. However, they do have a limitation of being relatively easy to override.
>
> Electrical interlocks are especially prone to this, so supervisors should keep a lookout to check that they are being used correctly. Where interlock fencing is used, it may be possible to simply climb over the barrier to avoid the interlocked gate altogether.
>
> Alternatively, someone could enter the danger zone and shut the gate behind them; the interlock does not 'know' if people are present or not, so without more forms of protection, it is still possible for a person to be injured.

Over-run protection

The interlock system may incorporate a time-delay device so that, for example, the door on a spin dryer will not open for 30 seconds once the cycle has finished, which will allow ample time for the drum to come to rest. This is just one type of what is known as over-run protection. Another example is train doors that incorporate motion-sensing devices and will not open until the train has stopped. Over-run devices may also be used in conjunction with automatic and trip guards, which we will now discuss.

Automatic guards

An automatic guard is sometimes referred to as a 'sweep guard'. It will push the operator aside as the machine operates. A power press is an example of this. The operator will start the machine by using a shrouded foot switch (shrouded to avoid accidental operation) to activate the press and the guard.

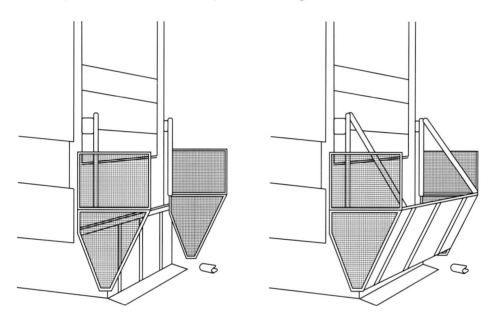

Figure 3: An automatic guard on a power press

Automatic guards are a less-favoured option than a fixed guard. This is mainly because the worker could overbalance and fall when being pushed aside. They are only really appropriate on slow-moving machinery; a rapidly descending press would cause the automatic guard to deploy equally rapidly, which could cause serious injury to the worker.

Protective devices

As well as guarding, there are other protective devices that can be used to protect workers when working with machinery. We will now look at some of these.

Two-handed devices

Two-handed controls require the operator to have both hands on the controls at the start and throughout the machine cycle. Such devices can protect the operator's hands if other types of guarding are impractical. They must be designed so that bridging the controls with one hand is not possible; both controls must be operated simultaneously. Clearly, a disadvantage of these devices is that they only protect the hands of the operator, leaving other people at risk. Such devices should only be used when rapid manoeuvrability of the work piece is required, the operator works alone on the machine (which is appropriately positioned in the workplace) and assessment shows that the overall risk is low.

Hold-to-run devices

As the name suggests, this is a control that is required to be held in a set position to allow the machinery to run. If released, the machinery will automatically come to a halt. As with two-handed devices, these devices only offer protection to the operator. However, for operations that require a limited range of movements and access to the area, these are useful.

Trip devices

There are various mechanical and photo-electric trip devices that cut power to the machine if the trip device is triggered. For example, a machine may be surrounded by a pressure-sensitive mat designed so that the machine will not operate if someone is standing on the mat. Another trip device is a 'light curtain'. These shine beams of light between sensors. If the beams are broken this will stop the machine from working.

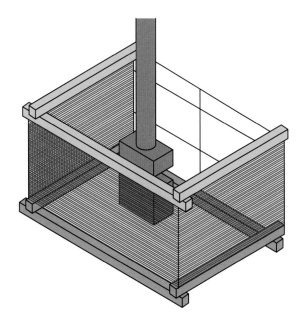

Figure 4: The light curtain can be seen on three sides of the machine

Trip systems have the obvious advantage of allowing clear access to the machine(s) but have the disadvantage of requiring regular skilled maintenance. Such systems must be checked at the start of each shift.

Emergency stop controls are discussed in 9.1.6: Safe use of work equipment and emergency controls.

Jigs, holders and push-sticks

These devices do not prevent access to the dangerous parts of the machine – they merely help to keep the operator's hands at a safe distance.

Jigs and holders are used to keep material in place. For example, they will hold the material close to a cutting blade while at the same time keeping the operator's hands away from the cutting area.

Push-sticks are used to help feed material through a piece of equipment; again these help to keep the operator's hands out of the 'danger area'.

Their effectiveness therefore depends on correct use by the operator, and it should be remembered that the other, more effective forms of guarding mentioned earlier should always be used in preference.

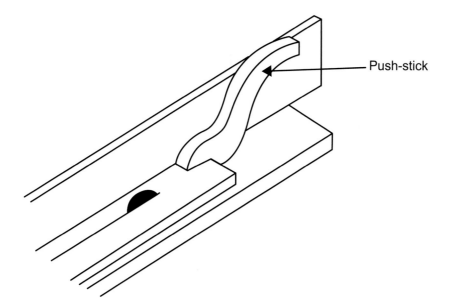

Figure 5: A push-stick being used to feed wood into a circular saw[2]

Information, instruction and training

Once everything possible has been done to eliminate risk through the choice of machine and fitting of appropriate mechanical safeguards, there may still be some residual risk that should be considered. There may be some cutting edges that cannot be guarded or the machine may throw out debris that could hit an unsuspecting person passing by. Therefore, adequate information and instruction to those working with, or in the vicinity of, the machinery must be provided. This could be in the form of working procedures.

Workers need to be provided with information on:

- how the equipment can be used safely;
- the general working environment required to operate the machine safely; and
- common problems that could happen when the machine is operated and how these can generally be resolved.

In addition, information can be conveyed via signage, for example signs instructing people passing by to keep out of areas where they may be hit by mechanical parts or struck by objects ejected from machines. Workers will also require specific instruction on how to operate the machinery. This can be done by producing written procedures or safe systems of work.

Another valuable source of information is manuals supplied by the machine manufacturer; all relevant information from these manuals should be included in written procedures/safe systems of work.

All information and instruction should be made available to relevant workers, whether they are temporary or permanent workers. In some cases, an experienced worker (supervisor) can convey instructions to other workers verbally. In these cases, the supervisor must observe the worker to ensure that they have understood the instructions and are working safely.

Workers will also need to be trained. Provision of information and instruction alone is not sufficient to ensure the machine is being operated safely. Therefore, all relevant workers must be trained; the level of training required will depend on the competence of the worker. For example, if they have used the same or a similar machine in other workplaces the worker may just need 'refresher' training, whereas someone who has not operated machinery before will need more intensive training.

Refresher training will need to be given to all relevant workers at various times to ensure that they are aware of the hazards and controls associated with the machinery. This is especially relevant if the machine has been modified or if the worker is returning to work after an extensive absence.

Personal protective equipment (PPE)

Some cases may require a combination of measures; reducing or eliminating some hazards and providing safeguarding and introducing safe working practices to help reduce risks from other hazards. In other cases, these measures will not be able to provide sufficient control and workers may be required to wear or use personal protective equipment (PPE).

In the case of machinery hazards, PPE could include hard hats and face shields to offer protection against being struck by moving parts or flying debris, gloves to protect against abrasive surfaces (remembering that gloves may increase the risk of entanglement) and boots to protect against objects dropped from the machinery.

To many managers, PPE may seem a cheap and easy option. However, it is not as cost-effective as it may seem. It does not eliminate the problem and, at best, mitigates it only for those who wear the PPE, while leaving others at risk. PPE is effective only if backed up by appropriate training and supervision, and these also cost money.

See further 9.3.2: Hazards associated with specific machines for how these different types of control arrangements are used on a range of different equipment.

9.4.2 Basic requirements for guards and safety devices

One of the first things to consider is that the guard must be compatible with the process. In other words, the guard must be suitable for the materials being machined and must not be weakened or otherwise adversely affected by lubricants, coolants and so on.

Guards also need to be strong enough to withstand impact from ejected materials and robust enough to tolerate prolonged and often quite harsh use.

To ensure that a guard continues to offer effective protection, it must be regularly and properly maintained, preferably while in position, since the need to remove the guard to maintain it may result in it not being refitted correctly, or even at all.

The guard, which is meant to protect workers, should not increase the risk. For instance, if the guard is solid metal and does not offer a view of the process, there may be a temptation not to use it. Alternatively, a guard that does not afford adequate ventilation in the way that, say, a wire mesh guard would, may lead to overheating of the machine or process.

Finally, it should not be easy to bypass a guard. Inventive workers will often find a way to operate a machine without the guard in place, so a combination of good design and effective supervision is essential.

9.4 Control measures for machinery

KEY POINTS

- The basic principle of machinery control measures is that employers should take effective measures to prevent access to dangerous parts of machinery or stop their movement before any part of a person enters a danger zone.
- There is a hierarchy that can be applied when selecting guarding options.
- Different types of fixed guards, interlock guards, automatic guards and trip devices all have advantages and limitations.
- Other types of protective device, information and training, and PPE can also help to protect people from machinery hazards.
- Whatever guarding measure is used, it must be compatible with the process, regularly and properly maintained and not easy to bypass.

References

[1] HSE, *Safe use of work equipment. Provision and Use of Work Equipment Regulations 1998. Approved Code of Practice and guidance* (L22, 4th edition, 2014) (www.hse.gov.uk)

[2] Adapted from HSE L22 (see note 1)

ELEMENT 10

FIRE

10.1: Fire principles

> **Syllabus outline**
>
> In this section, you will develop an awareness of the following:
> - The fire triangle: sources of ignition; fuel and oxygen in a typical workplace; oxidising materials
> - Classification of fires: A, B, C, D, F and electrical fires
> - Principles of heat transmission and fire spread: convection, conduction, radiation, direct burning
> - Common causes and consequences of fires in workplaces

Fire is a major risk to any organisation. It doesn't matter whether this is a low-risk office-based organisation or a high-risk chemical manufacturer. The consequences of a fire are devastating; some organisations never recover from the effects of fire and cease trading. When organisations can recover, it can take years to get back to the same state that the organisation was in prior to the fire.

To illustrate this, many people will remember the fire that destroyed part of Notre-Dame Cathedral in Paris in April 2019. It took less than 90 minutes from the alarm being raised for the spire to collapse. Around 400 fire-fighters were engaged in trying to control the fire. Fortunately, no one was killed but three emergency workers were injured. It is estimated that it will take between 10 and 20 years to return the building to its former glory.

It is therefore vital that organisations understand the risk of fire and what they need to do to control these risks. However, to understand how to control fire risks it is first necessary to understand the principles of fire. In this section we will be looking at fire classifications, how fires can start and, once they have started, how heat can be transmitted to spread the fire. Finally, we will look at some of the common causes of fire and the consequences of fire.

> **DEFINITION**
>
> *Fire is a chemical reaction giving off heat, light and (most of the time) smoke and is the visible effect of the process of combustion.*

10.1.1 The fire triangle

There are three essential elements required for a fire to start and to continue to burn. These are heat, fuel and oxygen; these are known collectively as the 'fire triangle'.

Remove any one of these components and a fire will not start, or it will be extinguished if it is already burning. We will look at this concept later when looking at the principles of fire prevention (stopping the fire from starting) and protection (mitigating the effects of fire).

Figure 1: The fire triangle

Sources of ignition (heat)

Heat is not the only thing that might cause a fire to start. It is probably better to describe this side of the fire triangle as 'sources of ignition' rather than just heat. A source of ignition is anything that has the capability of igniting materials and starting the combustion process.

Other forms of energy, such as a spark, can also contribute to a fire starting. Some common workplace ignition sources include:

- friction, for example from worn parts of machinery;
- hot surfaces;
- faulty electric supplies;
- faulty electrical equipment;
- static electricity;
- tools that can cause sparks by friction, such as grinding wheels;
- open flames, for example from a blowtorch or welding activities;
- smoking materials (matches and unextinguished cigarettes);
- lightning strikes;
- radiant heat from the sun; and
- faulty heating systems, including hot air blowers or electric bar heaters.

> **TIP**
>
> **Combustion**
>
> For a material to burn, there must be an energy (heat) source. There first must be sufficient energy applied to the material to raise it to a temperature that will generate vapour that must then be ignited. The heat component in the fire triangle therefore relates to the heat required to vaporise the fuel **plus** the heat to provide the means of ignition.
>
> Close inspection of any burning liquid or solid, for example ethanol or a piece of wood, reveals that the flame is not actually in contact with the liquid or the solid but is a small distance above it; fire is therefore a gaseous reaction – it is the vapour that is burning.

Sources of fuel

For a fire to start, there must be the material to burn (the fuel). Fuel is any material that will combust given the right set of conditions. Most things will burn; it is simply a matter of applying sufficient energy to them. Fires are categorised by the fuel involved (this will be discussed later).

Fire

Typical sources of fuel include:

- wood;
- paper and cardboard;
- fabrics;
- flammable liquids, including oils and solvents;
- flammable gases;
- flammable metals;
- foam; and
- rubber and plastics.

Sources of oxygen

The most obvious source of oxygen is in the air that surrounds us. The typical concentration of oxygen in the air that we breathe is 21%; this is more than enough to allow a fire to start and to continue to burn.

However, there are other sources of oxygen that must also be considered. For example:

- oxygen bottles (used for welding or medical purposes);
- ventilation systems; and
- air-conditioning systems.

These sources of oxygen can enrich the normal oxygen levels in the air, which will lead to a more intense fire.

Oxidising materials

An oxidising material is a chemical that can easily decompose to release oxygen or an oxidising substance. The oxygen is chemically bound in a material until it is freed by heat and/or a chemical reaction. These materials can add to a fire and cause it to spread. Oxidising materials should always be kept away from sources of fuel. Oxidising materials can ignite sources of fuel without the presence of an ignition source.

Figure 2: The oxidising symbol[1]

These materials are easily identified by their Globally Harmonized System (GHS) hazard pictogram. The pictogram for oxidising material is shown in Figure 2.

Examples of oxidising materials include:

- hydrogen peroxide, which is sometimes used in hairdressing salons as a bleaching agent; and
- sodium chlorate, sometimes used in weedkillers.

APPLICATION

Draw the fire triangle and then look around your current location. Identify the sources of ignition, fuel and oxygen and add these to your fire triangle.

10.1.2 Classification of fires

Fires are classified according to the fuel source. The fire classification will influence the type of extinguishing media that will be needed (see 10.3: Fire alarms and fire-fighting).

Different fire classification systems are used in different parts of the world, but here we will look at the UK classifications. There are five UK fire classifications as in Table 1.

Classification	Types of material
A	General combustibles, such as paper, wood, textiles, rubber and other organic carbon-based compounds
B	Flammable liquids
C	Flammable gases
D	Metals
F	Cooking oils

Table 1: UK fire classifications

The classification jumps from 'D' to 'F' as the 'E' classification is generally taken to refer to electrical fires. This is not considered to be a formal classification as electricity itself does not burn. However, firefighters do need to be aware of electrical hazards when attending an incident as the majority of buildings have an electricity supply.

10.1.3 Principles of heat transmission and fire spread

A basic principle of fire science is that heat energy moves along a high-to-low gradient; that is, heat will always move from a hot area to a cooler one. This transfer of heat can occur by one or more of the following:

- conduction – the transfer of heat in a static material;
- convection – by molecular movement in a liquid or gas;
- radiation – the transfer of heat through a gas or vacuum; and
- direct burning – the transfer of heat from one material's surface to another.

Conduction

Conduction happens when there is a difference in temperature between two parts of the same material; the heat will either travel along or through this material. The heat is 'passed' via microscopic particles in the material (molecules, atoms or electrons, depending on the material); the heated section's particles will vibrate quicker than in the cooler section. The particles will collide with their neighbour, which will pass the energy (heat) to the cooler area.

Materials differ greatly in their heat conducting ability. Metals are usually good heat conductors, while building materials such as brick and stone are poor conductors. If a building has a steel frame with untreated or uninsulated girders, the girders can conduct heat from one area to another and initiate secondary fires.

Convection

This is the process by which heat is transferred by movement of a heated fluid such as air or water. As the air is heated, its density decreases and it rises, being replaced by cooler fluid that is drawn in from below. The hot fluid can travel both vertically and horizontally through gaps in the building fabric, along air vents and roof spaces, spreading the fire to other parts of the building.

Radiation

All objects transmit radiant heat all the time. In the case of, for example, a brick wall at 25°C, the amount of radiant heat being given out is very low. However, in the case of a burning curtain, the amount of radiant heat emitted can be strong enough to initiate fires some distance away. Escape routes can become compromised or effectively 'blocked' by radiated heat. Heat radiated from a thick layer of smoke at ceiling level can set fire to the contents of the room above.

> **TIP**
>
> There is a basic physical principle known as the inverse square law. This says that at double the distance from a source of light or radiant heat, the intensity will reduce fourfold.
>
> This is true for what is known as a 'point source' of light or radiant heat: one that is radiating equally in all directions. However, when the source is a massive wall of fire, moving, say, 10 to 20 metres away will not make much difference. This is due to the colossal amount of radiant heat being given off, largely in one direction.
>
> An example of this occurred in the UK in 1985 when a wooden stand at the Bradford City Football Stadium caught fire. Tragically, 56 people lost their lives and hundreds more were injured during that incident.

Direct burning

This is the easiest way for fire to spread. It is the direct application of a flame to a material that it moves along or through. For example, if the corner of a piece of paper catches fire, the flame will spread across the paper. Burning material dropping onto something else is also another instance of direct burning. An example of this could be bits of a burning curtain dropping onto a carpet, setting the carpet on fire.

Fire spread

There are many things that could cause a fire to spread in a building. Figure 3 shows the different types of heat transmission and how these contribute to fire spread.

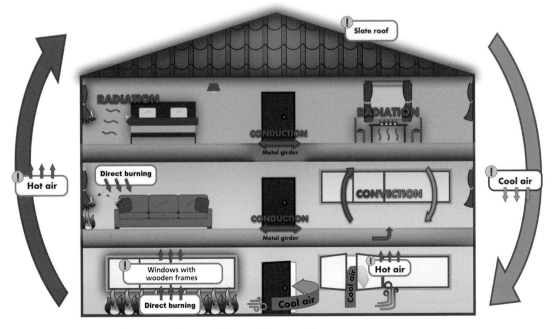

Figure 3: How types of heat transfer contribute to fire spread

Other things that could affect fire spread include:

- delayed discovery of/dealing with a small fire, allowing it to develop into a major fire;
- quantities of sources of fuel in a building; for example, storing instead of disposing of cardboard boxes that could further fuel a fire;
- incorrect disposal of some wastes, such as oily rags that could spontaneously combust given the right conditions;
- presence of dusts, gases or vapours that could explode and add to a fire;
- flammable material stored incorrectly or in unsuitable containers that could easily 'add' to a fire;
- poor building design combined with lack of appropriate fire-stopping measures; and
- use of incorrect extinguishing methods; for example, using a water fire extinguisher on an oil-based fire will spread rather than extinguish the fire (this will be explored further in 10.3: Fire alarms and fire-fighting).

APPLICATION 1

A small single-storey hardware shop (built of brick with a metal frame) displays goods on metal shelves. Goods such as nails are placed in cardboard boxes on the shelves. The shop has a large window in which it displays some of its goods on wooden shelves that are fixed in place with metal brackets. The floor is covered by hardwearing industrial carpet. The main door into the building is a sliding glass door; this is kept open most of the time. At the back of the shop is a small stockroom that has small windows around the top of the outer wall; these windows are open during the shop's trading hours.

What are the main ways that heat could be transferred around this building if a fire broke out during the shop's trading hours?

10.1.4 Common causes and consequences of fire

Common causes of fire

There are many potential causes of fire in workplaces. These reasons can be grouped in four broad categories: people, equipment/electrical faults, work activities and other conditions.

People

People can cause fires by making mistakes, having lapses of concentration or by just being careless. For example, throwing a smouldering match or cigarette into a waste bin, believing it to be fully extinguished, could be a mistake or because someone is careless. Starting to fry food, walking away and forgetting about it is a lapse of concentration. People may also start fires deliberately (arson).

Arson

Arson is when fires are set deliberately. Arson can occur for a variety of reasons that include retribution, such as a former worker setting fire to an organisation's premises as revenge for (what they believe to be) a wrongful dismissal. Other reasons include covering up another crime, insurance fraud, vandalism and pyromania.

> **TIP**
>
> **Pyromania** is a type of impulse control disorder where people are unable to resist starting fires. People with pyromania know that setting fires is harmful; however, setting fires is the only way they can relieve built-up tension/anxiety.

> **ADDITIONAL INFORMATION**
>
> Statistics from the National Fire Chiefs Council show that approximately 50% of fires in the UK are caused deliberately.

Equipment/electrical faults

Equipment used in the workplace can be the cause of fires if it is used when it is faulty. For example, if a fuse is bypassed in a plug, the equipment could overheat. Overloading circuits is another common cause of fires.

Figure 4 shows how this can happen. The extension itself, as well as the plugs, does not appear to be in a good state of repair. Adding this number of plugs to an already faulty extension can easily cause an overload.

Figure 4: Faulty or overloaded plugs can cause fires

Overloads, and potentially fires, can be caused by 'daisy chaining' electric extension cords when there are not enough available plug sockets or by using damaged extension cords. Poorly maintained equipment can also cause fires. For example, if worn parts of a machine are not replaced, the friction from moving parts could cause sparks.

Work activities

These might include working with open flames, such as using welding equipment or blowtorches, and other hot work activities, such as soldering, or any activity that involves using open flames or excessive heat. Cooking activities can also cause fires (especially those involving oils and/or open flames); this risk applies to both industrial/commercial and home cooking.

Other conditions

These include static electricity or natural conditions. Static electricity is a stationary electric charge on the surface of an object; it is typically caused by friction. This can cause sparks or crackling and will attract dust and hair. This is a particular problem in workplaces where there may be flammable atmospheres, such as mines and sewers, or where hazardous substances are routinely handled.

Natural conditions cause fires in several ways. The first is when rays from sunlight are focused on flammable materials and the flow of energy becomes concentrated enough to ignite the material. An example is when rays pass through glass and are focused on paper (you may remember doing this as a child, using a magnifying glass and a piece of paper).

The other natural condition is lightning. Lightning is an electrical discharge caused by imbalances between storm clouds and the ground, or in the clouds themselves. Each bolt can contain up to one billion volts of electricity and is extremely hot; lightning can heat the air around it to five times the temperature of the sun's surface. Given that lightning is often accompanied by strong winds, a lightning strike can rapidly form a fire.

Once a fire starts, it can develop and spread rapidly; it will create enormous amounts of heat energy and thick black smoke.

Consequences of fire

The consequences of fire are wide-ranging. For example, a small fire could cause minor or superficial burns to a small number of people, whereas a major fire could cause major (life-changing) injuries ranging from burns to internal organ damage. In a worst-case scenario, a fire can cause multiple deaths. Consideration also needs to be given to the fact that not only are the organisation's workers at risk from fire but so are emergency workers such as firefighters and ambulance crews when tackling the fire.

Property damage

This can range from minor smoke damage through to a total loss of the property. The design of the building, the degree of fire protection, the alertness of workers and the effectiveness of detection and alarm systems all play a part in how serious the consequences of fire are.

Financial consequences

Many organisations will be unable to recover from a catastrophic fire or it will take many years for them to get back to the same state they were in before the fire. In many cases, it will be difficult for an organisation to operate if its property has been destroyed; this means that the organisation will experience a drop in income, which may lead to redundancies and financial hardship for some workers.

Equipment and stock

As well as damage to the building itself, fire will either seriously damage or destroy equipment and/or stock that is in the building. Again, this will have an impact on how quickly an organisation recovers from a fire.

Reputation

The organisation's reputation is also at risk. For example, fire and the after-effects of fire (such as fire run-off water) can cause widespread environmental damage. If it is

Fire

later established that a fire happened due to the organisation cutting corners on, for example, fire protection measures, the organisation's reputation could be damaged. With the technology available in today's world, news organisations and social media are usually very quick to bring this type of news to society's attention. This could have a negative impact on the organisation's ability to recover from a catastrophic fire.

> **APPLICATION 2**
>
> Note some potential causes of fires in your current workplace.
>
> What are the consequences of a fire likely to be?

KEY POINTS

- The three sides (components) of the 'fire triangle' are heat, fuel and oxygen. A fire will not form (or will be quickly extinguished) without all three components.
- Oxidising materials are chemicals that can add to a fire and cause it to spread.
- There are five UK fire classifications. They influence what should be used to extinguish a fire.
- The four principles of heat transmission are convection, conduction, radiation and direct burning. These contribute to how fire can spread if left uncontrolled. Other things that affect fire spread include building design and the presence of fuel, waste or flammable material, dusts, gases and vapours.
- Workplace fires are commonly caused by people (including deliberately), equipment and electrical faults, certain work activities and natural conditions.
- Fires can result in injuries and fatalities, property and environmental damage, financial losses and reputational harm.

Reference

[1] Source: HSE, 'Hazard pictograms (symbols)' (www.hse.gov.uk)

10.2: Preventing fire and fire spread

> **Syllabus outline**
>
> In this section, you will develop an awareness of the following:
>
> - Control measures to minimise the risk of fire starting in a workplace:
>   - eliminate/reduce quantities of flammable and combustible materials used or stored
>   - control ignition sources, including suitable electrical equipment in flammable atmospheres
>   - use good systems of work
>   - good housekeeping
> - Storage of flammable liquids in workrooms and other locations
> - Structural measures to prevent the spread of fire and smoke: properties of common building materials (including fire doors); compartmentation; protection of openings and voids

Uncontrolled fire is one of the biggest risks to businesses. Not only can it kill or seriously injure people, it could also put an organisation out of business (as discussed in 10.1: Fire principles). It is therefore essential that organisations have fire safety arrangements in place that are suitable for the type of work activities carried out at their premises.

These arrangements are a legal requirement in the UK (see Tip box) and the precautions can be categorised in two ways:

- fire prevention measures that are put in place to help prevent fires from starting; and
- fire protection measures that are in place to mitigate (reduce) the consequences of a fire.

Here we will be looking at both practical and administrative controls that can be used to prevent fires.

The correct storage of flammable substances (no matter the quantity) is important; things can go badly wrong if these substances are not handled correctly. The Health and Safety Executive (HSE) has produced guidance on this issue that can be considered best practice and we will therefore look at how it can be applied in most workplaces.

The building structure is also really important and is usually the first line of defence if a fire does start. We will explore the various elements of a building that can help to prevent fire or smoke spread.

> **TIP**
>
> The main pieces of legislation relating to fire safety in the UK are:
>
> 1 Regulatory Reform (Fire Safety) Order 2005 – covers England and Wales;
> 2 Fire and Rescue Services (Northern Ireland) Order 2006;
> 3 Fire Safety Regulations (Northern Ireland) 2010;
> 4 Fire (Scotland) Act 2005; and
> 5 Fire Safety (Scotland) Regulations 2006.

> The Regulations that place particular duties on employers and workers outside England and Wales are items 3 and 5 in this list.
>
> Fire safety in the UK is regulated and enforced by the local fire and rescue authorities. In addition, HSE and the Health and Safety Executive for Northern Ireland (HSENI) enforce fire legislation for construction projects.
>
> As well as general fire safety legislation, there is also legislation that covers higher hazard workplaces where there are specific risks from dangerous substances that are capable of forming an explosive atmosphere. Here, we will look specifically at the:
>
> - Dangerous Substances and Explosive Atmospheres Regulations 2002 (DSEAR) – covers England, Wales and Scotland; and
> - Dangerous Substances and Explosive Atmospheres (Northern Ireland) Regulations 2003 (DSEAR NI).
>
> The HSE has produced the following guidance to supplement the information in the Regulations:
>
> - *Dangerous substances and explosive atmospheres. Dangerous Substances and Explosive Atmospheres Regulations 2002. Approved Code of Practice and guidance* (L138).[1]
>
> HSENI has approved L138 for use in Northern Ireland.

10.2.1 Control measures to minimise the risk of fire

In 10.1: Fire principles we looked at how all three parts of the fire triangle are needed for a fire to start or to continue to burn. In terms of control measures, simply removing one side of the fire triangle will either prevent the fire from starting or, if it has already started, will put it out. Therefore, the basic principle to keep in mind when controlling the risk of fire is simply to keep fuel and ignition sources separate from each other (there is not much you can do about the oxygen in the air).

There are some key techniques for the control of fire risks in workplaces. If used together (see Figure 1), they can significantly reduce the chance of a fire breaking out.

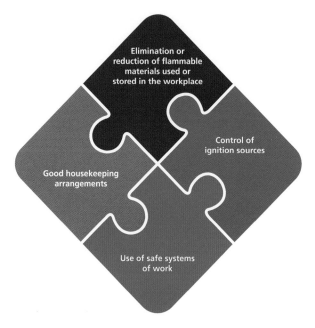

Figure 1: Controls working together to reduce the chance of fire

Elimination or reduction of flammable materials used or stored in the workplace

One of the easiest ways to reduce the risk of fire starting is to stop using flammable materials or decrease the amount of these on site at any one time. The greater the amount of flammable materials held, the greater the amount of fuel available for a fire; this is known as the 'fire load'.

Some things to consider include:

- Do you need that material or substance? Elimination of a fuel source is obviously the best course of action.
- If it is needed, could it be substituted with an alternative that may be less flammable? For example, a solvent-based cleaning product could be replaced by one that is water based.
- How much of the material or substance is required at any one time? Consider only ordering what is needed for the short term rather than, for example, storing materials for a year before they are used.

Particular arrangements are necessary for the storage of flammable liquids; this will be discussed later.

Control of ignition sources

There are a great number of potential ignition sources in any workplace. It is therefore extremely important that control measures for them are put in place. Methods of controlling ignition sources in a general workplace include the following:

- Separate ignition sources from fuel sources. This is one of the most obvious things to do; for example, do not have a waste cardboard storage point near a designated smoking area.
- Remove unnecessary sources of heat from the workplace or replace them with safer alternatives. An example would be removing a paraffin heater and replacing it with an oil-filled alternative or electric space heater.
- Install and use machinery and equipment that has been designed to minimise the risk of fire and explosion.
- Have good housekeeping arrangements.
- Ensure that all electrical fuses and circuit breakers are of the correct rating so that the potential for overheating is minimised (also see 11: Electricity).
- Maintain electrical equipment to prevent short circuits and overheating.
- Operate a permit-to-work system to control hot work.
- Enforce rules on smoking, such as use of signs to indicate where a designated smoking area is located. This area must have appropriate receptacles for extinguished cigarettes and matches.
- Make appropriate arrangements for the storage and disposal of oil-soaked rags to prevent spontaneous ignition.
- Take precautions to prevent arson, such as: ensuring good site security to prevent trespassers and vandals; storing sources of fuel (such as pallets, cardboard boxes) away from buildings and perimeter fences.

Control measures for ignition sources in higher hazard areas include all the above, but in addition you should consider the following:

- correct storage of flammable liquids and gases;
- using equipment and/or vehicles designed specifically for high hazard areas, known as 'intrinsically safe' equipment or 'Ex' equipment;

Fire

- earthing of all plant and equipment; and
- lightning protection.

This is not an exhaustive list of controls and others may be required depending on the type of activities carried out in the organisation.

In addition to this, suitable training should be provided for relevant workers to teach them about the hazard and the controls that are in place.

APPLICATION 1

Thinking of your organisation, or looking around your current location, consider whether the ignition sources are being controlled correctly. If they are not, what other controls would you recommend?

Electrical equipment for use in flammable areas

Consider 'electrical equipment' as a source of ignition and a 'flammable atmosphere' as a fuel. Where there is a sufficient mix of fuel in the air, there will be a potential for a fire or explosion should an item of electrical equipment emit a spark.

Vapour, such as petrol vapour, or flammable dust can create flammable atmospheres. Electrical equipment will need to be designed or protected, depending on which type of atmosphere it is used in.

Examples of flammable atmospheres include:

- confined spaces, such as tanks that have previously contained a flammable liquid; and
- flour mills where there is a large amount of dust suspended in the air.

Ideally, using electrical equipment in these areas should be avoided. However, this is not always practical. In areas where the atmosphere may become flammable, stringent precautions will be needed. Only appropriate equipment should be used (the terms often used are 'intrinsically safe' or 'explosion protected'). Equipment is categorised according to the hazardous zone (see Tip box) where it will be used.

Whenever possible, the risk of fire and explosion should be avoided by locating electrical equipment and wiring outside the flammable zone.

For example, an electrically powered motor could be located in a safe area and a pump located in the flammable zone. These would be connected by means of a drive shaft extended through a sealed gland, avoiding the use of electrical equipment in the flammable zone.

TIP

Flammable atmospheres are usually categorised by zones. The conditions for each zone will help to decide what type of flammable atmosphere could be present and what equipment can be used in them.

Zoning is required by law under DSEAR; L138 sets out zones for both vapours/mists/gases and dusts.[2] These are as follows.

Condition	Gases, vapours or mists	Dusts	Equipment category*
Explosive atmosphere is continuously present or present for long periods	Zone 0	Zone 20	1

10.2 Preventing fire and fire spread

Condition	Gases, vapours or mists	Dusts	Equipment category*
Explosive atmosphere is likely to occur in normal operation	Zone 1	Zone 21	**1 and 2**
Explosive atmosphere is not likely to occur in normal operation, but if it does, it will only be present for short periods	Zone 2	Zone 22	**1, 2 and 3**

* Equipment category descriptions
Category 1: Intrinsically safe equipment
Category 2: Flameproof, increased safety and pressurised equipment
Category 3: Other equipment such as powder-filled or non-sparking equipment

Marking of equipment

All certified equipment for use in potentially explosive atmospheres should be marked with some or all of the following information:

- manufacturer's name, address and (if relevant) its trademark;
- details of the conformity standard used (for example, CE marking for European standards);
- product type and serial number;
- temperature classification;
- certification number;
- product marking, including equipment group, equipment category (see previous Tip box) and a G (for use in a gas atmosphere) or D (for use in a dust atmosphere). The 'Ex' symbol (see Figure 2) must also be applied; and
- any other relevant information.

Figure 2: Ex mark for safe equipment

Use of safe systems of work

Safe systems of work were covered in 3.6: Safe systems of work for general work activities. Any working procedure that helps to reduce the chances of fire may be seen as a safe system of work. Safe systems of work can come in many forms. They can be written, formal procedures or verbal instructions via training. For example, most people will have gone through a fire evacuation drill at some stage. A specific safe system of work is the permit-to-work (PTW) system. This is used for more hazardous (high risk) activities where more control is required. Examples include welding activities in any area other than a welding shop, soldering, brazing or any activity that requires working with an open flame. PTW systems can also be used for other work activities, such as confined space working.

Permits for these types of activities are normally referred to as a 'hot work permit'.

A safe system of work for welding operations may feature the following fire safety precautions:

- use a PTW document;
- move or cover over any combustible material;
- damp the area down, especially if the floor is wooden;
- have a suitable portable firefighting extinguisher readily available;
- ensure that a 'fire watcher' is in position for at least 30 minutes after the end of the hot work;

- (where safe to do so) close mechanical ventilation ducts and windows; and
- have suitable evacuation procedures.

However, it is important to understand that, even if there is a PTW system in place, things can still go wrong. A safe system of work is just one control that must be in place for high hazard activities. PTWs are discussed generally in 3.7: Permit-to-work systems.

Housekeeping

Good housekeeping is one of the simplest and most effective safety measures. Not only is it instrumental in reducing instances of slips, trips and falls, it is also essential for good standards of fire safety.

Over time any organisation can find that a relatively large amount of flammable material can start to build up.

Housekeeping arrangements will help ensure that waste material is removed from the workplace periodically and is, therefore, not allowed to build up. It will also help to ensure that everything is in its place and can be found easily. Good housekeeping arrangements will make it much less likely that combustible materials will come into contact with a source of ignition.

10.2.2 Storage of flammable liquids in workrooms and other locations

> **TIP**
>
> The GHS is the Globally Harmonised System of Classification and Labelling of Chemicals. Prior to its introduction, countries had their own classification systems and associated regulations. Many of these systems had different hazard descriptions and pictograms. This made exporting chemicals to other countries difficult. For example, one country may have classified a chemical as highly flammable but other countries might not. The GHS was introduced to remove these obstacles and make it clear what the hazardous properties of chemicals were around the world.
>
> The GHS has been adopted into specific European legislation in the form of Regulation (EC) No 1272/2008 – classification, labelling and packaging of substances and mixtures (CLP). Post Brexit, Great Britain has retained the CLP Regulation, and this is now known as the GB CLP Regulation. The European legislation has been amended so that Great Britain can continue to adopt the GHS independently of the European Union. More information can be found on the HSE website.[3]
>
> The United Nations Economic Commission for Europe (UNECE) also provides guidance on the GHS and the different pictograms.[4]

Flammable liquids include solvents, fuel oils, varnishes, paints and innumerable other liquids that are used in industry. These liquids can be identified by the following GHS hazard pictogram (Figure 3) on packaging materials or in safety data sheets.

10.2 Preventing fire and fire spread

Figure 3: Flammable liquid symbol

> **TIP**
>
> Flammable liquids are usually classified according to their 'flashpoint' (the lowest temperature at which the application of an ignition source causes the vapours of a liquid to ignite under specified test conditions). The definition of 'flammable liquid' can vary but the GHS classifies it as a liquid with a flashpoint of not more than 93°C.
>
> In this are four categories:
>
> Category 1: Flashpoint <23°C and initial boiling point ≤35°C
>
> Category 2: Flashpoint <23°C and initial boiling point >35°C
>
> Category 3: Flashpoint ≥23°C and initial boiling point ≤60°C
>
> Category 4: Flashpoint >60°C and initial boiling point ≤93°C
>
> However, the CLP Regulation classifies a 'flammable liquid' as a liquid with a flashpoint of 60°C or below.[5]

Flammable liquids need to be handled with care. It is not the liquid that burns, but the vapour above the liquid. The HSE has produced guidance on the storage of flammable liquids in workrooms and other areas[6] and storage of flammable liquids in tanks.[7] We will refer to these as examples of good practice.

The first thing that needs to be considered is the quantity of flammable liquid that is stored. We usually distinguish between storage in bulk and storage in containers or packages.

Bulk storage in fixed tanks (quantities in excess of 1000 litres)

Bulk quantities of flammable liquids should be stored in fire-resisting tanks. These tanks can be above or below ground and should be built so that they have a bund (or secondary containment) that is big enough to hold 110% of the contents of the largest container stored in it.

A common problem with bunds is that rainwater will collect in them. There are two issues with this. Firstly, the rainwater will use up some of the capacity of the bund. If a tank was to develop a major leak there would be a good chance that the bund would not be able to hold all the liquid, causing it to spill into the surrounding environment.

The other issue is the contents of the tank. If the contents of the tank were to react with the rainwater in the bund, this could lead to the release of toxic vapours and/

or an explosion. It is therefore important that bunds undergo regular inspection and maintenance. Any damage to the bund must be repaired immediately. Checks should also be made so that rainwater does not build up in the bund over time; rainwater should be drained from the bund at regular intervals.

> **DEFINITION**
> A **bund** is a wall or dam (usually made of bricks or concrete) that completely surrounds a tank/container. If the tank/container leaks or ruptures, the bund will contain the leak and stop the spill from spreading further.

Ideally, above-ground tanks should be sited away from:

- site boundaries and/or neighbouring properties;
- fixed ignition sources; and
- storage of other hazardous/dangerous substances.

Tanks will also need to have a suitable separation distance from each other. This distance will depend on the size of the tank. These distances are not designed to give complete protection from fire or explosion but to allow sufficient time for evacuation of the area and to give the fire service time to arrive on site.

Tanks should also have hazardous zones, as illustrated in Figure 4.

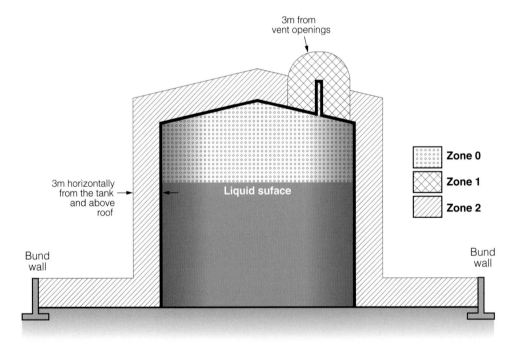

Figure 4: Above ground storage tank with hazardous zones[8]

Storage in containers/packages (quantities less than 1000 litres)

Bulk storage of flammable liquids should have designated storage areas away from other flammable/dangerous substances and ignition sources. Preferably, there should be a dedicated storage building or area that has plenty of ventilation, is away from the process area and is a suitable distance from other buildings (see Figure 5). If distance cannot be maintained, the wall that is closest to the storage area must be a fire wall. Buildings in the storage area must have suitable fire protection and prevention measures.

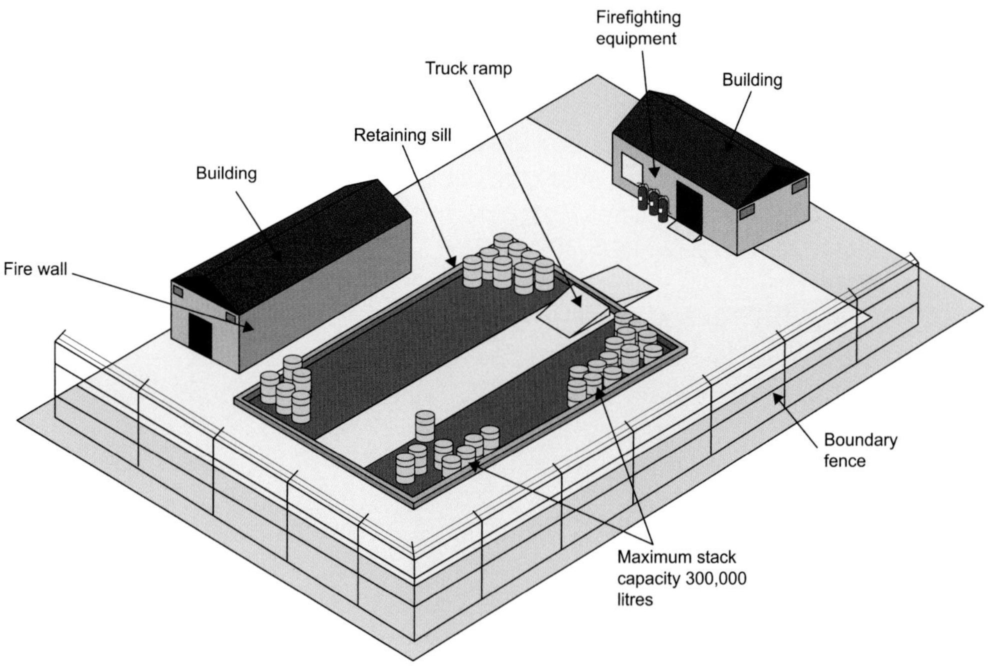

10.2 Preventing fire and fire spread

Figure 5: Example of an outside cylinder/drum storage area[9]

If stored in a building, conditions in the storage area chosen should not affect the construction of storage cupboards or containers. It is also important to consider the means of escape from the area, so it is vital that the storage area does not block evacuation routes.

Storage of small quantities in workrooms/areas

To reduce the risk from flammable substances, it is recommended that the maximum quantities stored in cabinets and bins are:

- 50 litres maximum for flammable liquids that are extremely/highly flammable or for liquids with a flashpoint below the maximum ambient temperature of the workroom/working area; and
- 250 litres maximum for other flammable liquids with a higher flashpoint of up to 60°C.[10]

Storage in these quantities should be considered good/safe practice rather than taken as absolute limits. There may be occasions when quantities over these limits are required. In these circumstances, a clear argument must be made for this and a risk assessment carried out.

The risk assessment should take into account:

- the properties of the materials to be stored;
- the size of the working area and the number of workers using it;
- the amount of flammable liquid being handled;
- the quantities of flammable liquid that may be accidentally released or spilled;
- ignition sources;
- how a fire could potentially spread;
- the ventilation in the work and storage areas;
- the storage area's fire protection; and
- emergency arrangements, which should include arrangements for closing the storage cupboard in the event of a fire as well as escape routes from both the storage and work areas.

Supplies needed for immediate use should be kept in small quantities in fire-resisting metal cupboards. These cupboards are available in different sizes and designs. The type required will depend on the type and quantities of flammable materials being stored.

Cupboards should be designed and built so that they provide a physical barrier between the fire and the flammable liquid. This will delay the liquids from adding to the fire and will allow workers sufficient time to safely evacuate the area.

Figure 6: Fire-resisting metal cupboards protect supplies

Every storeroom, cupboard, bin, tank and vessel used for storage should be clearly labelled so that all workers are aware of the flammability of the contents.

> **TIP**
>
> There is specific guidance on the construction of such storage cabinets. Some examples of this are:
>
> - the *DSEAR Approved Code of Practice and guidance* (L138);[11]
> - national or international standards, for example from the International Organization for Standardization (ISO);
> - insurance guidance; and
> - industry/best practice guidance.

Other good practice for storing or handling flammable substances includes the following:

- Storage should contain the liquid so that vapour does not escape during storage.
- Handle the liquid so that spillages do not occur; for example, by using funnels, careful decanting and use of transfer pipes.
- If spills do happen, have suitable equipment (trays and absorbent material) available to contain and dispose of the spill.
- Use correct waste disposal methods for spilled waste (remember that oil-soaked rags/cloths have been known to spontaneously combust given the right conditions).
- Provide ventilation so that any vapour that is created is dispersed.
- Control ignition sources in areas where flammable liquids are stored.

- Treat 'empty' containers as if they were full; they might be full of vapour, particularly if there is any heat involved. Many serious accidents have occurred in 'empty' storage drums and tanks, triggered by sources such as:
  - sparks from tools; and
  - heat from welding equipment being used to repair a tank.

Flammable gases

Flammable gases such as oxygen and acetylene will generally be contained in gas cylinders or tanks. If possible, the preferred option is to site the cylinders or tanks outside the building and to pipe the gas through fixed pipework to where it is needed. This will, of course, not be possible in all situations. For example, in mobile welding it will be necessary to take the gas cylinders to the place of work.

The same principles for storing flammable liquids can be applied to flammable gases.

As well as storage, it is important for those using flammable gases to be trained in their correct use. Specifically, they will need to know:

- how to change cylinders safely;
- how to check valves and connectors;
- how to test for leaks;
- how to follow safe working procedures; and
- that empty cylinders must be returned to storage as soon as possible.

It is also important to understand that only the minimum amount of flammable gases (such as LPG) should be kept in the workplace at any one time. It is a good idea to establish site rules banning the keeping of additional or unnecessary full or empty cylinders in the workplace.

It is also essential that where a process is liable to cause leaks or spills there are arrangements in place to contain it, or to immediately drain it off to a suitable container or safe place, or to render it harmless.

Areas where flammable liquids or gases are stored must be away from all possible ignition sources (when possible), including use of suitable electrical equipment. Smoking must be banned in and around these storage areas. Finally, reasonable steps should also be taken to minimise the likelihood of any vapours getting into general work areas.

10.2.3 Structural measures to prevent the spread of fire and smoke

Building structures need to be designed to withstand direct fire and to minimise the spread of both smoke and flames. Clearly, no building will withstand fire indefinitely. However, buildings should be designed and built to allow sufficient time for people to escape and reach a place of safety. This type of construction will also give the fire service the opportunity to put the fire out without there being a total loss of the structure.

Properties of common building materials

Materials commonly used in the construction of buildings behave differently when exposed to intense heat. The main building materials commonly encountered include steel, timber, concrete and brick.

Steel

Since steel is a metal, it will be affected by heat. It will expand when exposed to high temperatures; this expansion may affect the alignment of the structure, for example in a steel-framed building. As horizontal steel girders expand, they force the supporting girders away from vertical. This causes the load to be unevenly distributed, which can contribute to the eventual collapse of the building. In extreme temperatures, steel can begin to melt, which will severely compromise the structural integrity of the building. To avoid this, structural steelwork is routinely encased in insulating materials that give the building a greater chance of survival in the event of a fire.

Timber

Timber supports will char on the outside when exposed to direct flame, but it takes much longer for the central core of the timber to be weakened to the point of collapse. Other timber components such as floorboards and wood panelling are thinner and will burn much more readily. This therefore significantly adds to the 'fire load', which creates a more intense fire and assists in fire spread.

Concrete

Concrete can withstand fire well as its components are chemically inert; this makes it virtually non-combustible. Concrete also has low thermal conductivity, which means that heat cannot easily pass along or through it. When set, it is also non-toxic, which makes it one of the safest building materials. Exposure to direct heat can cause limited expansion of the concrete, which can lead to 'spalling', when the surface of the concrete cracks, peels or crumbles. In some cases, the concrete can chip, causing small chunks of concrete to fly off the surface.

Brick

Bricks will also withstand direct heat up to 1200°C. Bricks have been shown to retain their structural integrity even when exposed to extremes of heat.

> **APPLICATION 2**
>
> Thinking of the building materials discussed, how do you think these either contributed to the Notre-Dame Cathedral fire in Paris or helped protect the structure?
>
> News stories provide more details of this fire, how it spread and the damage it caused.[12]

Compartmentation and protection of openings and voids

If you imagine a building as a completely open space, you can probably imagine that if a fire started in one corner it would eventually spread to the opposite corner of the building, destroying everything in its path. This is obviously undesirable because it will inevitably result in a total loss (building, equipment, stock, etc).

Where appropriate, the building structure can be divided into separate 'boxes', known as compartments. Clearly, this can only be partially done in buildings like warehouses, which by their nature need to be open. However, it is an important design technique in other types of building such as offices, hospitals, schools, shopping centres and so on.

10.2 Preventing fire and fire spread

Figure 7: How compartments work in a building structure

As Figure 7 shows, the basic compartment is created by the walls, floors and ceilings, which should create a 'box' inside which a fire can be contained. The compartments are separated from one another by robust, fire-resisting building materials. However, it is not practical to leave these as self-contained boxes; services (gas, electricity, water, internet, etc) and ventilation pipes are required in the majority of circumstances. Even a simple storeroom will require electricity for lighting if there are no windows and/or for use during the evening or when conditions outside are gloomy.

These services will require cables or pipes to pass through the structure of the building. This will obviously mean drilling holes through the compartments. Any of these holes that are left unsealed after the work has been completed could allow smoke and flames to spread (horizontally and vertically) more easily through the building should a fire break out. It is therefore essential that any gaps/openings between compartments are effectively plugged. An example of this can be seen in Figure 7. Areas where services are required to run through the building have been fire-stopped (indicated by the red outlines).

Fire-stopping material is a form of passive fire protection. Materials often used for this include intumescent substances that swell when exposed to heat, such as intumescent strips in fire doors, mortar/cement, silicone and mineral fibres.

Fire doors are also an important component in maintaining the integrity of any compartment.

It is important to consider fire-stopping measures not only in new buildings but also during maintenance or renovation of existing buildings. When conducting a fire safety inspection you need to check voids and riser cupboards to make sure the integrity of compartments is maintained. If you can see into the next compartment, or look up to the floor above or down to the level below, then you need to install fire-stopping material.

APPLICATION 3

Take a few minutes to walk around the building you are in and consider how it has been compartmentalised. Are there any sections of the building that you think would need to be fire-stopped?

KEY POINTS

- Control measures required to minimise the risk of fire starting in a workplace include eliminating or reducing the quantities of flammable and combustible materials used or stored in the workplace, controlling ignition sources, using safe systems of work and using good housekeeping measures.
- Controlling ignition sources includes the selection of suitable electrical equipment for use in flammable atmospheres.
- It is important to consider the quantities of flammable liquid stored in workrooms and other locations and what should be done to reduce the likelihood of these substances starting or spreading a fire, including siting tanks and dedicated storage areas away from other buildings.
- Structural measures that can be used to prevent the spread of fire and smoke if a fire does start include understanding how common building materials will act in a fire, the principles of compartmentation and the importance of protecting openings and voids.

References

[1] HSE, *Dangerous substances and explosive atmospheres. Dangerous Substances and Explosive Atmosphere Regulations 2002. Approved Code of Practice and guidance* (L138, 2nd edition, 2013) (www.hse.gov.uk)

[2] See note 1

[3] HSE, 'The GB CLP Regulation' (www.hse.gov.uk)

[4] See UNECE, 'GHS pictograms' (https://unece.org)

[5] See note 3

[6] HSE, 'Storage of flammable liquids in process areas, workrooms, laboratories and similar working areas' (www.hse.gov.uk); and HSE, *Storage of flammable liquids in containers* (HSG51 3rd edition, 2015) (www.hse.gov.uk)

[7] HSE, *Storage of flammable liquids in tanks* (HSG176, 2nd edition, 2015) (www.hse.gov.uk)

[8] Source: HSE, HSG176 (see note 7)

[9] Adapted from HSE, HSG51 (see note 6)

[10] See note 6

[11] See note 1

[12] NBC News, 'An icon in flames: Notre Dame fire: what was damaged' (www.nbcnews.com)

10.3: Fire alarms and fire-fighting

> **Syllabus outline**
>
> In this section, you will develop an awareness of the following:
> - Common fire-detection and alarm systems
> - Portable fire-fighting equipment: siting, maintenance and training requirements
> - Extinguishing media: water, foam, dry powder, carbon dioxide, wet chemical; advantages and limitations
> - Access for fire and rescue services and vehicles

Ultimately, you want to stop fire breaking out in the first place. However, if the worst does happen, there are things that can be done to give an early warning of fire and equipment is available that can be used to extinguish fires (hopefully before the fire becomes serious or out of control). You will be able to see how the fire classifications discussed in 10.1: Fire principles will influence the type of extinguishing media required.

It is a legal requirement in the UK to provide both early warning systems and firefighting equipment. In England and Wales this duty falls under reg 13 of the Regulatory Reform (Fire Safety) Order 2005. Similar duties exist under the Fire Safety (Scotland) Regulations 2006 and the Fire Safety Regulations (Northern Ireland) 2010.

Firefighting media, such as portable extinguishers, are usually the first form of firefighting in a premises. In a survey carried out it was found that 75% of the incidents reported "*did not require the attendance and resources of the fire service*".[1] This statistic goes to show the significant contribution that portable firefighting equipment makes to the prevention of serious fires.

Should a fire become serious enough, the fire and rescue service (FRS) will be required. We will discuss the importance of having arrangements in place in this event.

> **TIP**
>
> The main pieces of legislation relating to fire safety in the UK are:
>
> 1 Regulatory Reform (Fire Safety) Order 2005 – covers England and Wales;
> 2 Fire and Rescue Services (Northern Ireland) Order 2006;
> 3 Fire Safety Regulations (Northern Ireland) 2010;
> 4 Fire (Scotland) Act 2005; and
> 5 Fire Safety (Scotland) Regulations 2006.
>
> The Regulations that place particular duties on employers and workers outside England and Wales are items 3 and 5 from the above list.
>
> Fire safety in the UK is regulated and enforced by the local fire and rescue authorities. The Health and Safety Executive (HSE) and Health and Safety Executive for Northern Ireland (HSENI) also enforce fire legislation for construction projects.

> In addition to the specific fire Regulations, the UK also has the following pieces of legislation that require consideration of fire when designing and constructing buildings.
>
> - Building Regulations 2010, Schedule 1 (applicable to England and Wales);
> - Building Regulations (Northern Ireland) 2000, Part E; and
> - Building (Scotland) Regulations 2004.
>
> The UK government (including devolved governments) has also produced specific guidance on fire safety during building design and construction. These documents cover the requirements for buildings to have a means of warning and escape, ways to prevent or mitigate fire spread, and access and facilities for the fire service. The main guidance is:
>
> - *The Building Regulations 2010, Approved Document B, Fire safety, Volume 2: Buildings other than dwellings;*[2]
> - *Building Regulations (Northern Ireland) 2012: Technical booklet E – Fire Safety* and the *Amendments Booklet* (applicable in Northern Ireland);[3] and
> - *Building standards 2022 technical handbook: non-domestic buildings* – Section 2 Fire (applicable in Scotland).[4]

10.3.1 Common fire-detection and alarm systems

As stated in the introduction, there is legislation that means organisations must put in place fire-detection and alarm systems. This legislation is high level, meaning that it says what must be done rather than how it is done.

To support this legislation, the UK administrations have produced further guidance that gives more information on how to put these systems in place. These documents often reference other standards that give very specific information. For example, in Approved Document B, BS 5839 is referenced in relation to categories of detection systems. See the Tip box for a list of relevant guidance documents.

We will now look at the details of various fire detectors.

Detectors

The purpose of fire detectors is to identify an outbreak of fire or smouldering materials in the early stages. This is done by sensing one or more of the following:

- heat (a rise in temperature);
- flames; or
- smoke.

Heat detectors

Heat detectors monitor the ambient air temperature and will sound an alarm when there is a change. Heat detectors do not usually sound false alarms. They are useful in areas where smoke detectors cannot be placed, such as kitchens, or in places where high flames/intense heat could happen; for example, flammable liquid storage areas. Heat detectors fall into two main categories:

- Rate-of-rise detectors: these sound an alarm when a sudden, rapid rise in temperature is detected. This is based on how fast and by how much the temperature changes rather than the temperature itself. The alarm will normally

be triggered when there is a change in temperature in an air chamber in the detector. The chamber also contains a differential pressure switch that will be tripped when there is a significant rise in temperature in one minute.
- Fixed temperature heat detectors: these monitor the ambient air temperatures against a preset level. A bimetallic strip is used, which expands when heated; this makes contact with an open electrical circuit to complete the circuit and cause the alarm to sound.

Flame detectors

Flame detectors are prone to false alarms, but they usually respond quicker than a heat or smoke detector. They are especially useful in high hazard environments such as hydrogen stations or around natural gas lines. They are usually linked to a fire suppression system as well as an alarm. Ultraviolet or infrared beams, or a combination of these, are usually used in flame detectors. They generally transmit a beam across a protected area that flames will interrupt. This is detected by a receiving unit and the alarm is sounded or the suppression unit activated.

Smoke detectors

Smoke detectors are early warning devices that sound an alarm when the presence of smoke is detected. Smoke detectors are very sensitive and can often give false alarms (you will probably have set off your smoke alarm at home when you have burnt some toast). Smoke detectors generally fall into three categories:

Figure 1: Smoke detectors are very sensitive

- Ionisation detectors: these use a radioactive source (kept in a chamber that has open vents to allow smoke to enter) to ionise the air between two electrodes (one with a negative charge and one with a positive charge). This creates a small current inside the chamber. When smoke enters the chamber it will change the current inside the chamber. When a sufficient change is detected, the alarm will activate.
- Light scatter detectors (also known as photoelectric smoke detectors): these contain a photoelectric cell fitted in a chamber at right angles to a light source. Smoke entering the chamber scatters the light and the resulting disturbance triggers an alarm.
- Obscuration detectors: these work on the opposite basis to the light scattering principle. A pulsing infrared light beam is sent from a transmitter to an analysing receiver (a photoelectric cell). When the intensity of the light is reduced by smoke for a period of around five seconds, the alarm is triggered.

APPLICATION 1

Take a walk around your workplace or building. Where are the smoke detectors located? Why do you think these locations were chosen?

Fire alarms

There are two main methods of raising a fire alarm – manually or electronically. Whichever method is used, the 'alarm' must be easily recognisable to all workers.

In small, low-risk premises, manual alarms may be the most suitable method, such as using bells, gongs, whistles or loud hailers.

The most common fire alarm systems are electronic. When selecting an alarm, consideration must be given to how the alarm will be recognised. Audible alarms are the most commonly used. However, other systems may also be required if you have workers or visitors who have hearing issues or if you work in a very noisy environment. For example, the alarm could be recognised by flashing lights or by hand-held devices that vibrate when the alarm goes off.

Fire alarms must have distinctive sounds and be distinguishable from other types of alarm in the building such as intruder alarms. The alarm must be able to be heard (or seen) in all parts of the premises.

> **TIP**
>
> **Activating fire alarms**
>
> Fire alarms can be activated automatically or manually.
>
> Manual alarm points are usually 'break glass' boxes; the alarm activates as soon as the glass is broken. Ideally, these should be fitted 1.4 metres above the floor and sited so that people do not have to travel more than 45 metres to reach one. These alarm points are normally sited on exit routes.

Figure 2: Fire activation point

Some systems will link directly to smoke, flame or heat detectors that will automatically set off the alarm when required.

> **CASE STUDY**
>
> **The costs of having inadequate fire prevention and protection measures**
>
> A building owner in England was prosecuted for major breaches of fire safety law following a significant fire. It took 70 firefighters and 10 fire engines to control the blaze that destroyed half of a four-storey building. Two women were rescued from the building and treated for smoke inhalation.
>
> Fire investigators found that the building owner had failed to carry out a fire risk assessment; the fire protection measures in the building were inadequate and there were no fire prevention controls (such as fire alarms or smoke detection systems).
>
> The building owner pleaded guilty; they received a suspended custodial sentence, a fine of £20,000 and were also ordered to pay prosecution costs of over £10,000.

> An FRS representative said: "Inspectors found failings in the building that showed a disregard of any proper fire safety measures and therefore a disregard for the safety of the tenants. Failings in fire safety measures also pose a heightened risk to firefighters who are already doing a dangerous job. There's no excuse for leaving people's safety to chance".[5]

10.3.2 Portable fire-fighting equipment

As discussed earlier, fires are classified by their fuel type. There are various portable extinguishing media that can be used to put out each type of fire.

The simplest methods are those used to smother fires, such as a fire blanket or sand. In 10.1: Fire principles we discussed the fire triangle; this method 'removes' the oxygen from the area of the fire, which will extinguish the flames. Fire blankets are commonly found in kitchens and can be used to put out cooking oil fires; they can also be used to extinguish small wastepaper bin or engine fires. Larger blankets can be used to protect people, either to extinguish flames or to protect them while escaping.

The most common portable firefighting method is the fire extinguisher. It is a legal requirement to have extinguishers in commercial premises. There are five main types of extinguisher:

- water;
- carbon dioxide;
- powder;
- foam; and
- wet chemical.

The electrical conducting properties of some extinguishing agents mean they must not be used for fires involving live electrics ('electrical fires'). For such fires, non-conducting agents such as powder and carbon dioxide must be used. If the electrical supply can be disconnected, then the fire can be extinguished according to its fire classification.

Siting, maintenance and training requirements

Fire extinguishers have limited capacity and so are only suitable for use on small fires (such as those in an office wastepaper basket). Therefore, for an extinguisher to be effective it must be used as quickly as possible after a fire starts. It makes sense, then, that the fire extinguisher should be positioned as close as possible to the likely source of a fire. Ideally, workers should not be required to travel more than 45 metres to reach one.

Fire extinguishers should be positioned along fire escape routes and next to fire doors. However, the location must not impede a fire evacuation or become an obstruction during the normal working day. The placement of fire extinguishers should not make them an additional workplace hazard that workers might trip or fall over, for example. Correctly locating extinguishers will also protect them from damage.

Ideally, extinguishers should be wall mounted; if this is not possible, they should be placed on purpose-built floor stands. The location of the extinguishers must be clearly signed so that the position can be seen from a distance. This is especially important when lighting levels are low.

As with any other item of equipment, fire extinguishers will need to be periodically inspected and maintained to ensure that they will work when needed. Suppliers of extinguishers often provide service contracts where a technician will visit at agreed intervals to carry out the necessary inspections.

Between formal inspections, workers should consider the condition of extinguishers during site inspections and any defects should be reported so that repairs can be made.

> **APPLICATION 2**
>
> Take a walk around your workplace or building. Where are the fire extinguishers located? Why do you think these locations were chosen?

Although fire extinguishers are in themselves easy to operate (usually by pulling out the pin and squeezing the handle), specific techniques need to be used when fighting fires. It is therefore necessary to ensure that anyone who might need to use a fire extinguisher, such as a fire marshal, undergoes appropriate training. This will help to ensure that the right type of extinguisher is used. It is also a good idea to periodically refresh the training given to fire marshals, as this knowledge is not something they will use on a daily basis.

Records for both the maintenance of extinguishers and training of users must be kept.

10.3.3 Extinguishing media and their uses

All fire extinguishers are coloured red. However, this can lead to confusion as, for example, a water extinguisher is the same size and shape as a foam or dry powder extinguisher. This could mean that the wrong extinguisher is chosen, which might result in the fire intensifying rather than being extinguished. To overcome this issue extinguishers have a coloured strip or panel to show what type of extinguisher it is.

Figure 3: Types of fire extinguisher

Table 1 shows the extinguishers that are commonly used.

Type	Label colour	Fire classification	Information
Water	Red (with white lettering)	A	Must not be used on electrical, metal or oil-based fires
Carbon dioxide (CO_2)	Black	B and electrical fires	Must not be used in enclosed spaces as it displaces oxygen in the area
Dry powder	Blue	A, B, C	Must not be used in enclosed spaces as the powder is easily inhaled
Foam	Cream	A, B	Water based so must not be used on electrical, metal or oil-based fires
Wet chemical	Yellow	A, F	Can be used on class A fires although it is more usual to use water and/or foam

Table 1: Fire extinguishers for different classifications

Special powders are available for Class D metal fires; such extinguishers will be needed in environments where metals such as manganese and aluminium are formed and processed.

Advantages and limitations of fire extinguishers

Extinguishers have both advantages and disadvantages. We have already touched on some of these.

Advantages
The biggest advantage is that the extinguishing media is immediately available; this can be the difference between stopping a fire or letting it take hold.

Another advantage is that the actual operation of the fire extinguisher is relatively easy (by pulling out the pin and pressing the trigger) and the training is simple and straightforward. This obviously has financial advantages for organisations, as they will not have to release workers from their 'day jobs' for very long or pay for expensive training courses. However, this is only the case for low- to medium-risk organisations. Those working on high-hazard sites will require much more extensive training, but that falls outside the scope of this qualification.

Disadvantages
Organisations must understand that extinguishers are only the first line of defence. In some cases, other fire suppression systems (such as sprinklers) may also be appropriate. The organisation's fire risk assessment should specify the types of firefighting equipment required.

Some types of extinguishers could have specific health effects on the users.

Fire extinguishers are readily accessible in the majority of buildings. There is, therefore, a danger of the extinguishers being overlooked as essential firefighting equipment. For example, they could be used for activities that they were not meant for (as doorstops, for example) so they may not be available when they are required. They could also be used when people are playing practical jokes, or just generally misused. This is a danger to the people in the vicinity (because those using the extinguishers are not trained in their use and some extinguisher contents can cause health conditions, such as burns from a CO_2 extinguisher). It could also mean that an empty extinguisher is returned to its original location. This would either vastly reduce or completely exhaust the extinguishing media available for use during a fire.

One of the biggest disadvantages is that if untrained workers incorrectly use an extinguisher, or use the wrong extinguisher, this could spread the fire instead of stopping it.

There may also be secondary effects. For example, wet chemical extinguishers are good for use on oil-based fires but they can also cause metals in the area to corrode.

Advantages and limitations by type

We have already noted the types of extinguisher available, so we will now look at the specific advantages and disadvantages for each type of extinguisher. These are given in Table 2.

Type	Advantages	Disadvantages
Standard water	• Cools the fire quickly/absorbs heat well • Long-range jet keeps user away from the fire • Heated water will evaporate, meaning there will be less long-term damage/quicker to clean up	• Good conductor of electricity so has limited use (Class A fires only) • Must not be used on electrical, metal or oil-based fires
Carbon dioxide (CO_2)	• Can be used on electrical equipment as it does not conduct electricity • No residue/clean-up costs	• Oxygen depleting so must not be used in enclosed spaces • Limited cooling properties • No protection against reignition of the fire • Possible burns from contact with uninsulated extinguisher horn
Dry powder	• Forms a barrier to exclude oxygen • Prevents reignition • Can be used in a wide range of temperature environments (eg 20°C to 60°C) • Not as vulnerable to frost as water-based extinguishers • Can be used on a wide range of fuel sources	• Must not be used in enclosed spaces as the powder is easily inhaled • Leaves lots of residue that can be difficult to clean up • Reduced visibility when extinguisher is in operation • Limited cooling properties • Can cause corrosion on electrical equipment
Foam	• Produces a foam blanket to smother the fire and leaves a film over the burning surface • Helps to prevent reignition	• Water based so must not be used on electrical, metal or oil-based fires

10.3 Fire alarms and fire-fighting

Type	Advantages	Disadvantages
Wet chemical	• Quickly puts out oil fires • Low chance of reignition after use • Clean-up can be done relatively quickly and cheaply • Low-pressure extinguisher, cutting down the chance of burning oils splashing and causing injury to the user • Can be used on Class A fires although it is more usual to use water and/or foam	• Can cause corrosion of some metals • Known irritant to skin and eyes • Not suitable for electrical fires

Table 2: Advantages and disadvantages of extinguisher types

10.3.4 Access for fire and rescue services

Site owners and operators must ensure that adequate access is provided for the fire and rescue services (FRS) so that they can deal with the fire as quickly as possible. This may require an additional entrance so that the FRS are not hampered by traffic entering or leaving the site.

Access routes should always be kept clear of vehicles and other obstructions such as waste bins/skips. These entrances should be manned, if relevant, so that the FRS can be guided to the fire as quickly as possible.

In some installations, such as oil and gas storage areas, the site will be laid out so that there is sufficient space to enable several firefighting vehicles to position themselves close to the fire. In higher-risk installations, the organisation will be required to liaise with the FRS; for example, they will be required to share their emergency plans with the FRS.

KEY POINTS

- It is a legal requirement in the UK to provide both early warning systems and firefighting equipment.

- Fire safety in the UK is regulated and enforced by the local fire and rescue authorities. The Health and Safety Executive (HSE) and Health and Safety Executive for Northern Ireland (HSENI) also enforce fire legislation for construction projects and there is additional legislation that requires fire to be considered when designing and constructing buildings.

- Common fire-detection systems identify a fire or smouldering materials in their early stages by sensing heat, flames or smoke and can be linked to alarm systems.

- Fire alarms must be easily recognised by all workers, have distinctive sounds and be heard (or seen, for example in a noisy workplace or if workers or visitors have hearing issues) throughout the premises. They can be activated automatically or manually.

- The requirements for siting, maintaining and training on the various forms of portable firefighting equipment include understanding the different fire classifications that each type of extinguisher (water, foam, dry powder, carbon dioxide and wet chemical) can be used for. There are advantages and disadvantages in using each of the extinguishing media.
- It is necessary to allow the fire and rescue services immediate and unimpeded access to a site in the event of a fire breaking out.

References

[1] FETA (Fire Extinguishing Trades Association) and IFEDA (Independent Fire Engineering & Distributors Association), 'Report on a survey into portable fire extinguishers and their use in the United Kingdom and other member countries of EUROFEU' (2015) (htttps://ifeda.org)

[2] HM Government, *The Building Regulations 2010: Approved Document B, Fire safety, Volume 2: Buildings other than dwellings* (2019, updated 2020) (www.gov.uk)

[3] Department of Finance and Personnel, *Building Regulations (Northern Ireland) 2012: Guidance Technical booklet E – Fire safety* (2012) and Department of Finance, *Amendments Booklet – AMD 7* (2022) (www.finance-ni.gov.uk)

[4] Scottish Government, *Building standards 2022 technical handbook: non-domestic* (2022) (www.gov.scot)

[5] Adapted from 'Serious fire safety law breaches at London flat lead to £20k fine and prison sentence', *Safety and Health Practitioner* (15 October 2021) (www.shponline.co.uk)

10.4: Fire evacuation

> **Syllabus outline**
>
> In this section, you will develop an awareness of the following:
> - Means of escape: travel distances, stairs, passageways, doors, emergency lighting, exit and directional signs, assembly points
> - Emergency evacuation procedures
> - Role and appointment of fire marshals
> - The purpose of fire drills, including roll call
> - Provisions for people with disabilities
> - Emergency escape routes to be recorded in building plans

We have looked at the contributors to a fire starting (sources of fuel, ignition and oxygen) and prevention and protection measures. However, no matter how well prepared you are, there is always the possibility of a fire starting. So it is essential that all organisations are prepared for this prospect. Having evacuation procedures in place and workers trained in their execution can ultimately save lives.

Here we will look at all things relating to evacuation should a fire break out.

Evacuation arrangements are essential for all organisations. We will illustrate this using the 2017 Grenfell Tower fire in London. A fire broke out and rapidly spread throughout the 24-storey tower block. There was a lot of confusion over evacuation, with many residents staying in their flats due to 'stay put' notices that were placed throughout the building and mobile phone advice from the fire service to residents.

The 'stay put' policy was then cancelled and a general evacuation took place. However, by that time it was too late for the majority of residents to evacuate; a later enquiry found that the 'stay put' policy should have been abandoned 80 minutes before it eventually was. Sadly, 72 people lost their lives in this fire and around 70 others were injured.

This example illustrates the importance of having evacuation procedures in place and, more importantly, why they need to be practised. We will therefore look at the requirements for evacuation routes that take people to a place of safety. We will also look at the contents of evacuation procedures, together with the roles of people in an evacuation, such as fire marshals.

Finally, we will discuss the special evacuation requirements for those with disabilities. Taken together, this should help you to understand what your organisation must have in place to ensure safe evacuation of the workplace if there is a fire.

> **TIP**
>
> The main pieces of legislation relating to fire safety in the UK are:
>
> 1. Regulatory Reform (Fire Safety) Order 2005 – covers England and Wales;
> 2. Fire and Rescue Services (Northern Ireland) Order 2006;
> 3. Fire Safety Regulations (Northern Ireland) 2010;

> **4** Fire (Scotland) Act 2005; and
> **5** Fire Safety (Scotland) Regulations 2006.
>
> The Regulations that place particular duties on employers and workers outside of England and Wales are items 3 and 5 on this list.
>
> Fire safety in the UK is regulated and enforced by the local fire and rescue authorities.
>
> In addition to the specific fire Regulations, the UK has the following pieces of legislation that require consideration of fire when designing and constructing buildings.
>
> - Building Regulations 2010, Schedule 1 (applicable to England and Wales);
> - Building Regulations (Northern Ireland) 2000, Part E; and
> - Building (Scotland) Regulations 2004.
>
> The government (including devolved governments) has also produced specific guidance on fire safety during building design and construction. These documents cover the requirements for buildings to have a means of warning and escape, ways to prevent or mitigate fire spread, and access and facilities for the fire service.
>
> - *The Building Regulations 2010, Approved Document B, Fire Safety, Volume 2: Buildings other than dwellings;*[1]
> - *Building Regulations (Northern Ireland) 2012: Technical booklet E – Fire safety* and the *Amendments Booklet* (applicable in Northern Ireland);[2] and
> - *Building standards 2022 technical handbook: non-domestic buildings* – Section 2 Fire (applicable in Scotland).[3]

10.4.1 Means of escape

A means of escape allows you to turn away from a fire and walk to a place of safety. The means of escape should lead directly to an outside open space away from the building. This means of escape should not include mechanical or other aids such as chutes, ladders, harnesses, lifts and other lowering devices. The exception to this rule is if you have to evacuate workers with mobility problems.

Means of escape is fairly complex, with lots of things to take into consideration. Some of these considerations are:

- construction of the building, including the fire resistance of different types of materials used in the escape routes;
- travel distances;
- stairs and passageways;
- fire doors;
- final exit doors;
- emergency lighting;
- directional and exit signage; and
- assembly points.

> **TIP**
>
> A means of escape should ideally:
>
> - allow all people to escape to a place of safety without external assistance;
> - have suitably located escape routes that are sufficient in number for the expected capacity of users;
> - be sufficiently protected from the effects of fire and smoke;
> - have adequate lighting and exits that are suitably signed; and
> - have provisions to limit the ingress of smoke to the escape routes and to restrict the spread of fire and remove smoke.[4]

Building construction

We have already looked at the properties of common building materials and how they are likely to behave in a fire. Both the building structure and the escape routes (such as corridors leading to exit doors) should be capable of withstanding the effects of fire for long enough to stop the building collapsing prematurely and to allow:

- time for all occupants to escape the building; and
- the emergency services the chance to bring the fire under control.

It is a legal requirement for a building to have fire-resisting products as part of its construction. Buildings should have external walls and a roof that will resist the spread of fire to neighbouring buildings; they should ideally be built from fire-resisting materials to help avoid early collapse in the event of a major fire. The same applies internally; linings on walls and ceilings must have a resistance to heat that will slow down the spread of flames over the surface of the materials. This is especially important for load-bearing parts of the structure.

Travel distances

This relates to the distances that building occupants will have to travel to reach an emergency exit that leads to a place of safety. This should ideally be no more than 45 metres (for a floor that has more than one exit) but this does depend on the type of building and the activities carried out in the building. The distance is measured from the furthest point on a floor to the exit.

Shorter travel distances will be required in high-risk buildings than in low-risk buildings. Distances are not measured in straight lines but must take into account the additional distance needed to walk around obstacles such as racking and shelving.

The UK fire guidance documents set the maximum travel distances to a fire exit in new buildings.[5] In reality, it is unlikely that you will be able to add fire exits to existing buildings without incurring considerable time and expense.

Stairs and passageways

Stairs and passageways should also use materials that are fire resistant and prevent smoke ingress (see the earlier section on building construction for details).

Stairs must be enclosed and wide enough (usually between 0.8 metre and 1 metre, depending on the type of building) to allow people to exit safely and must not narrow at any point; they should also be at least as wide as an exit giving access to the stairs.

The stairs will usually be connected to protected passageways. Dead ends must be avoided. Passageways should ideally be entered via self-closing fire doors that have automatic release mechanisms that will operate in the event of a fire.

Both stairways and passageways should be kept clear and free of obstructions. Organisations frequently use passageways and landings as additional storage space. This practice should be discouraged because it creates obstructions and adds to the fire load.

Fire doors

Fire doors are one of the most important links in fire safety. Fire doors should be constructed from fire-resistant materials and should be fitted with positive self-closing devices. Providing they are closed they will hold back heat and smoke for considerable periods. Fire doors usually offer protection for between 30 and 120 minutes. You may see that doors in your organisation have a rating shown on them – a rating of FD60 would offer protection for 60 minutes. Certified fire doors will usually have a label affixed to the top of the door confirming the fire rating. You may also see a coloured plug inserted in the door frame to indicate the fire rating.

APPLICATION 1

Take a look around your current location at the fire doors in place. Make a note of the fire ratings on the doors. Try to decide if the fire doors in the area have the appropriate fire rating.

Another important feature of fire doors is the intumescent seal. This is a strip of material (usually white) on the inside of the door. When it is exposed to heat the seal will expand, closing the space between the door and frame. This will stop the spread of fire and/or smoke (depending on the type of seal) beyond the door.

Figure 1: Example of an internal fire door

Fire doors must carry a 'fire door' sign to indicate that they should be kept closed. Automatic doors should also carry a warning for people to keep clear.

Fire doors should always be kept closed, except when there are practical reasons to keep them open, such as in hotels and elderly people's residential homes, where the initiating of the fire alarm automatically closes them. Doors should ideally open in the direction of escape and should be fitted with a vision panel so that people can see what is on the other side of the door.

Final exit doors

Final exit doors should open more than 90° so that those escaping will be able to fan out when leaving the building; this allows for a speedier evacuation. These external fire doors must be marked as appropriate, for example: 'PUSH BAR TO OPEN'. They must never be obstructed; a sign should be placed on the exterior of the door to warn people and vehicles to keep clear. External fire doors should never be locked when the building is occupied.

Emergency lighting

During a fire it is likely that normal lighting in a building will fail. It is therefore essential that buildings are fitted with emergency lighting to aid in a building evacuation. Emergency lighting should:

- indicate escape routes;
- provide sufficient illumination along these routes to permit safe escape; and
- ensure that fire alarms and firefighting equipment situated along the escape route can be easily located.

Generally, there are two types of emergency lighting:

1 maintained emergency lighting (a system where the lighting is in operation at all times); and
2 non-maintained emergency lighting (that only operates when the normal lighting fails).

There is also a third type of emergency lighting, generally known as 'standby lighting'. This is usually in critical areas of a building that require continuous lighting during a failure of the main supply, such as a hospital operating theatre.

Exit and directional signage

Exit (escape) routes should have easily recognisable signage. This is usually a green colour with a pictogram of a person running, along with a direction arrow. These signs should be placed at regular intervals and at all corridor junctions. Fire exit signage should be clear and unambiguous and signs should not contradict each other.

Figure 2: Escape routes should be clearly signed

Assembly points

These should be far enough away from the building so that people are not likely to be harmed by the fire and to allow those exiting the building sufficient space to escape. Care should be taken in selecting assembly points. Ideally they should be in the open air away from the fire; enclosed yards, alleyways and similar should not be used. For example, some schools may choose a playing field that is well away from the buildings; large warehouse complexes may locate their assembly point at the furthest point away from the buildings in their car park.

Ideally, assembly points should not be located next to the entrance that will be used by fire and rescue services. All workers must be aware of the assembly point locations, which must be clearly signed/marked.

As the signs will be outdoors, regular checks should be made on them to make sure that they have not been damaged, are not obscured by other items (especially vegetation in the area) and are still clear and legible. The signs will fade from weathering over time so should be replaced when the colouring/lettering starts to fade.

10.4.2 Emergency evacuation procedures

Whatever the nature of the building, clear procedures need to be in place for emergency evacuation. These procedures must be made known to all workers and must be regularly practised. Visitors to the site must also be made aware of emergency evacuation procedures.

Some of the things to consider in your evacuation procedure include:

- who will raise the alarm, and how;
- who will call the fire and rescue services;
- location of assembly points;
- details of competent people to take charge of the incident;
- any equipment required, for example flashlights;
- key people required, including first-aiders, incident controllers, people with technical or specific knowledge about the site; and
- essential equipment to be shut down if possible.

10.4.3 Role and appointment of fire marshals

Fire marshals should be appointed to assist with an evacuation. Fire marshals can also be known as fire wardens, incident officers, responsible persons, etc. Workers take on this role as well as their day-to-day duties. Fire marshals will need to be relied on to respond appropriately in an emergency; they must also be authoritative so that workers will listen to the instructions given.

Each fire marshal is allocated an area of the building that they will check once the fire alarm is sounded. They will look to make sure that everyone leaves the area and will then report to a co-ordinator so that this information may be passed on to the emergency services on their arrival. They may also be asked to supervise workers at the assembly points.

Fire marshals do not only have responsibility during an emergency; they may also be asked to look out for potential fire hazards so that proactive measures can be taken to deal with problems before they become emergencies.

Arrangements should be put in place to ensure coverage for sickness absence and holidays.

10.4.4 The purpose of fire drills

The purpose of fire drills is to make sure that all workers understand how to react in a fire or when the fire alarm sounds. Holding regular fire drills will help to ensure that all workers are familiar with the escape routes, procedures and firefighting equipment. Understanding how people are likely to react in a fire is an important part of a fire drill.

> **CASE STUDY**
>
> **Woolworth store fire in Manchester in 1979**
>
> A fire broke out in the furniture department that was located next to the restaurant (this was when furniture was still made from highly flammable materials). Although a member of staff shouted 'fire', people in the restaurant ignored this and carried on eating their meals. Because people could see no smoke, and none of the staff seemed concerned, the customers decided to ignore the warning. About five minutes after the first warning, a member of staff shouted 'fire, get out', but customers still ignored the warning and carried on eating. An eyewitness said that they thought "someone will get a fire extinguisher and put it out. It's only a little one".
>
> Elsewhere, around 50 people had made their way to the roof. A firefighter reached the people on the roof and tried to reassure them that they were in no danger. He started taking six people at a time from the roof to the ground using a hydraulic platform. However, each time he returned to the roof, the number of people remaining had reduced. Even though the firefighter had told them that the roof was the safest place, they had left this place of safety and gone back into the burning building.
>
> Sadly, 10 people lost their lives in this fire.[6]

Fire drills are different from fire alarm tests, which should take place weekly; these tests just check that the alarms work as expected when set off. Fire drills are used to help ensure that the emergency evacuation plans work in practice. After all, what you think will work on paper may not work in reality; there will always be things that are inadvertently overlooked or just do not work practically. Low-risk premises should aim to hold a fire drill at least once a year. Higher-risk buildings will require more frequent drills; the frequency will be guided by the type of building and/or activities carried out in the building.

Organisations may choose to do unannounced fire drills, which have the advantage of testing 'real' reactions to an emergency. Alternatively, some organisations may publicise the date and time of the fire drill in advance; the disadvantage of this is that some workers may decide not to be present. Whichever option is taken, worker reactions to the fire alarm should be carefully noted and any problems dealt with. Feedback should be provided to all workers to let them know how effective the evacuation was and to highlight any problem areas.

Depending on the arrangements that the employer has in place, a roll call could be taken during a fire evacuation to check that all people on site have left the building.

A fire drill also offers an ideal opportunity to test arrangements for the evacuation of disabled persons.

A record should be kept whenever a fire drill is performed. These records can help to improve your fire evacuation plan. They can also be used as evidence for the fire authority that the organisation is doing what is reasonable.

> **APPLICATION 2**
>
> Thinking about fire drills that you have taken part in, consider what went well, what did not go well and what you would do to improve the evacuation/fire drill.

10.4.5 Provisions for people with disabilities

Special consideration needs to be given to all sorts of disabilities – both physical and cognitive.

Disabled persons may need to have their own 'personal emergency evacuation plans' (PEEPs) to help ensure that they are able to safely evacuate the building.

Blind or partially sighted workers may require a 'buddy' to assist them out of the building. We discussed alarm arrangements for deaf workers previously; use flashing lights and/or vibrating devices to alert them once the alarm is activated.

Workers with cognitive disabilities may become confused and/or panic in emergency situations so may also require a 'buddy' to assist them.

People with reduced mobility may need to make their way to refuge areas (known as 'safe havens') that are constructed from fire-resisting materials. This will offer a place of safety until the person can be rescued. A refuge area must be contained in a protected zone, such as a stairwell or an adjacent protected lobby. If it is in a stairwell, the safe haven must allow sufficient space for other workers to use the stairs. In any event, it is essential that a two-way communication system is installed, which will keep people using the safe haven informed of the situation at all times and will allow them to inform the fire authorities of their location. Some organisations may also arrange for another worker to remain with the person in the safe haven to offer assurance and companionship while awaiting assistance.

10.4.6 Emergency escape routes to be recorded in building plans

It is important for escape routes to be recorded. This shows all workers and visitors to the building the quickest way out in the case of an emergency. In complex buildings or buildings where there are regularly large numbers of visitors, such as hospitals or hotels, there may be a plan indicating fire exits alongside fire evacuation information. In less complex buildings, a fire plan is not always necessary as evacuation routes are usually obvious from the evacuation signage.

Building plans should also be available to share with the fire and rescue service in the event of a real fire. If the roll call identifies that there are missing people, the plans will allow firefighters to go to the most obvious points to start looking for people.

KEY POINTS

- A means of escape enables people to turn away from a fire and get to a place of safety away from the building. Essential elements to consider are travel distances, stairs, passageways, doors, emergency lighting, exit and directional signs and assembly points.
- An emergency evacuation procedure should be known to all workers and visitors and practised regularly. It could include the appointment of fire marshals and other relevant people to specific duties during an evacuation.
- The purpose of fire drills is to check that workers know what to do if there is a fire or the fire alarm sounds. A drill may include a roll call to check everyone has left the building and making specific arrangements for the evacuation of people with different kinds of disability.
- It is important to include details of emergency escape routes in building plans so people know the quickest way out and also to help firefighters look for anyone who is missing.

References

[1] HM Government, *The Building Regulations 2010: Approved Document B, Fire safety, Volume 2: Buildings other than dwellings* (2019, updated 2020) (www.gov.uk)

[2] Department of Finance and Personnel, *Building Regulations (Northern Ireland) 2012: Technical booklet E – Fire safety (2012)* and Department of Finance, *Amendments Booklet – AMD 7* (2022) (www.finance-ni.gov.uk)

[3] Scottish Government, *Building standards 2022 technical handbook: non-domestic* (2022) (www.gov.scot)

[4] See note 1

[5] See notes 1, 2 and 3

[6] Adapted from Rob Williams, 'Inside the Manchester Woolworths blaze – the fire that claimed ten lives – and changed Britain', *Manchester Evening News* (7 May 2019); and 'What happened in the 1979 Manchester Woolworth fire?', BBC News (4 December 2012) (www.bbc.co.uk)

ELEMENT 11

ELECTRICITY

11.1: Hazards and risks

> **Syllabus outline**
>
> In this section, you will develop an awareness of the following:
> - Electric shock and its effects on the body; what affects severity: voltage, frequency, duration, resistance, current path; electrical burns (from direct and indirect contact with an electrical source)
> - Common causes of electrical fires, including portable devices overheating during charging
> - Workplace electrical equipment, including portable: what is likely to lead to accidents (unsuitable equipment; inadequate maintenance; use of defective/poorly maintained electrical equipment; use of electrical equipment in wet environments)
> - Secondary effects, including falls from height
> - Work near overhead power lines; contact with underground power cables during excavation work
> - Work on mains electricity supplies

Introduction

Here we will look at the fundamentals of electrical hazards and risks.

We will cover key terms such as voltage current and resistance and explore how these relate to each other. You will need to know a little about two different types of current: alternating and direct.

We will explore common electrical hazards and what affects the severity of these. We will look at the risks of electrical shock and how electrical fires and accidents commonly occur.

We will also consider the dangers of working near overhead power lines, excavating near underground power cables and working on mains electrical supplies.

11.1.1 Hazards and risks of electricity

Before we look at the hazards and risks of electricity, we will first introduce a few basic electrical terms and some basic principles.

Basic terms

You are probably familiar with electricity and electrical circuits that are most commonly encountered in the workplace and homes, such as switching on a light bulb/torch (mains or battery-operated) or using a power tool (mains or battery). In essence, electricity is simply the flow of electrons along or through a conducting medium such as copper wire. For electricity to flow and operate the tool or light, there must be a circuit and a source of electrical energy such as a battery or generator. A simple circuit is shown in Figure 1, made up of a battery, light bulb and connecting wires. The bulb converts the electrical energy into heat and light. In such circuits we often use terms such as voltage, current and resistance.

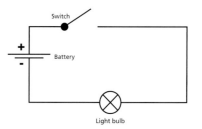

Figure 1: A simple electrical circuit

Voltage: this is supplied by the battery in the circuit shown in Figure 1. The voltage (or potential difference) 'pushes' electrons around the circuit. Voltage is measured in volts and uses the symbol 'V'. Examples of voltages commonly found include:

- mains electricity (230V in the UK);
- 1.5V in a battery; and
- 11,000V, 132,000V and higher in power transmission cables.

Current: this is the flow of electrons created when a voltage is applied across a circuit. Current is measured in amperes (or amps) and is signified by use of the symbol 'I'.

Resistance: this defines the ease (or otherwise) of flow of electricity. Resistance is measured in ohms and uses the Greek letter omega symbol (Ω). In our simple circuit shown in Figure 1, both the connecting wire and the light bulb have some resistance.

Ohm's law: in a simple circuit with a fixed resistance, the current (flow of electrons) increases as the voltage is increased. For a fixed voltage, the current decreases as we increase the resistance.

Current, voltage and resistance are related by Ohm's law: voltage in the circuit (V) = the circuit current (I) × the circuit resistance (R). This is usually written as:

$$V = I \times R \text{ or } I = V / R$$

This is a useful theory that underpins many of the technical controls used to limit the risks associated with electricity. The Application box gives an example of this theory in practice.

APPLICATION

Imagine a voltage source of 500 volts (V = 500). Imagine also that the resistance in the circuit is 50 Ω (R = 50). Using I = V / R, we can calculate the current. So in this example, the current I = 500 / 50 = 10 amps.

Now halve the voltage (to 250V) and keep the resistance the same. Using the same formula, I = V / R , we now get I = 250 / 50 = 5 amps. The voltage was halved, so the current halved as well.

Electricity

Types of electric current

There are two types of electric current in common use: alternating current (AC) and direct current (DC). DC involves the flow of electricity in one direction only; a battery, for example, will deliver DC. In AC the current repeatedly changes direction at a given frequency. In the UK, this frequency is 50 times per second (50Hz). The mains power system, which distributes power throughout the country for industrial, commercial and domestic purposes, uses AC. It is possible to convert AC to DC (using a 'rectifier') and DC to AC (using an 'inverter') for equipment that is incompatible with the originally supplied power type.

Both AC and DC can be dangerous, but generally you need a lower AC current than DC current for the same effect. Either can be fatal if the current level is great enough; both will cause people to be unable to let go of a conductor, although this happens at different levels (approximately 20mA for AC and 80mA for DC).

> **DEFINITION**
>
> **mA** stands for **milliamp** or one thousandth of an amp.

Now we will look at the dangers of electricity. Because AC is the most common type of current used in workplaces, most electrical incidents in the workplace involve AC. We will therefore mainly focus on the effects of AC.

Effects of electricity on the human body

Voltage and current

We have seen from Ohm's law that as voltage increases, so will current, for the same resistance. In electric shock, the human body becomes part of an electric circuit; the resistance of the body is a significant factor, but it varies a great deal, depending on several things such as the individual and the path the electric current takes through the body.

If you think of the current simply as energy, the more there is, the more harm will be caused. In general, for the average individual exposed to AC, the effects vary with current (see Table 1).

Current (mA)	Effect
2	Onset of sensation
8	Mild sensation
20	Contraction of muscles prevents victim releasing the live connection – very painful
40	Immediate resuscitation necessary to cope with muscular paralysis
80	Extreme distress and breathing difficulties
> 100	Onset of ventricular fibrillation
> 200	Breathing ceases, severe burns

Table 1: The effect on the body of varying electrical currents

The types of effect generally fall into the following three categories.

Ventricular fibrillation: this is an irregular twitching of the muscles in the walls of the heart. If it continues for more than about three minutes, the brain will become starved of oxygen.

Damage to the nervous system: an electrical shock can cause serious interference to the body's own electrically based brain/central nervous system.

Burns (direct and indirect): victims will sustain burns at the point of contact and may also exhibit 'exit wounds' where the electrical current finally leaves the body to get to earth. Surface burns to the skin are not uncommon and can be caused by direct contact or can be 'flash' burns, caused when the person gets close enough to a current source for the current to jump the gap (known as 'arcing'). When the victim has been exposed to higher levels of current, internal burns (especially through the chest cavity) invariably prove fatal.

Frequency

This applies only to AC. In general, low-frequency AC is more dangerous than high-frequency AC. The mains electrical distribution frequency in the UK is around 50Hz (meaning it changes direction 50 times per second) and this is considered the most dangerous frequency range.

Duration of contact

Thinking of electrical current as energy, the longer someone is connected to it, the more damage it will do. The damage sustained by the human body can also lower the body's electrical resistance, which increases the current further (Ohm's law).

Body resistance

Body resistance varies quite a lot between individuals. We know from Ohm's law that, for a fixed voltage, increasing the resistance reduces the current flow, whereas low resistance provides an ideal path for plenty of current to flow.

Current path

Resistance also varies with the route taken by the current from the point of contact to the point where the current leaves the body to get to earth. Probably the worst case is from fingertip to fingertip, which will involve the current flowing through the chest cavity, causing severe damage as it goes (such as burns).

11.1.2 Common causes of electrical fires

Electrical fires and explosions are mainly caused by following.

Cables and wiring overheating

This can result from:

- Overloading (carrying too much current): this may be because too many appliances are connected, or a blown fuse has been replaced with one of a higher rating (for example, a 5A fuse replaced with a 13A fuse).
- Damage: this could be either because of an accident or through misuse; for example, dragging the equipment by the cable or even traffic running over the cable can lead to weakened connections and thinning of the wire (so it is able to carry less current).

The heat from electrical equipment

Causes can include:

- flammable material such as clothes and curtains either falling on, or being too close to, an electric heater;
- hot electrical equipment, such as a soldering iron, being carelessly put down while still hot; and
- parts of equipment that overheat (for example, portable devices during charging) and ignite materials nearby.

Sparks and arcs from electrical equipment

While there is a technical difference, 'sparks' and 'arcs' are essentially the same thing; a spark can be considered a short-duration arc. An electrical system may contain a large amount of energy, and this may cause an electric arc to jump across quite wide gaps, causing burns, starting fires, directly igniting flammable atmospheres and so on.

11.1.3 Workplace electrical equipment

Workplace electrical equipment can be considered as either fixed or portable. Fixed equipment is permanently wired into the building's electrical system. It therefore forms part of the building's electrical circuit. As it is fixed in position, equipment of this type is much less prone to being damaged; it is not going to be moved, which is a prime cause of damage.

By far the bigger area of concern is portable electrical equipment.

This can be simply defined as any item of equipment with a plug that can be connected to and removed from an electrical socket. Examples range from an electric power drill through to a photocopier. Such equipment is much more prone to damage than fixed equipment. For example, trailing cables are prone to being crushed or otherwise damaged by chairs, traffic, people tripping on them, etc. Inappropriate handling and storage also increase the chances of damage; for example, equipment being frequently thrown into the back of a work van instead of being tidied away properly and placed in a suitable storage box.

What is likely to lead to accidents

The following are situations likely to lead to accidents with electrical equipment.

Selection and use of unsuitable equipment for the task or environment

This can include working in environments such as:

- Potentially flammable or explosive atmospheres: we have already seen that everyday electrical equipment may generate sparks or arcs or enough heat to start fires. For the same reason, they can ignite a flammable atmosphere. This kind of environment requires the use of special explosion-protected equipment. There are various designs for such equipment, including being sealed against ingress of the flammable atmosphere or reduced power (so that it does not have enough energy to ignite the flammable atmosphere). A widely used identification system for this type of equipment is the 'Ex' symbol marked on the equipment. This symbol will also be supplemented with other details such as the equipment category and whether it is designed for use with explosive gases or dusts (see Figure 2).

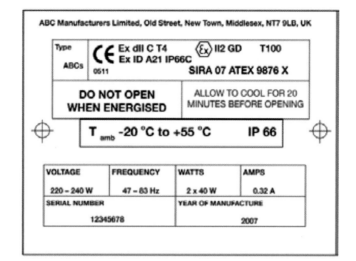

Figure 2: Example of Ex equipment marking

- Wet or humid environments: electricity and water do not generally mix, and the effect of introducing water into the equation is to reduce the resistance to the flow of electrical current. For example, ordinarily the casing around hand-held electrical equipment such as a drill (or, in a domestic setting, a hairdryer) stops the user from coming into contact with the live electrical conductors inside. In humid or wet environments, the water can get inside the casing and, in extreme cases, form a low-resistance path to the user holding it. Therefore, if equipment that is not designed for use in wet or humid environments is so used, the risk of serious injury or even death increases greatly. Using electrical equipment such as a hairdryer while taking a bath is therefore not a good idea!

Inadequate maintenance and inspection
Insufficient maintenance and inspection can result in defects going unnoticed and the defective equipment being used.

Substandard maintenance and repair
If maintenance is not carried out properly, such as twisting conductors together instead of using a proprietary joint, this could also mean defective equipment is used.

> **APPLICATION**
>
> Walk around your workplace and look out for the electrical equipment in use. Is it suitable for the environment in which it is used? Is it in good condition?

11.1.4 Secondary effects

These are indirect effects of electric shock. Even a relatively minor electrical shock may lead to a serious accident. A jolt from a faulty appliance can cause you to lose balance, lash out with your arm or lose control of the equipment. If a worker sustains such a shock when working at height, for example, they may fall and suffer serious injuries.

11.1.5 Work near overhead power lines and contact with underground power cables during excavation work

In essence, the major issue arising from overhead power lines is the fact that they are uninsulated. Although they often carry high voltages, insulation is not needed because safety is achieved by positioning the cable high in the air, so that contact becomes unlikely. Besides, insulating the full length of an overhead distribution cable would be prohibitively expensive and would make the cables far too heavy to suspend from pylons.

Figure 3: Overhead power lines are uninsulated

The fact that overhead lines are uninsulated, although positioned well out of the way of most people and plant, means that accidents do sometimes happen. In some cases, direct contact occurs, such as when a mobile crane or excavator inadvertently strikes an overhead cable on a construction site or farm. Similar things can happen with ladders and scaffold poles being carried. You do not even need direct contact; simply approaching too close can cause 'arcing' across the gap. Either form of contact can have fatal consequences. Therefore, the precautions for dealing with overhead lines mainly focus on keeping a safe distance away.

Buried power cables will, of course, be insulated as they are in contact with the ground in which they are buried. The difficulty with these arises from the fact that they are not visible and may therefore be struck when excavating, again leading to serious, often fatal, injuries if the insulation is breached.

11.1.6 Work on mains electricity supplies

Ideally, any work on electrical circuits should be done when the power has been switched off and the supply isolated. However, sometimes it is necessary to work 'live', which clearly carries a greater risk. This is more common when testing or fault-finding, so is a particular danger for maintenance personnel.

KEY POINTS

- Electricity is dangerous. The human body can all too easily become part of the electrical circuit and be electrocuted in the process.
- It is important to understand a little about the relationship between the voltage, current and resistance (in Ohm's law) and the factors that influence the severity of electric shock.
- Common causes of electrical fires include equipment overheating and accidents, such as using the wrong type of equipment for the environment.
- Not all electrical accidents are direct; some are secondary, for example an electric shock causing someone to lose their balance and fall when working at height.
- There are specific risks when working on mains electrical supplies and these are often encountered when doing construction work, working near overhead power lines or excavating near underground power cables.

11.2: Control measures

Syllabus outline

In this section, you will develop an awareness of the following:

- Protection of conductors
- Strength and capability of equipment
- Advantages and limitations of protective systems: fuses, earthing, isolation of supply, double insulation, residual current devices, reduced and low voltage systems
- Use of competent people
- Use of safe systems of work (no live working unless no other option; isolation; locating buried services; protection against overhead cables)
- Emergency procedures following an electrical incident
- Inspection and maintenance strategies: user checks; formal inspection and tests of the electrical installation and equipment; frequency of inspection and testing; records of inspection and testing; advantages and limitations of portable appliance testing (PAT)

Introduction

We have previously covered the hazards and risks associated with the use and misuse of electricity. Now we turn our attention to how these risks can be controlled.

We will look at the protection of potentially live electrical conductors, ensuring equipment is strong and capable, and the use of specific protective devices and systems such as fuses, earthing, supply isolation, double insulation, residual current devices and reduced and low voltage systems. We will also discuss the importance of using competent people and safe systems of work.

Finally, we will look at procedures for dealing with electrical emergencies and strategies for keeping electrical equipment in good condition (inspection and maintenance).

TIP

Great Britain has a range of guidance on electrical safety, including these from the Health and Safety Executive (HSE):

- *Electrical safety and you: A brief guide* (INDG231):[1] this provides a useful brief overview;
- *Electricity at Work Regulations 1989: Guidance on Regulations* (HSR25):[2] guidance on the legal requirements in the Electricity at Work Regulations 1989 and, in general, how you might comply with these;
- *Electricity at work: Safe working practices* (HSG85):[3] this gives specific guidance on safe systems of work (for example, when there is a need to work on live electrical systems);
- *Electrical test equipment for use on low voltage electrical equipment* (GS38):[4] this provides specific guidance on test equipment to use, for example to test if an electrical system is dead before work is carried out on it;

- *Avoiding danger from underground services* (HSG47):[5] this provides specific guidance on buried cables;
- *Avoiding danger from overhead power lines* (GS6):[6] this gives specific guidance on overhead cables; and
- *Maintaining portable electric equipment in low-risk environments* (INDG236):[7] this provides specific guidance on maintenance, inspection and testing approaches for portable equipment in offices and similar environments.

We will draw on relevant authoritative advice contained in the guidance, collected here to show you what is available should you wish to find out more.

Northern Ireland has similar provision,[8] which also frequently references the Health and Safety Executive (HSE) codes and guides applicable to Great Britain.

11.2.1 Protection of conductors

The main way in which conductors are protected is through insulation, which means the conductor is covered in a non-conductive coating or sleeve. Insulation can be damaged, so some forms of insulation incorporate protection against mechanical damage. For example, the use of armoured cable (where there is an extra layer of steel wire) when conductors are buried in the ground can protect from accidental contact while excavating.

Other examples include placing the insulated conductor inside trunking or a conduit. Conductors can also be 'protected by position'; for example, overhead power cables are usually not insulated and are instead placed high up, out of reach. Similarly, overhead pick-up wires for electric trains are not insulated (otherwise they would not work).

11.2.2 Strength and capability of equipment

Electrical equipment should be able to withstand the normal, overload and fault currents that might foreseeably run through it when it is in use. This includes having effective insulation. Equipment for use at work will normally need to be more robust than equipment that is used for home or hobby purposes because it is likely to be used a lot more and is therefore more prone to wear and tear.

Strength and capability will usually already have been considered by the equipment manufacturer or supplier, who will design their equipment for particular conditions of use. The user simply has to select the equipment suitable for the intended use at work.

You need to be especially careful when selecting the equipment you want to use in adverse conditions, such as damp/humid conditions, extreme temperatures, corrosive atmospheres, explosive atmospheres and extreme weather. Check the equipment is specified for use in such conditions when you select it.

11.2.3 Advantages and limitations of protective systems

A piece of electrical equipment (such as a circular saw, fridge, drill, food mixer, fan or heater), when connected to power and switched on, uses electrical energy in the circuit and converts it to something useful (such as motion or heat), as shown in Figure 1.

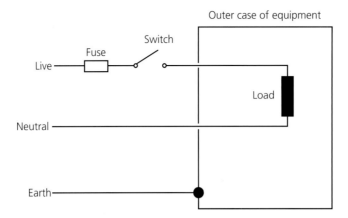

Figure 1: Overview of a basic electrical circuit

A circuit such as this will be protected by one or more devices or systems. We will look at some common ones now.

Fuses

A fuse is a deliberate 'weak link' in the circuit that protects the equipment in the event of a fault by automatically disconnecting the supply. It is used in many electrical systems and/or equipment.

The most common type of fuse contains a thin filament wire conductor. They are most commonly found incorporated into the plugs of mains-powered portable electrical equipment. As the circuit current flows through the conductor, it will generate a heating effect. The fuse size is chosen so that under normal conditions the heat generated is relatively small. However, if a fault develops, more current flows, heating the fuse wire to melting point. This breaks the circuit, thus cutting off the flow of current and rendering the equipment safe. For example, if a food mixer jams, the load will increase and the current being drawn will rise suddenly (imagine you are the mixer's motor – if you hit an obstruction, you will need more energy and will have to push a lot harder to make the mixer blades go around). The increased current demand will cause the thin filament of wire in the fuse to heat up and melt, thus breaking the circuit.

Fuses have an advantage in that they are relatively cheap and easy to install. Their limitation lies in the fact that they take time to 'blow', meaning that they do not protect people from electric shock – they only protect the circuit or equipment from further damage.

Furthermore, fuses can be replaced with incorrectly rated fuses (being rated for too high a current rating) or anything of a suitable size that conducts electricity. Using the wrong fuse, or bypassing the fuse entirely, means that the circuit will be able to draw too much current under fault conditions, which will cause a massive heating effect that goes beyond the capability of the appliance and could easily lead to a fire.

Earthing

In the UK, the electrical power distribution supply is deliberately connected to earth (the ground of our planet, Earth). This is usually done at substations and is called 'earth referencing'. It allows detection of certain types of fault (called 'earth faults'), triggering the operation of automatic disconnection devices, such as fuses.

Earthing is useful when equipment has exposed metal conductive parts. Equipment often has metal covers or cases to keep people away from the live electrical

11.2 Control measures

components and other dangerous parts inside. These metal covers, although electrically conductive, would not normally be live unless there was a fault. If these became live, for example by a current-carrying wire coming loose and touching the metal cover, and anyone were to touch it, it could cause an electric shock.

But if the metal cover is also connected to earth, it can detect the fault and, if a device such as a fuse is fitted in the path, can cut off the power automatically. This protection by earthing is based on the principle that electrical current in a system or piece of equipment will try to find its way to earth by the easiest possible route – literally the path of least resistance. If that is through a person, then they will probably be seriously injured; an earth connection provides an alternative, low-resistance route to earth, down which the current will flow, thus significantly reducing the current flow through the person. Planet Earth is so large that it is capable of absorbing any flow of electricity, provided there is a good electrical contact all the way into the earth.

A major limitation of earthing as a protective measure is that the earth line may be damaged and not offer a route to earth under fault conditions. The operation of the equipment will not be affected, so the only way to tell if there is an earth fault is to conduct regular inspection and testing.

APPLICATION

Imagine that, in addition to the fault with the food mixer described earlier, the earth connection is also faulty; perhaps it was severed at the same time as the line was damaged.

What might determine if the person with their hand on the metal casing of the mixer receives an electric shock? Imagine further that there is a metal sink next to the mixer; what might be the significance of this?

Following on from the Application example, perhaps the person is wearing rubber gloves and thick rubber shoes, giving a high total resistance, in which case only a very small current may flow through them. However, perhaps they have wet hands and are standing barefoot on a stone floor; the overall resistance (mixer–hands–body–feet–floor–earth) is now low, providing an easy path for the electric current to take to get to where it wants to be, the earth. In such a case, a considerable current may flow through the person on its way to the earth.

Even worse, with one hand on the live metal case of the mixer, the person reaches over and touches the metal sink – even thick rubber shoes are of no help now and a large current flows from hand to hand via their chest and heart. The fuse blows and the food mixer is left in a repairable state; once rewired and tested, it could be ready for use the next day. However, the person will not be so lucky and may have been killed or at least seriously injured.

Even in this worst-case situation, the victim's life might have been saved if there was a residual current device (an RCD, discussed later) in the circuit supplying the food mixer.

Isolation of supply

Firstly, it is important to distinguish between switching off a piece of equipment and isolating it.

Switching off will be achieved by direct manual operation or via a 'relay' (an electrically controlled switch). This will be the procedure for normal day-to-day operation, but if the electrical circuit/equipment is to be worked on, more security will be needed to protect the maintenance engineer(s).

Just switching off an appliance does not mean that it is isolated from the supply, nor does it prevent the equipment from simply being switched back on, thereby exposing anyone working on it to risk. If the equipment is to be worked on and more security is needed, it will need to be isolated so that the circuit cannot inadvertently be reconnected. This is an effective method of protection and can be achieved by 'locking off'. An isolator switch (which is capable of being locked off but cannot be locked in the 'on' position) is normally used for this purpose. Alternatively, parts of the circuit (such as a fuse) can be removed and locked away so that they cannot be put back without proper authorisation.

But clearly isolation may not be practical with equipment that is itself a source of electrical power. And in some cases it may not be desirable or even possible to isolate the equipment. For instance, the circuit may need to be kept live to conduct fault-finding – so-called 'live working'.

Figure 2: Locking off isolating switches can stop inadvertent reconnection

Double insulation

There are many pieces of electrical equipment in use that do not need to have any of their exposed metalwork earthed. Such equipment is instead double insulated (and will display a symbol to say so). This means the insulation on the wiring in the equipment is supported by an independent second level of insulation.

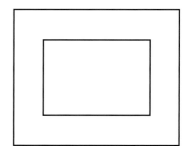

Figure 3: Double insulation symbol

Double insulation is commonly used for certain types of portable equipment, especially electrically driven motor tools, such as drills. This extra layer of insulation has clear advantages; since it is not part of the electrical system as such, it is not prone to the same types of failure. A breach of the double insulation, however, will leave the worker exposed to the risk of electrocution as there is no earth connection to provide a back-up.

Residual current devices

In some electrical fault conditions, there may be a risk of electric shock. During the time that it takes the fuse to melt, the victim may have received a severe or fatal shock, so there is a need for additional protection in the form of a residual current device (RCD). The RCD is considerably more than a 'fast fuse'; it does operate much faster than a fuse, but it also provides protection in certain fault situations where a fuse might provide no protection at all. However, in certain fault situations the fuse may be the crucial source of protection and so fuses and RCDs should be seen as complementary.

In essence, the RCD is designed to react in milliseconds (thousandths of a second) to any unexpected variation in the electric current being used by the equipment. This may be as trivial as a light bulb blowing in the kitchen or it may be a serious fault, such as the wiring of a food mixer having been badly damaged.

RCDs have a significant advantage over fuses because they operate much more quickly and at currents well below those normally associated with serious injury. However, they do have moving parts and so need to be regularly tested to ensure continued operation. Some may also trip repeatedly, leading to workers not using them because of the 'nuisance factor'.

Reduced and low voltage systems

Ohm's law tells us that if we reduce the voltage, the current will also reduce. Therefore reduced voltage systems tend to present a lower risk of injury.

Powered equipment such as drills and saws can be designed to operate effectively at lower voltages, and certain equipment, particularly lighting, can operate at voltages as low as 20V.

The demanding conditions of construction work, including extremes of weather, vehicle and material movement, changing workforce and heavy use, dictate the use of low voltage equipment whenever possible. The power for the low voltage tools will need to be provided by transformers that reduce the normal mains voltage (if it is higher) to the voltage required by the low voltage tools; such tools will have special connectors to ensure that they can only be connected to the correct supply. Note that, even when transformed to lower voltage, this can still entail a current large enough to cause harmful effects and so additional RCD protection will also be needed.

When using low voltage supplies, the distance to the transformer should be as short as possible and the leads from both the input (from the mains supply) and the lower voltage output should be protected as appropriate from physical damage, weather, chemicals and so on.

11.2.4 Use of competent people

Anyone working on electrical equipment or systems needs to be competent. This means having enough technical knowledge and experience to do the job safely. The more demanding the job, the more technical knowledge and experience will be required. The UK guidance suggests that a competent person should at least have:[9]

- adequate knowledge of electricity;
- adequate experience of the electrical work being carried out;

- adequate understanding of the system to be worked on and practical experience of that class of system;
- understanding of the hazards that may arise during the work and the precautions that need to be taken; and
- the ability to recognise at all times whether it is safe for work to continue.

There is also a recognition that someone who lacks all of this could still work on a job provided they are under the supervision of someone who does have it (how else does someone learn?). In practice this will probably mean using a qualified, experienced electrician for anything other than simple, low-risk jobs. This is especially important when working on live electrical equipment is considered necessary.

11.2.5 Use of safe systems of work

No live working unless no other option

Working on electrically powered equipment that is 'live' (connected to power, or energised) is clearly a lot more dangerous than working on the same equipment that is 'dead' (disconnected, or de-energised). The default should therefore be to work only on electrical equipment when it is 'dead', if possible. Working on 'dead' electrical systems may well be controlled using a safe system of work. This is especially true when working on high voltage systems, which can use a specific type of permit-to-work (see 3.7: Permit-to-work systems) called an 'electrical permit-to-work' to make sure that the electrical systems are disconnected, isolated, locked off and proved 'dead' before work commences.

Any suggestion of live working being necessary must be justified carefully. It is not enough for management to say that they did not want to stop production; there must be absolutely compelling reasons why live working has to be carried out, plus suitable precautions must be taken. Common reasons for working on live systems include testing (maybe as part of normal maintenance) or fault-finding.

> **TIP**
>
> Once this need for live working has been established, the engineers working on the live system must be protected by an appropriate system of work (which may well be controlled under a general permit-to-work system) that covers the following:
>
> - Keep exposed live parts to a minimum, both in terms of the time of exposure and the actual amount of live material that is exposed; for example, insulators can be placed over live conductors to minimise the chances of inadvertent contact.
> - Use protective equipment such as insulated tools, protective clothing (including gloves and footwear) and other protective devices.
> - Make sure there is enough working space and good lighting, to help reduce the chances of accidental contact with live conductors.
> - Use accurate circuit diagrams and information.
> - Restrict access to only those authorised to be there – keep everybody else out.
> - Those working on electrical equipment must be competent (have the necessary knowledge and experience or be under appropriate supervision) and must be familiar with the safe system of work being used.
> - A competent person on standby may also be required.

Isolation

Unless working on live equipment, the equipment should not only be switched off but isolated as well (where this is possible). Secure isolation (when equipment can be locked in the 'off' position, but not locked in the 'on' position) is important to remove the risk of accidental reconnection (for example, by another person) when someone might be working on the equipment.

To make sure everyone knows the equipment has been isolated, you should also post a 'danger' notice or tag. These are often combined into a system called Lock Out, Tag Out (LOTO). You should also test the equipment to make sure it really is 'dead'. This is because, in complex systems, you may have inadvertently isolated the wrong circuit.

Locating buried services

Hidden or buried electrical cables need to be identified and located to help prevent accidental contact when excavating. One of the main sources of information will be drawings or plans from those who either own or are responsible for the services (such as the electricity distribution company). But these plans alone cannot be relied on; they may not be accurate enough or reference points above ground may have changed over the years. You should also physically survey the area (things like lamp posts and inspection covers give clues) and use cable-detection tools (there are several different types, but all require skilled operators for correct and accurate interpretation). You may also need to carefully dig trial holes in the ground to verify your findings. Once located, the cables' direction line can be marked on the surface, using paint, for example.

Protection against overhead cables

In many cases, it is quite safe to work near to or underneath overhead power cables. That is because they are so high up there is no risk of accidentally getting too close or being in contact with them. The danger greatly increases, however, when you do work that means you might get close (such as working at height underneath cables) or you operate or carry certain equipment capable of accidental contact (such as cranes, excavators, metal scaffold poles or ladders).

The obvious first step is to avoid carrying out such work anywhere near live overhead power lines if you can. Remember that, due to the high voltage usually associated with power transmission cables, you do not even have to touch them; just getting too close is enough. It may also be possible to temporarily switch off the power at specific times when you or vehicles need to pass beneath, or close to, the cables.

If none of this is possible and the cables need to remain live, then you will need a safe system of work. The risk depends on all sorts of things, including whether you are just working near to the cables or you (and any tall machinery) need to pass directly beneath them.

If you do not need to pass beneath the cables, you can simply erect barriers at ground level to keep people and vehicles away. In general, these should be placed at least 6 metres horizontally from the cables to give a sufficient 'exclusion zone'. These barriers need to be highly visible to warn of the danger, so should be reflective/high visibility, including being lit at night. You need to be especially careful when using long-reach equipment, such as cranes or excavators, which, although positioned outside the barrier, may accidentally reach with their jibs or booms into the danger zone. In these cases, additional, higher-level visual warnings can be installed, such as a line of coloured flags ('bunting') following the line of the barrier but 3 to 6 metres vertically above ground. You may also extend the 'exclusion zone' much further, so that there is no possibility of accidental contact.

If equipment needs to pass beneath the cables, you can create a sort of passageway through the barriers on either side. This is commonly achieved with non-conducting 'goalposts' and warning signs (about the maximum clearance height) at either end of the passageway and additional bunting/barriers lining the passageway

Figure 4: 'Goalposts' provide a passageway under cables[10]

You also need to warn drivers to lower the jibs/booms of their mobile machinery before travelling through the passageway; jibs may also be physically restrained before entering the passageway. Good working practices also include making sure that any long-reach equipment, like metal ladders, is always carried horizontally, close to the ground.

Sometimes it may also be possible to cover exposed cables with temporary insulating material. This is more commonly the case when overhead power lines cross over, near to or connected to an existing building, so that someone working on the building could be at much greater risk of contact with the power cables.

11.2.6 Emergency procedures following an electrical incident

Despite the precautions in place, accidents involving contact with live electrical conductors (leading to electric shock, burns, etc) do sometimes happen. If you discover an apparent case of electrocution or cable strike, these are the main points to remember.

- Always assume the equipment, cables, etc are still live:
  - Do not go near or touch the equipment, etc.
  - If the victim is still in contact with the equipment, do not be tempted to pull them free, because you too could be electrocuted when you touch them.

- In the case of overhead power cables, if they are damaged and touching the ground, the ground itself where it touches the cables may also be live, so you should not approach it. If the cables were damaged by a vehicle that is still in contact, the driver should either stay inside the vehicle or jump out of it – not climb out, because they will potentially be making a path to earth through their own body if they set foot on the ground while still holding onto the vehicle.
- Call for help (such as a trained first-aider or the emergency services) if possible. Otherwise, if something happens to you as well, no one will come and rescue you.
- Switch off the electricity supply if possible by removing the plug (if safe to do so), switching off at the main fuse box, etc. If this is not possible, the victim must be moved as quickly as possible away from the source of power and this must be achieved without endangering anyone else. The victim should be pulled away from the source by means of a non-conducting implement such as a wooden broom handle (dry) or a dry sheet used as a lasso. If high voltages are involved, such improvisation may be dangerous to the 'rescuers'. For example, if the victim is found slumped over equipment in an electricity substation, extreme care needs to be exercised.
- If the victim is unconscious and has stopped breathing, artificial resuscitation must be started immediately and continued, even if the victim appears to be dead.

11.2.7 Inspection and maintenance strategies

A general rule is that all electrical systems, equipment and tools should be maintained in good condition, in efficient working order and in a state of good repair. This of course means that, among other things, they must be electrically safe.

Fixed electrical systems should be installed to a suitable standard in the first place. These standards are contained in *British Standard 7671: Requirements for Electrical Installations*, co-published with the Institution of Engineering and Technology (IET); they are usually referred to as 'The IET Wiring Regulations'.[11] In the same way, electrical equipment (portable and non-portable) should also be designed and built to a suitable standard by the manufacturer and, in the case of non-portable equipment, installed (or permanently connected) properly. Once in use, these systems and equipment need to be maintained to the same standards to prevent danger. There are three basic ways of doing this: user checks, formal visual inspections and combined inspections and tests.

User checks

Many electrical issues can be identified easily by users through a visual check before using an appliance or connecting to the electrical system, etc. This does not require any special expertise but simply an awareness of the signs that something is not right; for example, broken or cracked electrical sockets, loose or damaged cables/insulation, bent plug pins, scorch marks. If such signs are noted, the equipment should not be used but instead passed to a competent person for further testing, defect repair and, if necessary, replacement.

Formal inspection and tests of the electrical installation and equipment

User visual checks will pick up obvious defects (which account for the majority of issues), but deeper checks are also needed to pick up the less obvious issues. These checks should be carried out by someone with specific training and experience and include:

- Formal visual inspection: this is done less frequently than a user check and by a person with a little more training. It could include, for example, removal of the plug top (if possible) to check the internal plug wiring and whether a correctly rated fuse has been installed.
- Combined inspection and test: this is the least frequently conducted of the three methods and involves using test equipment, for example to check the continuity of the earth lead (for appliances that are not double insulated) and the integrity of the insulation (for all appliances and installations).

Frequency of inspection and testing

In general, there are no firm rules on the frequency of inspection or testing – this tends to be left to organisations to decide, based on their assessment of the risk. For electrical installations/systems, every five years or so is common. Many portable appliances, however, are likely to need more frequent checks. This is because they are handled much more and are therefore more likely to suffer damage of the sort that leads to danger. But again, this depends on the risk.

For example, an appliance that sits on an office desk in a non-hazardous environment and is seldom, if ever, moved may only require a formal visual examination once a year and may never need to be tested. On the other hand, a portable power tool that is being transported from site to site and is used in harsh conditions is likely to sustain damage, so might be checked by the user every day before use and formally tested every three months. INDG236 (see earlier Tip box) includes a table with suggestions for inspections and tests for some common portable electrical equipment, such as computers and kettles. But these are only suggestions and not legal requirements.

Records of inspection and testing

Electrical inspections and testing (and the results of these) should be recorded (in a spreadsheet, for example). And, especially for portable equipment, often a label is attached to the equipment to give reassurance to workers and also to act as a marker for the next date of checking.

Advantages and limitations of portable appliance testing

As already mentioned, portable equipment needs a lot more frequent attention and, as a result, you often find that electricians can have a business dedicated to portable appliance testing (PAT), offering these services to all organisations.

PAT has notable advantages, the main one being that it will detect circuit defects that go unnoticed in a visual inspection. This will allow for the item to be removed from service before it has the chance to cause injury. Regular PAT will also enable records to be kept that show a gradual deterioration in the earth continuity or in the efficiency of the insulation.

Limitations of PAT lie in the frequency with which it is carried out. The test is only valid at the time it is conducted; conceivably, a defect could arise a day later but not be detected until the next test, which could be many months away. In addition, if the tester does not record the readings, but instead just records 'Pass' every time the item is tested, then there will be no way to see whether the equipment is degrading over time.